The N

The following review of the complex science of entomology is predicated on the assumption that a text should emphasize general principles over systematic and applied details. This approach seems appropriate in view of my limited objective, *viz.*, summarization of material salient to beginning aspects of the subject. I touch on what I regard as the most important aspects; I largely ignore insect classification and control except to the extent that they bear on comprehension of principles; and I adopt the minimal terminology and the fewest scientific names necessary for expressing insect complexities.

This guide to the insects may be used as the sole text of a short course in entomology. Because of its relative brevity, it also lends itself to use by students enrolled in a formal entomological course should they wish to supplement their lecture and laboratory experiences or prepare for examinations; it lends itself to use by those needing a review of the "essentials," whether or not they are enrolled in a formal insect course; and it lends itself to use by agriculture and veterinary students, by teachers at the elementary and secondary school level, and by others unable to avail themselves of the opportunity of intensive, formal study of the subject.

A knowledge of general zoology is assumed background.

Entomology

by

S. K. Gangwere

First Page Publications
12103 Merriman • Livonia • MI • 48150
1-800-343-3034 • Fax 734-525-4420
www.firstpagepublications.com

© 2005 S.K. Gangwere
All Rights Reserved

Cover Design: Kimberly Franzen
Editorial Liaison: Sarah Hart

First Page Publications
12103 Merriman • Livonia • MI • 48150
1-800-343-3034 • Fax 734-525-4420
www.firstpagepublications.com

Library of Congress Cataloging-in-Publication Data

Gangwere, S. K.
  Entomology / by S.K. Gangwere.
    p. cm.
    Summary: Overview of principle components entomological study.
    Includes bibliographical references.
  ISBN 1-928623-26-3
  1. Insects. I. Title.
  QL463.G325 2005
  595.7--dc22
        Library of Congress Control Number: 2004117394

This book is affectionately dedicated to

Jacqueline Dawn Gangwere

*whose love, help, and encouragement have furthered this
entomological project and others that,
through the years, I have undertaken*

Table of Contents

Preface .... *ix*
Acknowledgments .... *xi*
Introduction
    Chapter 1 1
External Body Structure
    Chapter 2 23
Internal Body Structure and Physiology
    Chapter 3: Skeletointegumentary System. 51
    Chapter 4: Nervous System. 63
    Chapter 5: Muscular System. 79
    Chapter 6: Endocrine System. 91
    Chapter 7: Digestivoexcretory System. 103
    Chapter 8: Circulatory System. 121
    Chapter 9: Respiratory System. 131
    Chapter 10: Reproductive System. 145
Development
    Chapter 11: Metamorphosis. 153
Distribution
    Chapter 12: Insect Natural History. 161
    Chapter 13: Insect Habitats. 179
Behavior
    Chapter 14: Insect Feeding Behavior. 191
    Chapter 15: Insect Locomotion. 209
    Chapter 16: Insect Mating and Reproduction. 225
    Chapter 17: Insect Aggregations and Societies. 239
Systematics and Evolution
    Chapter 18: Nomenclature, Taxonomy, and Systematics. 253
    Chapter 19: Classification and Major Insect Groups. 265
Bibliography 313
Glossary 321
Index 339

Preface

Fix thy corporeal and internal eye
On the young gnat or new-engendered fly.
Laying their eggs, they evidently prove
The genial power, and full effect of love.
Each then has organs to digest his food,
One to beget and one to receive the brood;
Has limbs and sinews, blood and heart, and brain,
Life and her proper functions to sustain;
Though the whole fabric smaller than a grain.
— Prior

Entomology is the science concerned with insects and the way in which they carry out their life processes in complex interaction with the environment. Entomologists, those who practice this discipline, have as their goal resolution of the maze of relationships that insects share with one another and with the physical and biotic environment. They have achieved notable successes. They have identified and named well over a million species as well as the major groups to which these entities belong; they have outlined the general features of insect life processes; and they now understand many of the complexities of insects' role in nature.

I am indebted to earlier authors. Virtually all information within this book is from the voluminous literature of entomology, and comparatively little is from my personal laboratory and field experiences which, in themselves, are biased, being weighted toward study of the distribution and behavior of the grasshoppers and related insects. As much as I might wish to acknowledge the source of each individual observation used, I find it a practical impossibility to do so at this pedagogical level even assuming that I could determine who first said what and when; I cannot. Most basics have been stated and restated so often that original credit is difficult to fix.

For purposes of brevity, I omit chapter outlines or resumes arguing that they are not essential in an abbreviated text. Information appropriate to a resume is readily conveyed by perusal of italicized words and chapter headings.

A few suggested references that one might wish to consult are provided at the end of each chapter, the full citations of which appear in the bibliography toward the end of the book.

Acknowledgments

I acknowledge the generous assistance of colleagues who kindly reviewed drafts of selected chapters within their areas of expertise. These specialists include Drs. Spencer J. Berry, Professor Emeritus, Wesleyan University, Middletown, CT; John Brown, Curator, USDA, c/o U. S. National Museum, Washington, DC; the late R. G. Chapman, Professor Emeritus, University of Arizona, Tucson, AZ; Donald G. Cochran, Professor Emeritus, Virginia Polytechnic Institute and State University, Blacksburg, VA; Fred Delcomyn, Professor, University of Illinois, Urbana, IL; Al B. Ewen, retired from Agriculture Canada, Saskatoon, Sask., Canada; Oliver S. Flint, Jr., Curator Emeritus, U. S. National Museum, Washington, DC; Darryl T. Gwynne, Professor, University of Toronto, Mississauga, Ontario, Canada; Thomas J. Henry, Curator, USDA, c/o U. S. National Museum, Washington, DC; Laurence Mound, CSIRO, Canberra, Australia; James L. Nation, Professor, University of Florida, Gainesville, FL; David A. Nickle, USDA, Beltsville, MD; Allen L. Norrbom, USDA, Beltsville, MD; Daniel Otte, Curator, Academy of Natural Sciences, Philadelphia, PA; M. Paul Pener, Professor, Hebrew University, Jerusalem, Israel; David R. Smith, Curator, USDA, c/o U. S. National Museum, Washington, DC; Courtenay N. Smithers, Curator, Australian Museum, Sydney, NSW, Australia; Thomas Smyth, Jr., Professor Emeritus, Pennsylvania State University, University Park, PA; Douglas W. Whitman, Professor, Illinois State University, Normal, IL; and Joseph Woodring, Professor, Louisiana State University, Baton Rouge, LA. I also thank Ms. Nancy Wilmes, Science Library, and her colleagues at Wayne State University, Detroit, MI, for generous access to essential library and other resources. Above all, I thank Dr. Roger G. Bland, Professor Emeritus, Central Michigan University, Mt. Pleasant, MI, and Dr. Leslie A. Mertz, of Kalkaska, MI, formerly of Wayne State University, Detroit, MI, both of whom kindly reviewed the entire manuscript. I offer heartfelt thanks to the above and to all others who assisted in one way or another for their valuable criticisms, suggestions, and technical help that enabled me to update and improve the text and avoid numerous pitfalls into which I might have fallen. Their assistance notwithstanding, I alone am responsible for all statements made.

Illustrations from various sources were modified and redrawn except for certain original figures or for figures for which permission to reprint was obtained. The source of each drawing used is indicated in the figure explanations.

Chapter 1

Introduction

The Nature of Insects

Insects are small-bodied creatures that, owing to their exceptional mobility, reproductive potential, and environmental adaptability, have become widespread virtually throughout the world. They far outstrip all other animals combined in numbers of species and individuals. Being both numerous and intimately involved in the ecological food chain, they play a critical role in nature. They inflict incalculable economic losses on mankind owing to their destructiveness to crops and to the innumerable diseases and health problems that they occasion.

The science of insect study is known as *entomology* (Gr. *entomon* = insect + *logos* = discourse). Insects are representatives of animal class *Insecta* (L. *insectus* = cut into parts) based on their obvious body division into segments. Thus, along with spiders, crabs, millipedes, centipedes, and various lesser-known animals of that ilk, insects belong to phylum *Arthropoda* (Gr. *arthros* = joint + *poda* = foot) based on their external skeleton.

By way of introduction, insects are arthropods with the body divided into three distinct regions: head, thorax, abdomen. The *head* is a fused, probably six-segmented structure bearing *biting-type mouthparts* (often modified into suctorial, lapping, or other subtypes as described in Chapter II), a pair of *antennae*, a pair of *compound eyes*, and usually two or three *simple eyes*. The *thorax* is three-segmented, each segment of which bears a pair of *walking legs* and the last two of which usually have a pair each of membranous or thickened *wings* with supportive *veins*. The *abdomen* consists of 11 or fewer segments generally devoid of appendages, except for those modified into, or associated with, the *genitalia (Table 1, Fig. 1-1)*.

Most insects are air breathers adapted for terrestrial life by possession of branched, internal air tubes, or *tracheae*, that open to the outside on most segments of the thorax and abdomen through pairs of valved apertures called *spiracles*.

Insects have separate sexes, the paired *gonads*, or sex organs, of which consist of multiple tubules leading into a Y-shaped sex duct *(Fig. 1-2)*. Insect development usually involves a complicated *metamorphosis*, or change in body form, featuring immature and adult stages separated by molts *(Fig. 1-7)*.

Chapter 1 / **INTRODUCTION**

TABLE 1
THE INSECT ORDERS IN BRIEF

Names	Habitats	Eyes & Antennae	Mouthparts	Wings	Tarsi	Abdomen	Development	Primary Characters
Protura Telsontails	Terrestrial, under debris	No eyes, no antennae	Entognathous sucking	Primitively wingless	1 sgmt & 1 unpaired claw per leg	9-12 sgmts w telson, 3 prs styli, no cerci	Anamorphosis	Wingless hexapod w fore legs bent up
Collembola Springtails	Usually terrestrial, often under debris	Eyeless or w few ommatidia, sgmt type antennae	Entognathous chewing	Primitively wingless	1 sgmt & 1 pr claws per leg	6 or fewer sgmts, no cerci	Ametabolous	Wingless hexapod w springing organ
Diplura Japygids, campodeids	Terrestrial, under debris	No eyes, sgmt antennae	Entognathous chewing	Primitively wingless	1 sgmt & 1 pr claws per leg	10 sgmts, 1 pr cerci/forceps	Ametabolous	Wingless hexapod w paired styli & cerci or forceps
Microcoryphia Bristletails	Terrestrial, under debris	Large cmpd eyes, ocelli, sgmt antennae	Unicondylic ectognathous chewing	Primitively wingless	Usually 3 sgmts & 1 pr claws per leg	11 sgmts, 1 pr cerci, & median caudal filament	Ametabolous	Wingless hexapod w many paired styli & 1 pr cerci shorter than median caudal filament
Thysanura Silverfish, firebrats	Terrestrial, under debris or in buildings	Small or no cmpd eyes, often no ocelli, & annulate antennae	Dicondylic ectognathous chewing	Primitively wingless	2-5 sgmt & 1 pr claws per leg	11 sgmts, 1 pr cerci, & median caudal filament	Ametabolous	Wingless hexapod w paired styli & 3 often subequal caudal appendages
Ephemeroptera Mayflies	Terrestrial adults near water, naiads aquatic	Cmpd eyes, 3 ocelli, & short annulate antennae	Ectognathous chewing in naiads; vestigial in adults	2 prs net-veined triangular wings	3-5 sgmt & 1 pr claws per leg	10 sgmts & 2-3 caudal filaments	Hemimetabolous	Net-veined, triangular wings, 2-3 caudal appendages
Odonata Dragonflies, damselflies	Terrestrial adults near water, naiads aquatic	Cmpd eyes, 3 ocelli, & short antennae	Hypognathous chewing	2 prs net-veined, long wings w costal nodus & stigma	3 sgmt & 1 pr claws per leg	10 sgmts & 1 pr short cerci	Hemimetabolous	Net-veined, long wings w costal nodus & stigma
Blattodea Cockroaches	Terrestrial, in cracks, under debris, or in buildings	Cmpd eyes, ocelli, often absent, long antennae	Opisthognathous chewing	Wings long, short, or absent	5 sgmt & 1 pr claws per leg	10 obvious sgmts & 1 pr short cerci	Hemimetabolous	Flat, ovoid, running insects w shield-like pronotum covering head

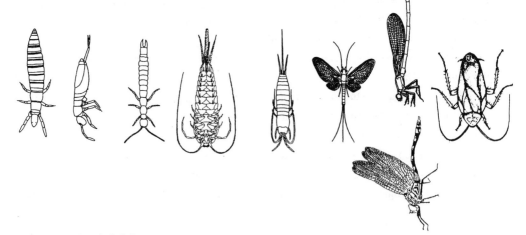

Table 1. The insect orders in brief.

The Nature of Insects

Names	Habitats	Eyes & Antennae	Mouthparts	Wings	Tarsi	Abdomen	Development	Primary Characters
Mantodea Mantises	Terrestrial plant-perchers	Cmpd eyes, 3 ocelli, slender antennae	Hypognathous chewing	Wings long, short, or absent	Fore legs raptorial, mid & hind legs not specialized, 5 sgmt	11 sgmt & 1 pr sgmt cerci	Hemimetabolous	Elongate predators w raptorial fore legs
Isoptera Termites	Terrestrial, in wood or ground	Cmpd eyes & 2 ocelli or none, long antennae	Prognathous chewing	2 prs similar wings w basal suture or wingless	4-5 sgmt & 1 pr claws per leg	10 visible sgmts & 1 pr cerci	Hemimetabolous	Polymorphic social insects w basal suture
Phasmatodea Stick insects	Elongate plant-perchers	Small cmpd eyes, no ocelli, long antennae	Prognathous chewing	Often wingless	5 sgmt & 1 pr claws per leg	11 sgmts & unsegmented cerci	Hemimetabolous	Stick-like or leaf-like insects of concealing habit
Orthoptera Grasshoppers, katydids, crickets	Usually terrestrial on ground or vegetation	Cmpd eyes, 2-3 ocelli, long or short antennae	Hypognathous chewing	Usually 2 prs, the fore pr tegmina & hind pr flight wings	1-4 sgmt, w 1 pr claws & arolium per leg	11 sgmts & 1 pr unsegmented cerci	Hemimetabolous	Veined tegmina, often saddle-like pronotum, & jumping hind legs
Grylloblattodea Ice crawlers	Terrestrial, under rocks	Cmpd eyes reduced or absent, no ocelli, long antennae	Prognathous chewing	Wingless	5 sgmt w 1 pr claws & no arolium	11 sgmts & 1 pr long, flexible cerci	Hemimetabolous	Long, slender, primitive, wingless insects
Dermaptera Earwigs	Terrestrial, under debris or on vegetation	Cmpd eyes & ocelli present or absent, long antennae	Prognathous chewing	Short tegmina & truncate, membranous, folded hind wings	3 sgmt & 1 pr claws per leg	11 sgmts & 1 pr forceps	Hemimetabolous	Veinless, short tegmina & 1 pr caudal forceps
Plecoptera Stoneflies	Terrestrial adults near water, naiads aquatic	Cmpd eyes, 2-3 ocelli, filiform antennae	Prognathous chewing or reduced	2 prs membranous wings, hind pr w ample anal area	3 sgmt, 1pr claws, & arolium per leg	10 visible sgmts & 1 pr long cerci	Hemimetabolous	Membranous wings, hind pr w ample anal area
Embioptera Web-spinners	Terrestrial, subsocial, in silk tunnels under debris or plants	Cmpd eyes, no ocelli, & filiform antennae	Prognathous chewing	2 prs membranous wings or wingless	3 sgmt & 1 pr claws per leg	10 sgmts & 1 pr short cerci	Hemimetabolous	Subequal thorax/abdomen, dilated fore tarsi
Zoraptera	Terrestrial, in small groups under bark or debris	Cmpd eyes & 3 ocelli or eyeless, moniliform antennae	Hypognathous chewing	2 prs wings w reduced venation or wingless	2 sgmt & 1 pr claws per leg	11 sgmts & 1 pr short cerci	Hemimetabolous	Gregarious & polymorphic, wings (if any) shed

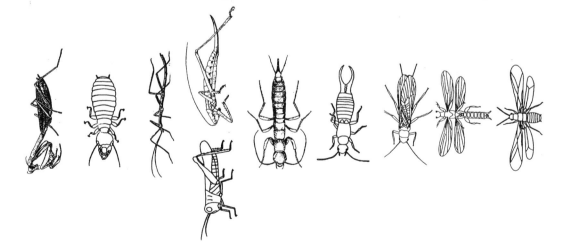

Chapter 1 / **INTRODUCTION**

Names	Habitats	Eyes & Antennae	Mouthparts	Wings	Tarsi	Abdomen	Development	Primary Characters
Psocoptera Bark lice, book lice	Terrestrial, under debris or bark or in buildings	Cmpd eyes & 3 or no ocelli, long filiform antennae	Hypognathous chewing	2 prs membranous wings w stigma & reduced venation or wingless	2-3 sgmt & 1 pr claws per leg	9 visible sgmts, no cerci	Hemimetabolous	Louse-like non-parasites, no cerci
Phthiraptera Lice	Terrestrial parasites	Cmpd eyes reduced or absent, no ocelli, short 3-5 sgmt antennae	Modified chewing or sucking	Wingless	Legs w un-sgmt or 2-sgmt tarsi, adapted for clinging to hair or feathers, 1-2 or no claws	7-10 sgmts, no cerci	Hemimetabolous	Wingless, depressed parasites of warm-blooded animals
Thysanoptera Thrips	Terrestrial, under debris or bark or on plants	Cmpd eyes, 3 or no ocelli, short antennae	Hypognathous asymmetrical rasping/sucking	Long- or short-winged or wingless; if former wings are strap-like w reduced venation & "hairs"	1-2 sgmt, 1-2 claws, & protrusible vesicle per leg	11 sgmts, no cerci	Holometabolous	Strap-like wings w "hairs," eversible vesicle on tarsi
Hemiptera Bugs, cicadas, hoppers, aphids, etc.	Terrestrial, on plants, some bugs aquatic	Cmpd eyes & 3 or no ocelli, long many sgmt antennae	Hypo-, pro-, or opisthognathous	2 prs wings, fore pr leathery w few veins, hind pr membranous; either or both may be absent	1-3 sgmt & usually 1 pr claws per leg	Varied, 9-11 sgmts, no cerci	Usually hemimetabolous but some holometabolous	Sucking beak btwn fore legs or on front of head, fore wings leathery & of uniform texture or hemelytral
Neuroptera Lacewings, antlions	Terrestrial, on plants	Cmpd eyes 3, 2, or no ocelli, long or short antennae	Adults pro- or hypognathous chewing, larvae suctorial	2 prs net-veined, subequal wings	5 sgmt & 1 pr claws per leg	10 sgmts, no cerci	Holometabolous	Complex, net-veined wings, veins much branched near margin, no costal nodus
Megaloptera Dobsonflies, alderflies	Terrestrial adults on plants near streams, larvae aquatic	Cmpd eyes & 3 or no ocelli, long antennae	Prognathous chewing	2 prs net-veined, subequal wings	5 sgmt & 1 pr claws per leg	10 sgmts, no cerci	Holometabolous	Complex, net-veined wings, no costal nodus

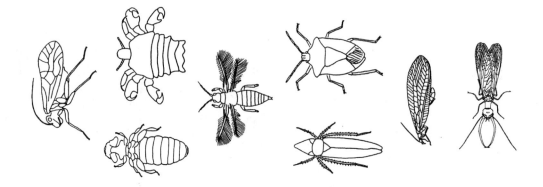

The Nature of Insects

Names	Habitats	Eyes & Antennae	Mouthparts	Wings	Tarsi	Abdomen	Development	Primary Characters
Raphidioptera Snakeflies	Terrestrial, adults on plants, larvae under bark	Cmpd eyes & 3 or no ocelli, usually filiform antennae	Prognathous chewing	2 prs net-veined subequal wings	5 sgmt & 1 pr small claws per leg	10 visible sgmts, no cerci	Holometabolous	Net-veined wings w costal stigma, no nodus; elongate "neck" or prothorax
Coleoptera Beetles, weevils	Terrestrial, under debris or on plants, some families aquatic	Cmpd eyes, no ocelli, 11-sgmt antennae of varied form	Usually prognathous chewing	Membranous hind wings complexly folded under elytral fore wings	3–5 sgmt & 1 pr claws per leg	5–8 visible sgmts, no cerci	Holometabolous	Horny or leathery elytra that conceal folded hind wings
Strepsiptera Stylopids	Terrestrial, free-living males & usually parasitic females	Males w & females without cmpd eyes, no ocelli, antennae flabellate to wanting	Reduced chewing in males & vestigial in parasitic females	Haltere-like fore wings & broad, membranous hind wings in males, wingless females	2–5 sgmt, w or without claws in males, parasitic females are legless	10 sgmts & no cerci in males, sgmts obscured in females	Holometabolous	Haltere-like fore wings in males; larviform parasitoid females
Mecoptera Scorpionflies	Terrestrial, on plants	Cmpd eyes & 3, 2, or no ocelli, filiform antennae	Small chewing mouthparts at apex of hypognathous beak	Usually 2 prs long, membranous wings, primitive venation	Usually 5 sgmt & 1 pr claws per leg	10 visible sgmts, 1 pr short cerci	Holometabolous	Elongate beak, primitive wings, male terminalia sometimes scorpion-like
Trichoptera Caddisflies	Adults terrestrial, larvae aquatic	Cmpd eyes & 3, 2, or no ocelli, filiform antennae	Hypognathous reduced chewing mouthparts	Usually 2 prs membranous wings, primitive venation, "hairy"	5 sgmt, 1 pr claws, & arolium per leg	10 visible sgmts, 1–2 sgmt cerci	Holometabolous	Moth-like, "hairy" wings, larvae often case-makers
Lepidoptera Moths, skippers, butterflies	Terrestrial on plants	Cmpd eyes & 2 ocelli or none, clubbed, recurved, or varied antennae	Hypognathous sucking mouthparts	2 prs dissimilar wings, simple venation, scale-covered	5 sgmt & 1 pr claws per leg	10 sgmts, no cerci	Holometabolous	Broad, membranous, scaly, wings & proboscis coiled beneath head

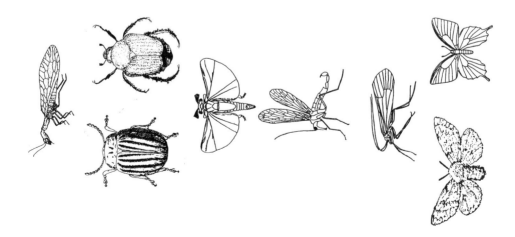

Chapter 1 / **INTRODUCTION**

Names	Habitats	Eyes & Antennae	Mouthparts	Wings	Tarsi	Abdomen	Development	Primary Characters
Diptera Flies	Terrestrial, on plants, occasionally aquatic	Cmpd eyes & 3 or no ocelli, varied antennae	Usually hypognathous adapted for piercing or sponging	Fore wings elongate, membranous, w reduced venation, hind wings are halteres	5 sgmt, 1 pr claws, 1 pr pulvilli, & sometimes empodium per leg	Up to 11 sgmts but sometimes only 4-5 visible, cerci fused	Holometabolous	1 pr fore wings & 1 pr halteres, antennae often aristate
Siphonaptera Fleas	Terrestrial mammal or sometimes bird parasites	Cmpd eyes reduced or absent, no ocelli, concealed short antennae	Hypognathous suctorial	Secondarily wingless	5 sgmt & 1 pr claws per leg	10 sgmts, vestigial cerci	Holometabolous	Wingless, laterally flattened parasites
Hymenoptera Sawflies, ants, wasps, bees, *etc.*	Terrestrial, within or on ground or plants, sometimes social	Usually both cmpd eyes & ocelli, varied antennae	Hypognathous chewing sometimes modified for lapping or sucking	Usually 2 dissimilar prs of elongate, membranous wings w reduced venation & often costal stigma	Usually 5 sgmt, 1 pr claws, & empodium per leg	10 sgmts, not all visible, 2nd sometimes constricted into petiole, & often 1 pr cerci	Holometabolous	Modified ovipositor, often "wasp waist," & membranous wings w few irregular veins & cells

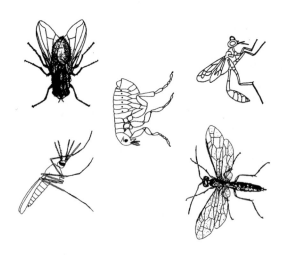

The Nature of Insects

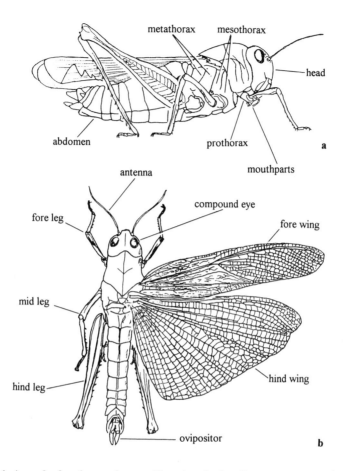

Fig. 1-1. (*a*) Lateral view of a female grasshopper. Note that the *head* bears antennae and compound eyes; the *thorax* (here divided into pro-, meso-, and metathorax) bears legs and wings; and the *abdomen* is terminated by an ovipositor used in egg deposition; (*b*) dorsal view of a female grasshopper (Figures are modified and redrawn from CSIRO, 1991).

Size

Insects have followed an evolutionary course favoring numerous individuals of small size rather than only a few of large size. The layman thinks of virtually any small, crawly, cold-blooded animal with legs as an insect. This impression is not entirely absurd. Many organisms of this description are indeed insects or at least arthropod relatives of insects, and most are small in comparison with vertebrates. Most insects measure from 2 – 40 mm in length, with the average probably less than 10 – 15 mm. A few wasps literally capable of creeping through the eye of a needle are dwarfed by some protozoans (*Fig. 1-3 f*).

Small size affords advantages in hiding and is reflected in the habitat occupancy of most insects. They have gone into, become specialized for, and now dominate practically every land and freshwater *microhabitat*. They live under stones and debris, within soil, beneath tree bark, in axils or flow-

Chapter 1 / **INTRODUCTION**

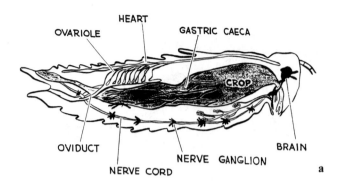

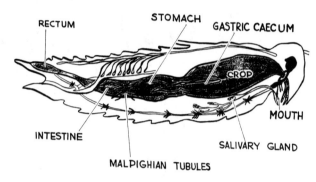

Fig. 1-2. (*a*) Internal structure of a female grasshopper showing especially nervous (brain, nerve cord, nerve ganglia), reproductive (oviduct, or female sex duct, and ovaries consisting of ovarioles), and circulatory organs (heart); (*b*) internal structure of a female grasshopper showing especially its digestivoexcretory organs (food storage crop, digestive stomach and caeca, absorptive and egestive intestine), and excretory Malpighian tubules.

ers of plants, on foliage, or they bore within plant or animal tissues or move about over soil, rocks, or vegetation.

Small size presents certain disadvantages, however, particularly with respect to potential water loss. Small objects have a proportionately greater surface area with respect to volume than do large objects, posing the possibility of heightened evaporation, which is a function of surface. Thus, small insects are highly subject to water loss offset, in part, by possession of valved respiratory apertures kept closed as much as possible and a wax- and cement-covered outer body wall largely impervious to water loss.

Contrasting with the small size of insects mentioned above is the comparatively large size of some others. The world's longest insects are certain walkingsticks, one species of which attains a body length of 330 mm. The world's heaviest insects are certain tropical scarab beetles, one of which has a body length of 120 mm and a bulk about 20 times that of its relatives in northeastern USA, fitting into one's hand like a tennis ball. Certain giant silkworm moths whose wings attain a length up to 242 mm are the extant insects with the greatest wingspan.

The Nature of Insects

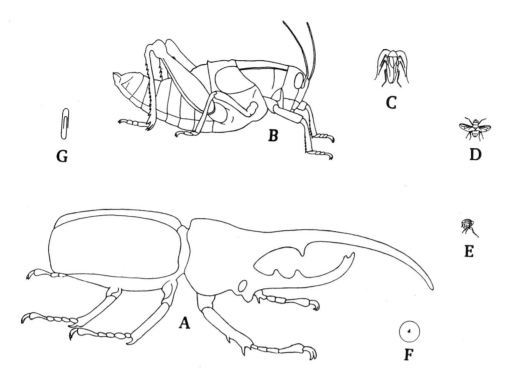

Fig. 1-3. Comparative sizes of (*a*) rhinoceros beetle, (*b*) lubber grasshoppper, (*c*) cockroach, (*d*) house fly, (*e*) flea, (*f*) chalcidid wasp, and (*g*) paper clip (Modified and redrawn from various sources, with the chalcidid encircled for purposes of illustration).

A large body size tends to isolate its possessors from the very microhabitats for whose occupancy most insects have become specialized and gives them an ecological role analogous to that of songbirds and other relatively small, mobile vertebrates. It exposes them to increased predation offset only to a degree by their armor-like outer body wall. It also brings them close to the functional limits of their respiratory and skeletointegumentary systems. Their respiratory system is a type that carries oxygen directly to the body tissues by ever-branching air tubes. This system functions less efficiently with a large body size, to the point where it becomes selected against in nature. Equally important is the fact that body weight increases by cubes and strength of support by squares. This means that the largest insects are obliged to have a disproportionately heavy outer body wall and appendages to support their increased weight. Thus, limits are imposed on insect size making it unlikely that ones larger than the ponderous beetles mentioned above will evolve in the future.

Wings

Insects differ from other living animals, except birds and bats, in possession of functional wings. Their powers of flight having been established about 350 million years BP (before present), insects

Chapter 1 / **INTRODUCTION**

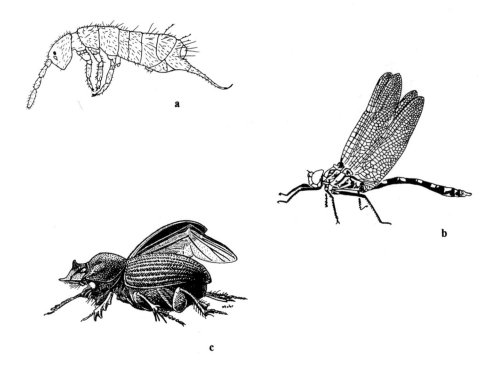

Fig. 1-4. (*a*) A springtail, representative of the wingless arthropod group *Apterygota*; (*b*) a dragonfly, representative of the non-wing-folding *Paleoptera*; and (*c*) a scarab beetle, representative of the wing-folding *Neoptera* (*a* is modified and redrawn from various sources, *b* is modified and redrawn from Kennedy, *c* is by courtesy of Illinois Natural History Survey).

have flown longer than any other kind of animal. Moreover, birds and bats simply modified their fore limbs into wings, but insects evolved totally new, flap-like extensions of the body wall that became hinged for use in flight. Hence, insects have not lost their total complement of legs through flight.

Moreover, the method that most insects use to produce flapping flight is unique. Birds and bats fly by using muscles that insert from the body directly onto the wing surface. Insects use indirect muscles that attach to the thorax, not to the wings (except for dragonflies and a few other insects that use direct wing muscles). Contraction of their indirect muscles alternately depresses and then elevates the thorax which, in turn, flaps the wings.

Insect classification is based, in part, on the presence or absence and on the kinds of wings, if any. Subclass *Apterygota* (*a* = wanting or without, *ptera* = wings) consists of primitive arthropods that do not have and presumably never did have functional wings. Included are: (1) *Protura* or telsontails, (2) *Collembola* or springtails (*Fig. 1-4 a*), (3) *Diplura* or japygids and campodeids, (4) *Microcoryphia* or bristletails, and (5) *Thysanura* or silverfish and firebrats. All, except Thysanura, are merely close relatives of insects, not true insects, though conventionally treated in courses in entomology. The second subclass, *Pterygota*, consists of true insects that are winged or at least have evolved from winged ancestors. Pterygotes include infraclasses Paleoptera and Neoptera. *Paleoptera*, such as mayflies and dragonflies

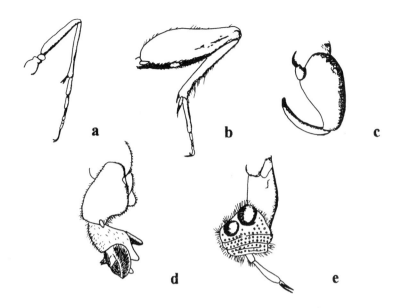

Fig. 1-5. Insect leg adaptations for (*a*) walking, (*b*) leaping, (*c*) grasping prey, (*d*) digging, and (*e*) clasping-sucking (*a* and *b* are legs of a cricket, *c* a bug, *d* a mole cricket, and *e* a diving beetle (Modified and redrawn from various sources).

(*Fig. 1-4 b*), are generalized insects unable to fold the wings backward over the abdomen at rest. *Neoptera* are more advanced insects than Paleoptera, having a wing articulation that permits backward flexion, sometimes of a complex type (*Fig. 1-4 c*).

Some orders of Neoptera are wingless despite having evolved from winged stock. Their secondary winglessness is adaptive. It is characteristic, in general, of many parasitic groups (*Fig. 19-12 h*), of particular taxonomic groups and species (*Fig. 19-7 d*), and of one or of the other sex (*Fig. 19-11 f*) in otherwise winged orders.

Part of the significance of insects' acquisition of flight lies in their improved survival. Flight assists them in eluding enemies and seeking shelter. It has helped them take over and dominate microhabitats throughout the world, except the depths of the ocean. It facilitates their long-range dispersal for reproduction and allows them to utilize specialized habits. For example, flight makes feasible adoption of dung-feeding habits by certain beetles because only wide-ranging animals can depend for their sustenance on such an irregularly distributed food source. Flight coupled with metamorphosis makes possible different habits during different life stages, as with certain herb-feeding caterpillars of one habitat that transform into nectar-feeding moths or butterflies in another, completely different habitat.

Structural Variability and Adaptability

Insects' basic structure is subject to almost endless variation according to species and taxonomic group. For example, their legs may be adapted for walking, running, hopping, swimming, clinging,

Chapter 1 / **INTRODUCTION**

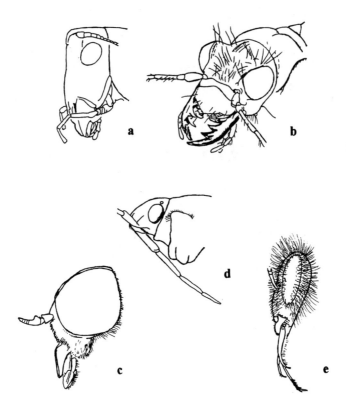

Fig. 1-6. Comparative structure of (*a*) a katydid's generalized *hypognathous head* (downward projecting mouthparts) with jaws adapted for chewing plant tissues, (*b*) a beetle's *prognathous head* (forward projecting mouthparts) with enlarged prey-seizing jaws, (*c*) a horse fly's hypognathous head, its proboscis concealing slashing blood-sucking stylets, (*d*) a bug's *opisthognathous head* (backward projecting mouthparts) with an elongate, segmented proboscis concealing slender piercing stylets for sucking plant juices, and (*e*) a bee's hypognathous head with chewing mandibles and a soft proboscis for lapping pollen and nectar (Modified and redrawn from Oldroyd, 1962).

catching prey, digging, producing sound and/or hearing, grooming, or other purposes (*Fig. 1-5*). Their wings are subject to variations mentioned earlier and may also serve in protection, species or sex attraction, or sound production as well as their primary function, which is flight. Their mouthparts may be adapted for, among other purposes, chewing, cutting, sucking, or sponging (*Fig. 1-6*) but, in a few instances, are vestigial organs quite useless in ingestion. Mouthparts may be used also for such non-feeding purposes as warning, fighting, species or sex attraction, grooming, or sound production.

The foregoing represents only part of the gamut of insects' variability. For example, the adult caste of termites and ants varies from wingless workers, to powerful-jawed, flightless soldiers, and to winged sexual forms (*Fig. 17-4*), all in the same species!

Most adult insects exhibit a pronounced *sexual dimorphism*. Females tend to be larger-bodied than males and to have a different form, behavior, and sometimes color or color pattern. Gypsy moths (*Fig. 16-3 b, c*) provide a good example.

The Nature of Insects

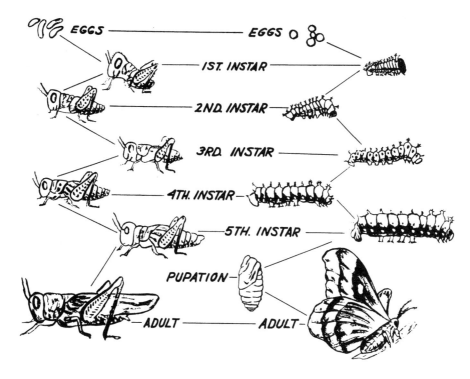

Fig. 1-7. *Hemimetabolous development* of a grasshopper (*left column*) compared to the *holometabolous development* of a moth (*right column*) from egg (*top line*), through juvenile stages (*middle lines*), to adult stages (*bottom line*) (Modified and redrawn from various sources).

Metamorphosis

Metamorphosis, or change of body form with life stage, adds another dimension to insect variability. A few primitive insects and their arthropodan relatives, such as the wingless Protura, Collembola, Diplura, Microcoryphia, and Thysanura, have a *direct development*, one in which the life stages resemble one another (*Fig. 11-1, a, b*). Theirs involves eggs which hatch into small, sexually immature, adult-like juveniles that eventually transform into larger, fertile *adults*.

Grasshoppers and their allies, termites, earwigs, bugs, and certain other insects exhibit a development involving relatively slight differences between the life stages in terms of size, body proportions, and the presence or absence of wings and genitalia. This second type of development is termed *hemimetabolous*. The eggs of such insects hatch into small, wingless juveniles, or nymphs, with a vague general resemblance to adults of their species, and the *nymphs*, in turn, eventually transform into larger, sexually mature, usually winged adults (*Fig. 1-7, left column*). Nymphal grasshoppers, for example, typically have short wings compared to adults (*Fig. 11-1 c, d*). The aquatic juveniles of mayflies, damselflies, dragonflies, and stoneflies are exceptions. Their young, often called *naiads*, do not resemble the winged, sexually mature terrestrial adults into which they eventually transform (*Fig. 11-1 e, f*).

Chapter 1 / INTRODUCTION

Finally, there are insects with a holometabolous development. Their complete type of development involves eggs that hatch into actively feeding, non-reproductive larvae that eventually molt into seemingly quiescent pupae and then into adults (Fig. 1-7, right column). Larvae represent the feeding stage and adults the dispersal and reproductive stage. This radical transformation can best be appreciated by comparison of a worm-like, foliage-feeding caterpillar with the winged, nectar-feeding adult butterfly or moth that it eventually becomes. The two extremes, as unlike as mammals are from fishes, could be taken as different classes of animal by the uninitiated, yet they belong to the same species!

Numbers of Insects

Numbers of Species

Insects far outstrip all other animals combined in their numbers of species. Approximately three of every four creatures crawling, walking, or flying over the earth, burrowing within it, or swimming through its waters are insects. Of those that are not, many are arthropodan relatives of insects (*Graph 1-1*).

The present system of scientific names dates to the 10th edition of Linnaeus' *Systema Naturae* (1758). Linnaeus, the Swedish botanist who originated the binomial system of designating plants and animals, recognized 4,379 kinds of animals, of which 1,937 were insects. Since Linnaeus' time, scientists have collected insects from some of the most inaccessible places on earth and have improved our knowledge of insect systematics to the point where many previously undetected species are known. An estimated 7,000 new species per year are added to the recognized total. Notwithstanding these burgeoning numbers, the proportion of insects to other animals has held at levels of about 75 per cent of the total fauna.

These data are estimates based on current lists of scientific names. Additional study invariably discloses that some recognized names are invalid, duplicate names for the same species. Synonyms result from recognition of color or other variants that do not represent true species, from study of insufficient samples, or from outright taxonomic error. Whatever the error, synonyms must be subtracted from the total recognized list. However, the number of names subtracted from the list owing to synonymy is dwarfed by the numbers of newly described species added to the list annually. Some authorities estimate that the insect species presently recognized represent but one-fifth or less of the total living insect fauna. Others feel that the present list is perhaps 75 per cent or more inclusive. Whichever estimate is right, the list of recognized insect species is destined to grow in the future notwithstanding the increased habitat and species loss occasioned by mankind's destructiveness to the environment.

Of all insects, *Coleoptera* (beetles) compose, by far, the largest group. Beetles contain over two-fifths of all known insect species, a total of over 370,000 species (*Graph 1-2*). Another two-fifths consists of *Lepidoptera* (butterflies and moths), *Hymenoptera* (sawflies, wasps, ants, and bees), and *Diptera* (flies), leaving the remaining one-fifth composed of the remainder of the insect class, some

Numbers of Insects

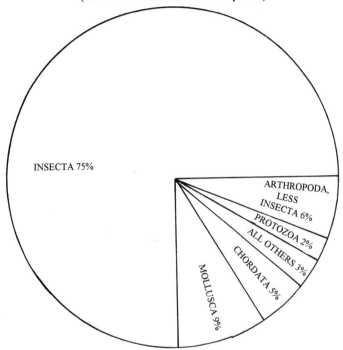

Graph 1-1. Comparative size of major animal groups in terms of numbers of species (Based on data from Arnett, 1985, Borror *et al.*, 1989, USDA, 1952, other standard works, and personal communication). Non-insect arthropods include, among others, spiders, mites, lobsters, crabs, centipedes, and millipedes; chordates include chiefly vertebrates; and molluscs include clams, snails, and octopods.

24 orders in all. The largest groups among the latter are *Hemiptera*: *Homoptera* (cicadas, aphids, leafhoppers, and their relatives), *Hemiptera*: *Heteroptera* (true bugs), and *Orthoptera* (grasshoppers and related orders).

Numbers of Individuals

There is no way to estimate accurately the numbers of individual insects because their world totals exceed present means of measurement and do not take into account the marked seasonal, locality, and other variations to which they are subject. Notwithstanding these imponderables, one authority has estimated the world population of insects at about 10 billion individuals per km^2 of land surface. The validity of estimates of this magnitude is difficult to establish. Perhaps it is sufficient merely to cite a few examples.

As many as 25,000 individual aphids have been recorded on a single tomato plant (*Fig. 1-8 a*); ants are often so abundant in our cities that one cannot walk a few blocks without crushing some

Chapter 1 / **INTRODUCTION**

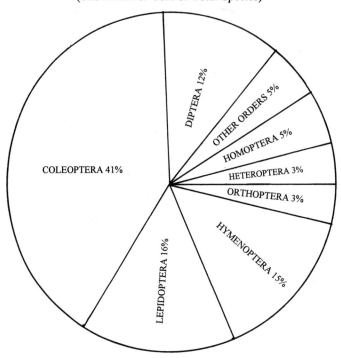

Graph 1-2. Comparative size of insect orders in terms of numbers of species (Based on data from Arnett, 1985, Borror *et al.*, 1989, USDA, 1952, other standard works, and personal communication).

beneath his feet; house flies crawl over people, food, dung, garbage, and other organic wastes worldwide; hordes of biting mosquitoes and black flies and of non-biting midges annoy us as they swarm over woodland, swamp, and nearby places notwithstanding treatment of their breeding grounds with powerful insecticides; and those who live in the vicinity of the Great Lakes cannot fail to be impressed by the myriads of newly emerging mayflies that alight everywhere along the shoreline and are crushed beneath our feet.

An estimate of social insect nest size is more readily made than is one of a general insect population, but even a nest census presents numbers so great as to make the task laborious and the result uncertain. For example, hives of the honey bee may contain 30,000 – 60,000 individuals; nests of certain ants range from a few dozen to several hundred thousand individuals; and nests of termites vary from a few hundred to over three million individuals in some tropical species that construct massive above-ground nests (*Fig. 17-3 b*).

These figures are based on average conditions in a given season and locality, but what of plague conditions? In the last century, swarms of the now-extinct Rocky Mountain Locust appeared in the

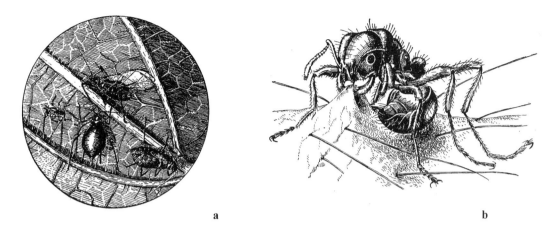

Fig. 1-8. (*a*) Aphids, including several wingless and a winged individual, perched on a leaf on which they are sucking; (*b*) a viciously biting fire ant (After USDA, 1952).

Great Plains of the USA and devastated the countryside over to the Mississippi River. One swarm was estimated at 85 km high, 170 km wide, and 510 km long and must have contained over 124 billion individuals! Similar outbreaks are known to occur today in the Old World Desert and Migratory Locusts, as well as in a number of other insect groups.

Reproductive Potential

Insects' abundance is a consequence of their unusually great fecundity, short life cycle, and sometimes multiple generations per year.

The number of eggs laid per female insect varies from a few to thousands. One tropical termite has been observed ovipositing at a rate of an egg every two seconds, for a total of 43,000 eggs per day! The average female insect probably lays 100 or more eggs in her lifetime, so many of which hatch and reach reproductive age that the total produced by the overall population becomes unbelievably great.

Most insects have a short life cycle during which they may be highly prolific. For example, the house fly, an insect with multiple generations per year, has a life span of about a month during which each female produces six or more lots of eggs of up to 150 eggs per lot.

Foods

Insects occur mostly in association with green plants that serve them, directly or indirectly, as food. They live on, in, or near these plants and make use of their tissues for food, shelter, and egg laying. Over half of all known insect species eat living vegetation. These plant eaters, or *herbivores*, con-

Chapter 1 / INTRODUCTION

sume all manner of vegetation, including so-called "lower plants," ferns, gymnosperms, herbaceous monocots and dicots, and woody dicots, and they attack all plant parts from roots, to stems and leaves, and to flowers and fruit. Practically nothing vegetable is immune from their ravages. There are also many insect *scavengers* that feed on dead plant tissues. Other insects are *carnivores* that live on, in, or near various animals and use their bodies, products, or wastes as food. Though these carnivores are directly associated with animals, not with plants, their animal hosts are themselves ecologically tied to particular plant assemblages.

Importance of Insects

Balance of Nature

A given proportion of producers, herbivores, carnivores, and decomposers is characteristic of biotic communities whose very perpetuation requires that an equilibrium, the so-called *Balance of Nature*, be struck. *Producers* constitute the first trophic level (*Fig. 12-6*). They consist of the herbs, trees, vines, and other green plants that, by photosynthesis, harness solar energy into the food stuffs on which the world's ecosystem depends. *Herbivores* such as grasshoppers and caterpillars are among the innumerable organisms that derive their energy of metabolism from eating the tissues of producers. Most insects belong to and dominate this second trophic level, both on land and in fresh-water. As such, they constitute much of the animal basis of the terrestrial food chain and are second only to plants in their ecological significance. Other insects belong to the third trophic level. They are *carnivores* that eat *herbivores*, especially other insects. Some of them, for example, certain ladybird beetles, eat noxious plant-feeding insects and, hence, are effective in biological control. Still other insects belong to the fourth trophic level. Cockroaches, granary weevils, dung beetles, skin beetles, and house flies are among these *scavengers*. They consume dead plant and animal tissues, exudates, and organic wastes of all manner and thereby assist bacteria and fungi in returning to soil the nitrogenous substances that the photosynthetic green plants require to make protein foods.

Insects play a critical role in nature. Anything that disturbs them alters the ecological energy flow and, if sufficiently severe, may cause breakdown within the overall environment. For example, an epidemic in a given herbivorous insect may deplete its numbers. This population decrease, while it exerts less forage pressure on the vegetation, reduces the available food supply of countless insectivorous amphibians, reptiles, birds, small mammals, and other carnivores which, consequently, may starve in absence of sufficient numbers of insect prey.

Pollination

The flowering plants that dominate today's landscape are pollinated by two primary agents: wind and animals. Most plants rely on the second, *animal method*, and most animal pollinators are insects. Among important pollinating insects are flower flies and other Diptera, butterflies and other

Lepidoptera, and social wasps, solitary and social bees, and other Hymenoptera. Bees alone account for 80 per cent or more of all pollination.

The plant-insect relationship is a mutually beneficial interaction between partners. One partner, the insect, obtains food as it pollinates the other partner, the plant. This has led to such close interdependence that these plants are unable to fruit in absence of their insect partners. This relationship is true of the majority of agricultural and horticultural plants including fruit trees, vegetables, and special crops (except grasses, which are wind pollinated). Pollen transfer from plant to plant is sometimes relatively simple, not involving specialized adaptations. Pollen grains, in this instance, merely adhere to the visiting insect's hair-like setae as it feeds until knocked loose at the next flower it visits. In other cases, both plant and insect are highly modified toward effecting the transfer.

Insect Products

Insects are useful to mankind's economy in countless ways aside from pollination, just a few of which bear mentioning. For example, the honey bee produces *honey* and *beeswax*; scale insects produce *lac*, a substance involved in preparation of shellac, varnish, and various natural dyes; and the silkworm moth produces *silk*. Insects also tend to be high in *food value*; hence, they provided nourishment for aboriginal man and continue to be eaten today in many parts of the world; and butterflies, beetles, dragonflies, and other colorful, aesthetically pleasing insects have become important in today's *art, decoration,* and *literature*.

Economic Entomology

On the opposite side of the ledger are the incalculable economic losses that insects cause. Insect plant pests significantly damage or destroy up to one third of every crop grown including grains, vegetables, fruits, and citrus. Their destructiveness to roots, stems, buds, leaves, flowers, and fruit causes a decreased yield and requires great outlays of money for spraying and other expensive control measures. Insects include a host of destructive livestock pests such as horse flies, bot flies, sucking lice, fleas, and others, as well as many pests of mankind's dwellings. Termites, powder-post beetles, and carpenter ants damage furniture and structural timber; clothes moths and carpet beetles ruin rugs, clothes, furs, and other fabrics; cockroaches, granary weevils, and other pantry pests are among the many insects that infest, tunnel in, eat, or otherwise contaminate mankind's food.

An entire field called *economic entomology* has been developed to study and treat the depredations of these insects of agricultural and veterinary importance (*Fig. 1-9 g - l*).

Medical Entomology

Another discipline, *medical entomology*, has arisen in response to mankind's insect-transmitted diseases and health problems (*Fig. 1-9 a - f*). This field has become an important subdivision of today's medical sciences.

From time immemorial, insects have shared mankind's lot and made his life miserable. They adversely affect his health by acting as vectors of disease. Filth-inhabiting insects such as house flies,

Chapter 1 / **INTRODUCTION**

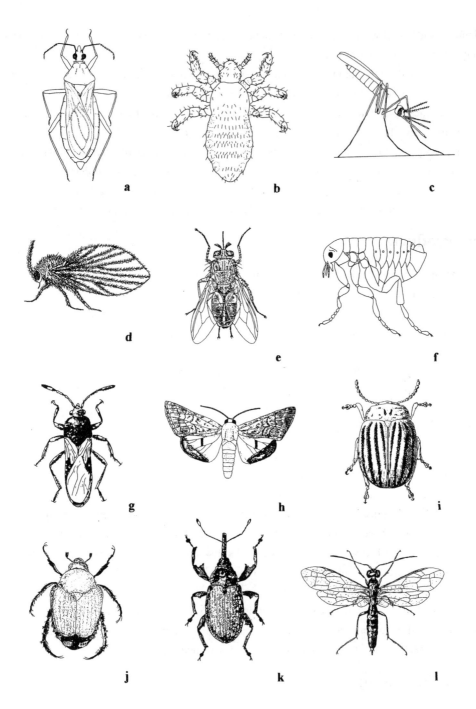

Fig. 1-9. Various insect pests of (*a – f*) medical and (*g – l*) agricultural/horticultural importance. The medical pests and the diseases/conditions that some of them transmit include (*a*) conenose bugs — Chagas' disease, (*b*) human body louse — epidemic typhus, (*c*) mosquitoes — malaria and yellow fever, (*d*) sandfleas — leishmaniasis, (*e*) tsetse flies —sleeping sickness, and (*f*) fleas — bubonic plague. The agricultural/horticultural pests include (*g*) chinch bug, (*h*) corn earworm, (*i*) Colorado potato beetle, (*j*) Japanese beetle, (*k*) cotton boll weevil, and (*l*) wheat stem sawfly (Modified and redrawn from various sources).

cockroaches, and other insect agents of disease pick up pathogens on their body as they walk over and feed upon feces, garbage, and other wastes, and, upon gaining access to mankind's food, contaminate it. Among the numerous insect vectors of disease are mosquitoes, some of them involved in transmitting malaria, yellow fever, and other serious diseases that affect fully one-sixth of the human population.

Many insects are important for their painful *bite* or *sting*, sometimes accompanied by severe swelling or bleeding (*Fig. 1-8 b*). Examples include the bed bug, body lice, Diptera such as horse flies, deer flies, and the stable fly, fleas, and the fire ant. The honey bee, wasps, and many other Hymenoptera are notable for their lance-like sting, a modified ovipositor, or egg-laying device, that can be thrust into an intruder's flesh forcing toxin into the wound and causing sharp pain and subsequent irritation. Necrosis of the wound, partial paralysis, shock, and even death may result in instances of especial victim sensitivity.

A few caterpillars, especially of the family Megalopygidae, have stinging "hairs" over the body. These structures are actually hollow, glandular setae that release poisonous secretions upon being broken off within the intruder's skin. They may cause severe irritation and even a dermatitis.

Suggested Additional Reading

Berenbaum (1995); Borror, Triplehorn, & Johnson (1989); Chapman (1998); CSIRO (1991); Davies (1988); Gullan & Cranston (2000); Kettle (1995); Metcalf & Metcalf (1993); Norris (1991); Richards & Davies, vol. 1 (1977); USDA (1952).

Chapter 2

External Body Structure

*For what is form, or what is face,
But the soul's index, or its case?*
— Cotton

General Organization of Insect Body

An arthropod consists of a tubular trunk supported by paired, articulated appendages (*Fig. 2-1 a*). The trunk, containing the viscera, is capped at the anterior and posterior ends by the acron and the periproct, respectively. The mouth is located at or beneath the *acron*, and the anus is on the *periproct*. The trunk is usually reinforced by a hard *exoskeleton* of calcium carbonate or sclerotin. This skeleton is not a continuous, inflexible tube, in which movement would be impossible, but it is divided lengthwise into a series of rigid, ring-like *segments* joined by soft, flexible *conjunctivae*, or intersegmental membrane, generally tucked in to permit bending (*Fig. 2-1 b*).

A single arthropod hypothetical segment can be likened to a box that articulates by flexible connections with similar boxes before and behind (*Fig. 2-2*). The top of each box is the *tergum*, the bottom the *sternum*, and the sides are *pleura* (sing., *pleuron*). Integument evaginates at the sternopleural junction on either side to form paired limbs that attach by flexible *arthrodial membrane*. Each limb, like the body itself, consists of a linear series of sclerotized tubes, or *articles*, connected by arthrodial membrane.

A series of segments placed end to end and capped, at one end, by the acron and, at the other, by the periproct composes the simplest possible combination. This worm-like arthropod with legs is not a realistic possibility. A mouth must perforate one end and an anus the other, and the two must connect lengthwise by a digestive tract. A nervous system and associated sensory structures must concentrate in the head region; and, concomitant with establishment of this head-tail orientation, there must occur further modification of the legged, worm-like animal through differential fusion, change in form, strengthening, and selected reduction of segments and of limbs.

Insects are advanced over their myriapod ancestors (centipedes, millipedes, *etc.*) in the degree to which they have fused their segments into distinct body regions, each associated with particular functions (*Fig. 2-1 b*). There are three such regions in insects: head, thorax, and abdomen. The *head*, involved in feeding, coordination, and environmental orientation, bears mouthparts, antennae, and eyes, all on a rigid head capsule derived in part from the acron. The *thorax*, specialized for locomotion, bears legs and usually wings. The *abdomen*, associated largely with elimination and reproduction, lacks developed legs but bears modified appendages called *cerci* (sing., *cercus*) and genital appendages.

Chapter 2 / EXTERNAL BODY STRUCTURE

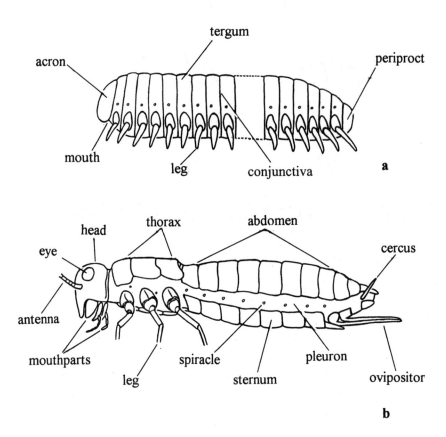

Fig. 2-1. Body plans of (*a*) hypothetical arthropod and (*b*) hypothetical wingless insect (Modified and redrawn from R. E. Snodgrass, 1935).

Head

The head is the most highly modified of the three body regions. It consists of 5 to 7 segments so fused and obscured that differences of opinion have arisen as to its exact composition. From anterior to posterior, it includes at least an acron (which is *not* a true segment) and antennal, intercalary, mandibular, maxillary, and labial segments to be discussed below.

Mouthpart Position

There are three basic mouthpart positions. Grasshoppers and their allies (*Fig. 2-3 a*), caterpillars, and other largely plant-feeding insects feature a vertical head and mouthparts that project downward, the so-called *hypognathous* adaptation. Most beetles (*Fig. 2-3 b*), earwigs, and certain other largely predacious insects have a horizontal head and mouthparts that project forward, the *prognathous* condition. Finally, cockroaches and leafhoppers, cicadas, and their allies (*Fig. 2-3 c*) feature an expanded

Head

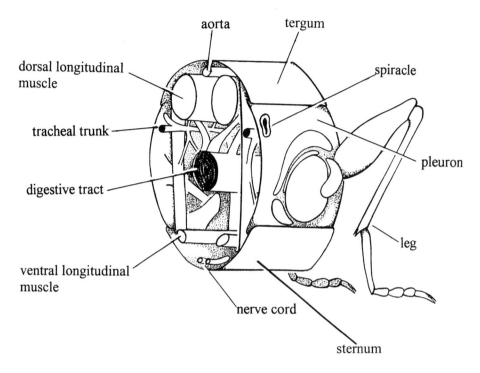

Fig. 2-2. An idealized insect thoracic segment (Modified and redrawn from R. M. & J. W. Fox, 1964).

facial region that deflects the mouthparts downward and backward to the point where they may lie between the fore legs, the *opisthognathous* condition. In absence of knowing with certainty which condition is primitive, hypognathy is arbitrarily chosen as standard for the topographic descriptions that follow, based on which the labrum is regarded as anterior, the labium posterior, with the laterally disposed mandibles and maxillae lying between them (*Fig. 2-4 b*).

Organization of Generalized Head

The head of a mature insect is a sclerotized capsule from which almost all semblance of segmentation is lost. It bears so-called "sutures" that are seldom indicators of intersegmental line. The usual kind of head suture is a *sulcus*, or groove, incidental to an internal strengthening ridge. However, the "epicranial suture" of authors lacks an internal groove or ridge, being merely a dorsal line of weakness along which the cuticle separates during molting. Lacking meaning as a morphological boundary, it is properly called the *ecdysial cleavage* line (*Fig. 2-4 a*).

The head usually bears laterally or dorsolaterally a pair of *compound eyes* and, between them, paired *ocelli* (discussed below). A median ocellus is often located beneath and between the latter, and paired *antennae* are located beneath or between the compound eyes. The head opens below at the mouthpart articulations and behind at the *occipital foramen*, an aperture by which the neck attaches

Chapter 2 / EXTERNAL BODY STRUCTURE

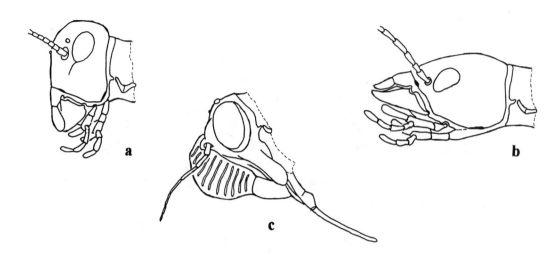

Fig. 2-3. Lateral views of (*a*) hypognathous, (*b*) prognathous, and (*c*) opisthognathous insect heads based on mouthpart position (Modified and redrawn from R. E. Snodgrass, 1935).

to the thorax and through which pass the ventral nerve cord, esophagus, aorta, salivary ducts, and a pair of tracheal trunks (*Fig. 2-5 a*).

An elaborate supportive endoskeleton called the *tentorium* is located within the head of all true insects (*Fig. 2-5 b*). It consists of four *apodemes*, or sclerotized cuticular invaginations, which usually fuse internally to make a strong brace for the head and its musculature. Though internal, the tentorium is visible externally at the paired anterior and posterior tentorial pits (*Fig. 2-4 a, b*). The former are minute indentations near the anterior articulation of the mandibles, and the latter are similar indentations near the ventral extremities of the postoccipital suture that separates occiput and postocciput. The tentorium acts as a brace, particularly between the mandibular articulations.

The head capsule, as noted, bears grooves, or sulci, known loosely as "sutures," among them the postoccipital, occipital, and frontoclypeal sutures (*Fig. 2-4 a, b*). These may well be lines of articulation along which adjoining segments have united. If so, the frontoclypeal and occipital grooves are not functional sutures but fused "obsolete sutures." However, the postoccipital suture immediately behind the occipital suture represents the true intersegmental line that separates the maxillary and labial segments. The occipital suture, where fully developed, extends around the head capsule and terminates at the posterior articulation of the mandibles (*Fig. 2-5 b*). The frontoclypeal suture that ends at the anterior tentorial pits possibly represents the posterior boundary of the acron. All other "head sutures," including the frontogenal, clypeolabral, and subgenal sutures, are secondary structures, though relatively constant throughout the Insecta.

Areas of head capsule set off by these sulci are given distinctive names. Several such intersutural areas, all unpaired, compose the anterior head surface. The *frons* is the facial area between the frontoclypeal suture ventrally, the two arms of the ecdysial cleavage line dorsolaterally, and the antennal bases laterally (*Fig. 2-4 a*). The frons bears the median ocellus when that simple eye is present. It is

Head

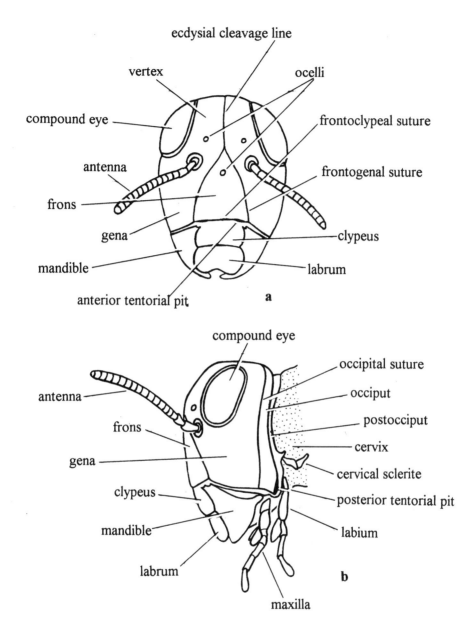

Fig. 2-4. (*a*) Frontal and (*b*) lateral views of a generalized insect head (Modified and redrawn from R. E. Snodgrass, 1935).

readily separated from the more ventral *clypeus* by the frontoclypeal suture or, in its absence, by an imaginary line between the anterior tentorial pits. A lip-like, movable flap called the *labrum* articulates with the ventral margin of the clypeus at the clypeolabral suture.

The dorsal surface of the head is the *epicranium*, and the anterior extension of epicranium between the compound eyes is the *vertex*. The vertex bears the paired ocelli and antennae but is not

Chapter 2 / EXTERNAL BODY STRUCTURE

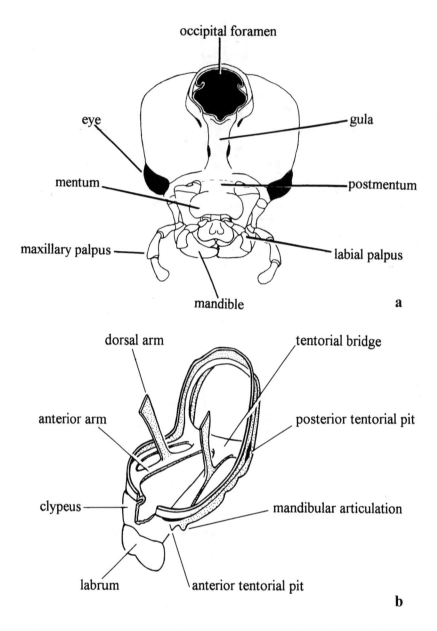

Fig. 2-5. (*a*) Ventral view of the head of a blister beetle showing a well-developed gula; (*b*) a cut-away view of an insect head showing the tentorium (Modified and redrawn from R. E. Snodgrass, 1935).

differentiated as a separate sclerite. The posterior portion of epicranium delineated by the occipital suture is the *occiput* (*Fig. 2-4 b*). It is not always a distinct sclerite but may be defined as a general area in insects with an obsolete occipital suture. The occiput itself is sometimes divided by a postoccipital suture, in which case a *postocciput* continuous with the neck may be recognized. The postocciput bears the occipital *condyles*, or points of articulation by which the head articulates with the cervical sclerites mentioned later.

The lateral surface of the head capsule below the compound eyes is composed principally of paired *genae* (*Fig. 2-4 b*). Their separation from the frons is indefinite in absence of frontogenal sutures, in which case they may be defined as general areas. Behind the genae are the *postgenae*. The latter are ventral extensions of the occipital arch set off from the genae by the occipital suture. When absent, as that suture often is, the postgenae may be defined arbitrarily as the caudolateral faces of the head capsule.

The ventral aspect of the head is completed by the mouthparts, except in instances of prognathy. Mouthparts are discussed later, but modifications that allow for prognathy may be mentioned here. The necessary forward projection of the head is achieved by lengthening the postgenal regions and interpolating below a new sclerite called the *gula* (*Fig. 2-5 a*). It is an unpaired ventral plate between the labial postmentum and the occipital foramen. Gular sutures form the gula's lateral boundaries.

Eyes

Insect eyes are of three types: *ocelli*, stemmata, and compound eyes. Ocelli are simple eyes visible as circular or subcircular convexities of clear cuticle undivided into facets. They occur in both nymphal and adult insects and are often three in number disposed in a triangle (*Fig. 2-6 c*), but the median ocellus of the three is sometimes lacking (*Fig. 2-6 b*). *Stemmata* are the simple eyes of adult springtails, silverfish, fleas, and holometabolous larvae (*Fig. 2-6 a*). They consist of 1 to 7 separate convexities on each side of the head in a position roughly corresponding to that of the missing compound eyes. Paired *compound eyes* are characteristic of most pterygote insects and of the jumping bristletails of the apterygote family Machilidae but are lacking in most holometabolous larvae. They usually appear as multifaceted convexities on the sides of the head (*Fig. 2-6 b, c*). They tend to be circular in outline but are sometimes subcircular or even irregular in shape. Each surface of a compound eye is generally divided into hundreds or thousands of hexagonal facets, each the external corneal layer of a single visual unit.

Antennae

Antennae, the modified, movable appendages of the second segment of the head, are often called "feelers," but they are much more than that because they bear smell, taste, and touch sensilla and often serve other specialized functions as well. They are of two types: *segmented antennae*, in which each *article*, or subsegment, is independently movable by intrinsic muscles arising from the preceding article (*Fig. 2-7 a*), and *annulated antennae*, in which only the *scape*, or basal subsegment, has intrinsic muscles, so each annulated antenna operates as a unit by contraction of extrinsic muscles from within the head (*Fig. 2-7 b*).

Protura (which are not true insects) lack antennae. All other hexapods, both adults and juveniles, have antennae, though they may be strongly reduced. Segmented antennae occur in the primitive apterygote orders Diplura, Collembola, and Microcoryphia, and annulated antennae are found in the Thysanura and throughout the Pterygota.

Chapter 2 / EXTERNAL BODY STRUCTURE

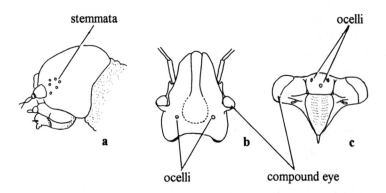

Fig. 2-6. (*a*) Lateral view of a caterpillar head, (*b*) dorsal view of a bug head, and (*c*) frontal view of a cicada head (Modified and redrawn from various sources).

Antennae arise from sockets situated between or beneath the compound eyes. These openings are often strengthened by a sclerite with a ventral condyle by which the appendage as a whole articulates with the head. An antenna includes the following articles: scape, which is a basal, often wide and/or elongate article that operates by extrinsic muscles from the tentorium; pedicel, a second article that, in annulated antennae, lacks intrinsic musculature; and flagellum, a third, apical article that is usually subdivided into a number of annuli subject to structural variation useful in taxonomic separation (Fig. 2-8). The antennae are moved by levator and depressor muscles arising from the tentorium and inserting into the scape and by extensors and flexors arising from within the scape and inserting into the pedicel (*Fig. 2-7 b*). The flagellum lacks muscles but has a sensory nerve. The musculature at the antennal base in Collembola, Diplura, and Microcoryphia is the same as in pterygotes, but each annulus of the flagellum has intrinsic musculature.

Biting Mouthparts

Insect mouthparts are adapted to the kind of food habitually eaten. The two most obvious mouthpart types are biting-chewing, or *mandibulate*, and sucking, or *haustellate*. Mouthparts in the non-insect hexapods (Collembola, Diplura, Protura) lie concealed within a pocket produced by downgrowth of the genae. This *entognathous* condition does not occur in true insects, in all of which the exposed mouthparts are said to be *ectognathous*.

Biting-chewing mouthparts are a generalized type consisting of two pairs of opposable jaws guarded in front and behind (using hypognathy as the descriptive standard) by the lip-like labrum and labium, respectively. Bristletails, dragonflies and damselflies, grasshoppers and their allies (*Fig. 2-4 b*), termites, earwigs, beetles, adult lacewings and their allies, most bird lice, and insects of certain other orders have biting mouthparts. Mouthparts of this type also occur in a reduced or vestigial condition in, among others, adult mayflies and in the larvae of many insects whose adult mouthparts are suctorial.

The labrum of biting-chewing insects is a flattened, unpaired lip suspended from the lower margin of the clypeus (*Figs. 2-4 a, 2-10 c*). It covers the anterior face of the mandibles and the preoral

Head

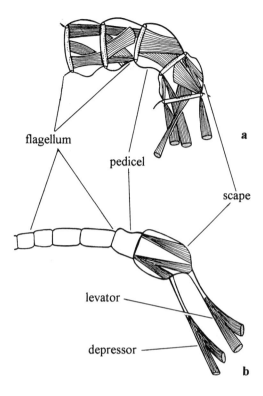

Fig. 2-7. Basal articles of (*a*) segmented-type apterygote antenna, as opposed to those of (*b*) annulated-type pterygote antenna (Modified and redrawn from A. D. Imms, 1939).

cavity and has a movable articulation with the clypeus, from which it is usually separated by a clypeolabral suture. The internal labral surface is continuous with the membranous wall of the clypeus and is provided with an *epipharynx*, a membranous organ with small tactile hairs and taste sensilla.

The paired mandibles located immediately behind the labrum constitute the first pair of jaws. They presumably represent remnants of the ambulatory appendages of the fourth primitive segment of the arthropod head. They take two distinctly different forms among biting insects. One type, characteristic of most apterygotes and of mayflies, is elongate, unicondylic (having a single articulation), and deeply excavated along the inner margin to accommodate insertion of a well-developed ventral muscle group (*Fig. 2-9 a*). The second type, found in Lepismatidae and in pterygotes as a whole, is relatively short and dicondylic (having two articulations) (*Fig. 2-9 b*). Dicondylic mandibles have an efficient hinge action, as opposed to the more primitive unicondylic mandibles, with their pivotal action.

A pterygote *mandible* is usually a sturdy, heavily sclerotized, unsegmented structure with a broad, triangular base. It has two points of articulation along its external basal margin (*Fig. 2-4 b*). One is a socket into which fits a process from the clypeus. The other is a rounded process that fits into a corresponding socket on the subgena. Each individual mandible is moved along its hinge line by contractions of a pair of antagonistic dorsal muscles, one a relatively weak abductor muscle that pulls the

Chapter 2 / EXTERNAL BODY STRUCTURE

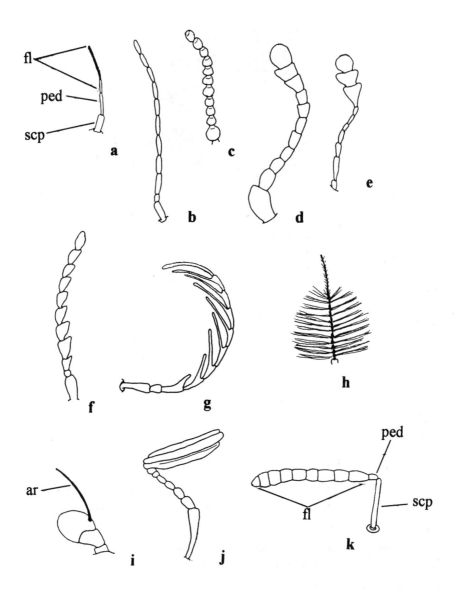

Fig. 2-8. Common variations in antennal form including (*a*) bristle-like, (*b*) thread-like, (*c*) bead-like, (*d*) club-like, (*e*) head-like, (*f*) saw-like, (*g*) comb-like, (*h*) plumose, (*i*) aristate, (*j*) leaf-like, and (*k*) elbowed types, with *ar* = arista, *fl* = flagellum, *ped* = pedicel, *scp* = scape (Redrawn from various sources).

jaw away from the opposing jaw, and the other a more massive adductor that brings the jaw forcibly against the opposing jaw (*Fig. 2-9 b*). Each mandible usually bears an apical incisor area for cutting food and a basal molar area for masticating it. The nature and extensiveness of these surfaces differ according to food habit, being flat, roughened, toothed, ridged, *etc.*, in different insects (*Graph 14-1*). The molar area may disappear altogether in carnivores. There is no feeler-like *palpus* in mandibles, in which respect they differ from the maxillae and labium.

Head

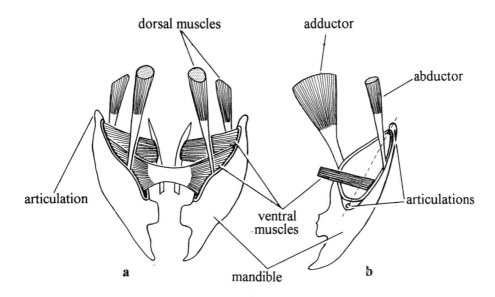

Fig. 2-9. (*a*) The apterygotan unicondylic *vs.* (*b*) the pterygote dicondylic mandible types dissected free to show their musculature (Modified and redrawn from R. E. Snodgrass, 1935).

The *maxillae*, or second pair of jaws, presumably represent the appendages of the fifth primitive head segment. They are located immediately behind the mandibles (*Fig. 2-4 b*) and have a flexible attachment to the head capsule at the inner margin of their bases. They function as accessory jaws, holding food for the mandibles to incise and masticate.

A maxilla commonly consists of the following parts (*Fig. 2-10 a*): cardo (pl., *cardines*) which is a proximal, sometimes laterally oriented hinge sclerite that attaches to the head capsule by a single condyle; *stipes* (pl., *stipites*), an elongate sclerite directed downward from the cardo and serving as basis of the distal *endite lobes* and maxillary palpi. The endite lobes include a mesal, often toothed and setose *lacinia* and a lateral lobose or flattened *galea*. The *maxillary palpus* is an antenna-like sensory appendage of 1 to 7 articles.

Immediately behind the paired maxillae is the lip-like *labium* that articulates either with the neck membrane or with the gula mentioned earlier (*Figs. 2-4 b, 2-5 a*). The labium appears single but actually consists of a pair of fused, maxilla-like organs, being serially homologous with the maxillae (*Fig. 2-10 b*). Its basal segment, or *postmentum*, is homologous with a pair of fused maxillary cardines; its prementum with a pair of stipites; its apical *glossae* and *paraglossae* with the laciniae and galeae, respectively; and its *labial palpi* (consisting of 1 to 4 articles) with the maxillary palpi. The postmentum is sometimes secondarily divided into a proximal *submentum* and a distal *mentum*.

Primitively, there is a median, unpaired tongue-like organ called the *hypopharynx* that projects forward from the internal surface of the labium at the back of the preoral cavity. It occupies a position within the preoral cavity roughly opposite that of the labral epipharynx. The mouth opens where the epipharynx becomes continuous with the inner surface of the clypeus, and the salivary duct aperture opens where the hypopharynx attaches to the labium.

Chapter 2 / EXTERNAL BODY STRUCTURE

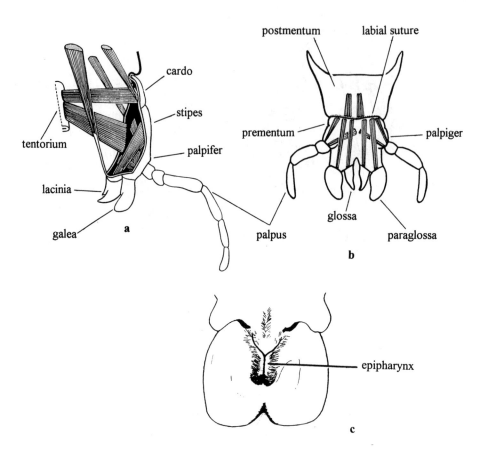

Fig. 2-10. (*a*) A generalized left maxilla and (*b*) a generalized labium, each showing its internal musculature; (*c*) an internal view of a grasshopper labrum showing the setose epipharynx (*a* and *b* are modified and redrawn from R. E. Snodgrass, 1935).

Suctorial Mouthparts

Suctorial mouthparts are homologous with, and have been derived from, biting-chewing mouthparts. Therefore, the same names are applied to them, though individual parts bearing the same name are structurally dissimilar. Sucking lice, aphids, leafhoppers, bugs, and their allies, butterflies and moths, true flies, fleas, and some lesser orders include insects with suctorial mouthparts, either piercing or non-piercing in function. The mouthparts of mosquitoes are representative of the piercing type.

Mouthparts of female mosquitoes consist of six needle-like, piercing *stylets* within an elongate labium (*Fig. 2-11 a, b*). The stylets include an anterior labrum-epipharynx, two mandibles, two maxillae, and a posterior hypopharynx. The labium is a sheath tipped by labella. Maxillary palpi are well-developed. The food channel is formed between the appressed labrum-epipharynx and hypopharynx, and the latter also contains the salivary channel. The preceding description does not apply to male mosquitoes whose mandibles and maxillae are reduced, rendering them into non-piercers able merely to dip the beak into liquids and imbibe.

Head

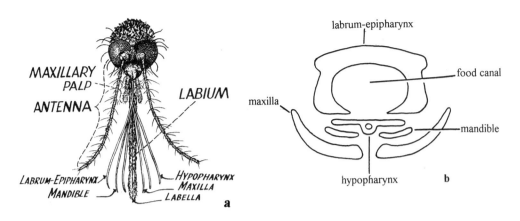

Fig. 2-11. (*a*) Head of a female mosquito showing the proboscis with exserted stylets and (*b*) cross-section of the mosquito stylets with the enclosing labium removed (Modified and redrawn from W. B. Herms, 1950).

The mouthparts of bugs are likewise piercing-sucking in adaptation but the stylets are reduced in number from the six of mosquitoes. Here, the beak consists of four piercing stylets within a flexible, segmented labium that attaches to the head capsule (*Figs. 2-12 a, 2-13 a*). There are no palpi. The stylets are arranged in overlapping fashion, with the two centrally located maxillae held together by a system of tongues and grooves, the two mandibles next, and the overlapping labium at the periphery. The maxillae fit together along the midline to form both the food and the salivary canals. The short labrum and the hypopharynx are at the base of the proboscis. Other parts characteristic of biting mouthparts are either lost or reduced.

Mouthparts of the stable fly and of certain other Diptera are likewise piercing in adaptation, but they are reduced to two stylets (*Fig. 2-12 c*). The proboscis, derived from the labium, consists of a short rostrum and a long, pointed, trough-like haustellum tipped by the sclerotized, toothed labella derived from the labial palpi. The two stylets include the anterior labrum-epipharynx and the posterior hypopharynx, both unpaired and enclosed within the groove of the haustellum. Mandibles are lacking, as are maxillae, except for the paired maxillary palpi. The food channel is formed between the opposing labrum-epipharynx and the hypopharynx, and the salivary channel courses within the latter.

The mouthparts of house flies and certain other advanced Diptera contain exactly the same elements as do those of the related stable flies but are adapted for sponging rather than piercing. They illustrate the non-piercing type. Their proboscis, or labium, consists of a conical rostrum, with paired maxillary palpi, and a cylindrical haustellum ending in soft, spongy labella (*Figs. 2-12 d, 2-13 c*). The haustellum is thrust downward and forward into food but, at non-feeding times, is kept folded against the rostrum. The two stylets (the anterior labrum-epipharynx and the posterior hypopharynx) lie within the labial groove and between them form the food channel. The salivary channel courses within the hypopharynx. Food is conveyed by capillarity into the food channel through numerous labellar grooves.

Of the many other types of suctorial mouthparts, I shall mention only one additional one, the siphon-type mouthparts of butterflies and moths (*Figs. 2-12 b, 2-13 b*). Their mouthparts consist of little more

Chapter 2 / EXTERNAL BODY STRUCTURE

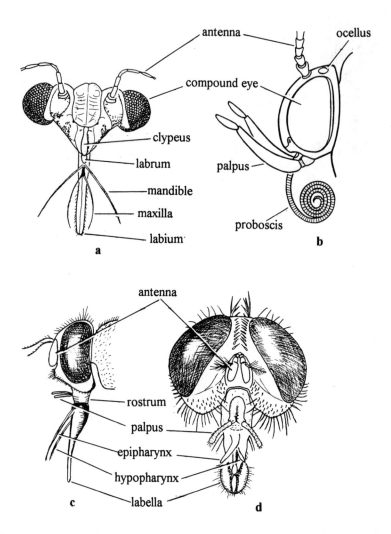

Fig. 2-12. (*a*) Frontal view of bug head, with four exserted stylets; (*b*) lateral view of moth head, with coiled proboscis derived from two appressed maxillary galeae; (*c*) lateral view of stable fly head, with two exserted stylets; and (*d*) frontal view of house fly head, with two exserted stylets (*b* is modified and redrawn from R. E. Snodgrass, 1935; *a*, *c*, and *d* are modified and redrawn from W. B. Herms, 1950).

than an elongate, coiled proboscis capable of great projection. This structure is formed of two appressed, interlocked maxillary galeae which, between them, make the food channel. The remaining parts of the maxillae are reduced or lost. The labium is a mere plate bearing prominent labial palpi, and the labrum is a similar transverse structure beneath the face. Mandibles are lacking or vestigial.

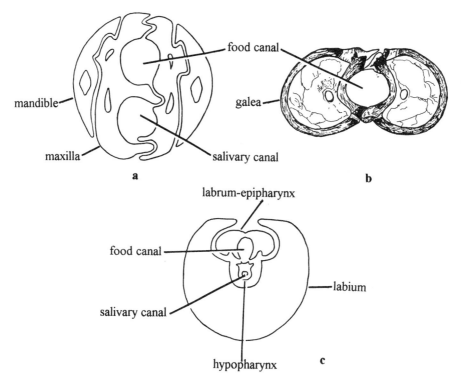

Fig. 2-13. Cross-sectional views of suctorial mouthparts of (*a*) bug (the ensheathing labium not shown), (*b*) moth, and (*c*) house fly (Redrawn from various sources).

Neck

The membranous neck is a composite area resulting from fusion of the dorsal labial and ventral prothoracic segments. It is usually strengthened by lateral *cervical sclerites*, one set on each side of the head (*Fig. 2-4 b*). They articulate, in front, with the postocciput's occipital condyles and, behind, with an excavation on the fore margin of the prothorax. Paired dorsal and ventral cervical sclerites occasionally supplement the lateral ones. These sclerites strengthen the head attachment and facilitate movement, especially lifting and depressing.

Thorax

The thorax is the body region between head and abdomen. It consists of an anterior *prothorax*, a middle *mesothothorax*, and a posterior *metathorax* which represent body segments seven, eight, and nine, respectively, according to classical theory. These segments differ from those of the head in being

Chapter 2 / EXTERNAL BODY STRUCTURE

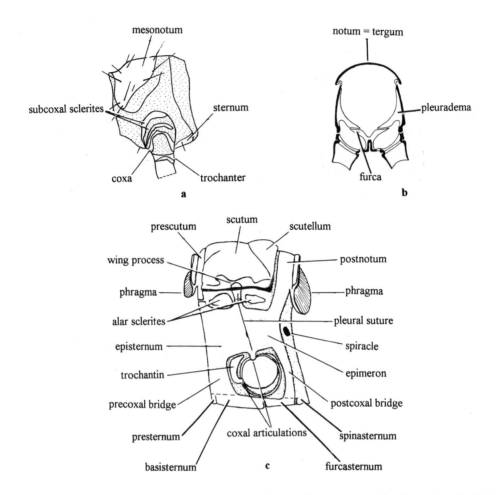

Fig. 2-14. (*a*) Lateral view of the largely membranous pleuron of an apterygote showing subcoxal sclerites and leg base, (*b*) cross-section of a generalized pterygote mesothoracic segment showing the endoskeletal pleuradermata and furca, and (*c*) lateral view of a generalized pterygote mesothoracic segment (*a* is modified and redrawn from A. Berlese, 1910; *b* and *c* are modified and redrawn from R. E. Snodgrass, 1935).

comparatively well-developed, uniform, and separated by distinct *conjunctivae*, and they differ from the abdominal segments in that each of them consists of all three sclerites, tergum (or notum), sternum, and pleura. The thorax is clearly more generalized than are either of the other two body regions.

The thorax usually bears legs for locomotion and wings for flight, and it contains the muscles and nerve centers involved in those activities. The thoracic segments are comparatively uniform in the primitively wingless apterygotes, but the wing-bearing second and often the third thoracic segments of pterygotes are sometimes grossly altered by addition of strengthening sclerites and the musculature necessary for flight. This alteration occurs even in those pterygotes that are secondarily wingless in response to adaptation for a parasitic or other specialized habit.

Thorax

Apterygote Thorax

Apterygotes are primitively wingless non-insect hexapods with a thorax that lacks the sclerotization and musculature of a flying insect. Apterygotes' similar thoracic segments are provided with largely membranous pleura (*Fig. 2-14 a*) whose leg insertion is strengthened by *subcoxal sclerites.*

Pterygote Thorax

The prothorax of pterygotes is often more generalized than are either the meso- or the metathorax, and it lacks functional wings in all insects, both fossil and extant. Its dorsal sclerite, the *tergum* (= notum), and its ventral sclerite, the *sternum*, are joined on each side by a well-developed *pleuron* presumably derived through expansion and coalescence of the subcoxal sclerites mentioned above.

The needs of flight place severe strain on the pterygote thorax. Its exoskeleton must be strengthened to accommodate the increased musculature needed to power the wings and avoid buckling upon their operation. Accordingly, the wing-bearing segments are more complex and more strongly sclerotized than are those of apterygotes, especially in the pleural region, and are provided with an internal arch-like endoskeletal brace consisting of the *pleurademata* and *furca*, formed by invaginations from the paired pleura and the sternum of their respective segment, either the meso- or the metathorax (*Fig. 2-14 b*).

Pterygotes' wing-bearing mesothorax is greatly altered. The *alinotum* (*notum* is the proper term for a tergum that bears wings) is enlarged and may or may not be similar to that of the metathorax. It articulates with the wing base on either side. It may be divided by transverse sutures into the following dorsal sclerites (from front to back): *prescutum, scutum, scutellum,* and *postnotum* (*Fig. 2-14 c*). The postnotum fuses laterally with the paired pleura to make a strong support between the alinotum and sternum.

The original divisions of the pterygote sternum have become obscured by evolutionary fusion, loss, and modification. Only recently have morphologists resolved its homologies. They now feel that it includes (from front to back): *presternum, basisternum, furcasternum* (sternellum), and *spinasternum*. These parts fuse into a plate-like ventral sclerite that connects with the pleura by *precoxal* and *postcoxal bridges* forming, on either side, the leg sockets. The presternum is usually obscured by fusion with the basisternum. The basisternal-furcasternal separation is visible in insects with a sternacostal suture extending transversely between the leg bases. The narrow spinasternum overlaps the true intersegmental line (*Fig. 5-6 b*).

The paired pleura apparently developed by expansion, fusion, and strengthening of the subcoxal sclerites mentioned earlier with respect to the apterygote thorax. Each wing-bearing pleuron has a *wing process* above (serving as the fulcrum for wing movement) and a pair of leg condyles, or *coxal articulations*, below (*Fig. 2-14 c*). Between the wing process and the dorsal coxal process courses the pleural suture. It is not a true suture but merely the external indication of a strong apodeme or cuticular ingrowth. It divides each pleuron into an anterior *episternum* and a posterior *epimeron*. Below each wing base and forward of the episternum is a spiracle, and nearby are the *alar sclerites* involved in wing articulation.

Chapter 2 / EXTERNAL BODY STRUCTURE

The preceding describes the thorax of generalized insects but is not uniformly applicable throughout the Pterygota. Stoneflies retain the generalized plan. In certain insects, the prothorax is reduced to a small annulus, but, in cockroaches and some other orthopteroids, it is developed into a strong dorsal shield. Insects of most other orders favor one of the two wing-bearing segments, with consequent reduction in the other. This trend involves displacement of some sclerites, fusion and reduction in others, and appearance of new "sutures" dividing existing ones. This alteration reaches the point where, in some insects, the meso- and metathorax fuse into a single functional unit called the *pterothorax*. Development of the latter is correlated with habits, as in strong-flying, specialized insects whose thoracic structure bears little resemblance to the generalized thorax. Accordingly, terms used to describe the pterothorax are complicated, and names adopted for use in the different orders may be dissimilar, causing confusion.

Thoracic Endoskeleton

The apodemes that constitute the internal thoracic skeleton include the phragmata, pleurademata, furcae, and spinae. The paired *phragmata* (sing., *phragma*) serve as attachments for the powerful longitudinal wing muscles (*Fig. 5-6 a*). They arise dorsally from the meso- and metathorax and the first abdominal segment, each by invagination along the antecostal suture and/or postnotum (*Fig. 2-14 c*). The *pleurademata* (sing., *pleuradema*) invaginate from the paired pleural sutures and join with the furca of their respective segment to form an almost continuous internal arch for support and muscle attachment (*Fig. 2-14 b*). The furcae arise from the sternacostal suture of their respective segment. The *spinae* are digitiform processes that invaginate from the spinasternum.

Legs

Arthropods stem from a precursor whose body was divided into many segments, each bearing paired appendages. In the head, these appendages became modified into antennae and mouthparts or were lost; in the thorax, they were retained as legs; and, in the abdomen, they were either lost or became cerci or genital structures.

There are six thoracic legs in insects, hence the alternative class name Hexapoda formerly applied to insects as a group. These legs are arranged in pairs associated with the pro-, meso-, and metathorax and are termed, respectively, *prothoracic* or fore-, *mesothoracic* or mid-, and *metathoracic* or hind legs. They are present in almost all stages of all taxonomic groups, except for the wormlike larvae of true flies and fleas, females of many scale insects, and certain other insects with a specialized mode of existence. The fore legs of Protura are held forward and upward, antenna-like, as if to replace the missing antennae, leaving only two pairs of functional legs (*Fig. 19-1 a*).

An insect leg is composed of the following articles: coxa, trochanter, femur, tibia, tarsus, and pretarsus with claws (*Fig. 2-15 a*).

The *coxa* is usually a short, stout article. It is the functional but not primitive basal attachment of the leg to the thoracic body wall. It articulates by flexible membrane, usually reinforced by two coxal condyles. These articulations restrict its movement largely to the horizontal plane.

Thorax

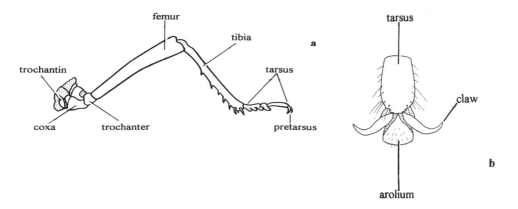

Fig. 2-15. (*a*) Mid leg of a grasshopper; (*b*) pretarsus and apical part of a tarsus of a cockroach (Modified and redrawn from R. E. Snodgrass, 1935).

The *trochanter* is a short article between the coxa and the femur. It is usually reduced and is almost always single. Its femoral attachment tends to be fixed and rigid but that with the coxa is a hinge joint permitting vertical movement. (The trochanter should not be confused with the trochantin, which is a ventral coxal process).

The *femur* is the section of leg between the trochanter and the tibia. It is generally the leg's most prominent, robust article. It may be armed with stout but immovable spines. Its bulk tends to correlate positively with development of the tibial muscles that it encloses, so it is massive in leaping insects such as grasshoppers.

The *tibia* attaches to the femur proximally and to the tarsus distally. Its femoral articulation permits vertical motion, so it is capable of being flexed against, and closely applied to, the femur (*Fig. 3-6 a*). It is generally long and slender and may equal or exceed the femur in length. It often bears one or more apical spurs and, along its length, many spines. *Spurs* have a movable articulation; *spines* are fixed.

The slender *tarsus*, or foot, attaches distally to the tibia. It is commonly divided into five or fewer articles, or *tarsomeres*, on the underside of which are often borne pad-like *plantulae*, but it is moved as a whole by muscles arising from the distal end of the tibia.

The *pretarsus* is often interpreted as the ultimate part of the tarsus but is actually a small, distinct article of its own. It usually bears a median outgrowth or lobe called the *arolium* and a pair of movable *claws*, the development of which varies in different taxonomic groups (*Fig. 2-15 b*). A pair of pads called *pulvilli* are located between the claws in certain flies, and a pad-like or spinous *empodium* replaces the arolium in those insects (*Fig. 15-1 d, e*).

Wings

Insect wings are flat, bag-like, paired evaginations from the dorsolateral body wall that originate from caudolateral folds of the tergal margin. Consequently, they are composed of double thicknesses of the

Chapter 2 / EXTERNAL BODY STRUCTURE

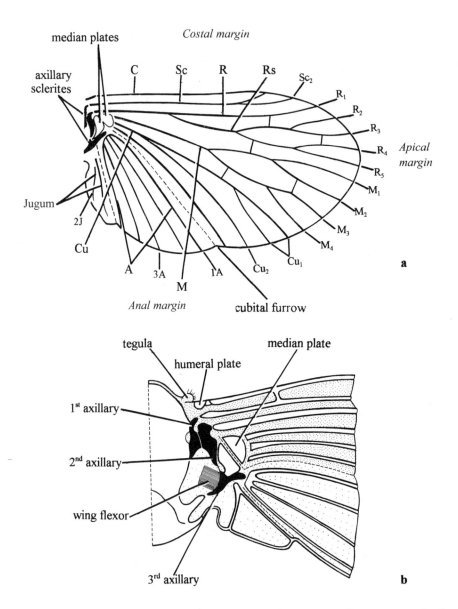

Fig. 2-16. (*a*) Wing venation according to the Comstock-Needham system; (*b*) a pterygote wing base showing the associated sclerites (Modified and redrawn from R. E. Snodgrass, 1935, and other sources).

same three layers as the body wall itself (cuticle, epidermis, and basement membrane) in mirror-image duplicates, with the basement membranes together. Each wing articulates above with the tergum that supports it and hinges below on the *wing process* on its side. Associated with this membranous articulation may be three or four *axillary sclerites* used in wing folding (hence, developed only in Neoptera), two *median plates*, an anterior *humeral plate*, and often a lobular *tegula* (*Fig. 2-16 a, b*). Two *alar sclerites* are located below the wing on either side of the wing process (*Fig. 2-14 c*).

Thorax

The wing margins are designated *axillary* where the wing articulates, *humeral* at the anterobasal angle, *costal* along the leading edge, *apical* at the tip, *anal* along the trailing edge, and occasionally *jugal* at the caudobasal angle of the wing. The central portion of the wing is the *disc*, along the length of which course a number of *veins*, and crosswise between the veins may extend *cross veins* (*Fig. 2-16 a*). The whole effect is of a thin, membranous fan supported, during extension, by a truss-like system of veins and cross veins. The fan folds lengthwise at given furrows and plaits during flexion and may also undergo transverse folding in its distal parts (*Fig. 1-4 c*). Such folding is automatic during repose owing to the nature of the structural pattern and vein elasticity.

Wing venation differs among the major groups of insects and is useful in separating many genera and species. Its specificity was apparent to the early entomologists who, on this basis, classified the insects of interest to them. A confusion of terms arose. Then, in this century, upon appearance of the Comstock-Needham Scheme based on wing ontogeny, there became available a universal system of venational nomenclature more consistent with homology. This system remains in general use today. However, recent comparative morphologic/paleontologic studies of primitive venation have led to somewhat modified interpretations from those that follow.

The principal *veins* extend longitudinally from wing base to apex in the following sequence (from the costal to the anal margins, respectively): costa, subcosta, radius, media, cubitus, anal, and jugal. Their respective abbreviations (always starting with a capital letter) are: C, Sc, R, M, Cu, A, and J. Branches, if any, are indicated by numbers that increase in magnitude from anterior to posterior.

The unbranched *costa* is marginal. *Subcosta*, associated with the first axillary sclerite, tends to be short and to divide once into branches Sc_1 and Sc_2. Radius, which is often the strongest vein, is associated with the second axillary sclerite. It divides once into R_1 and radial sector or R_s, the latter of which, in turn, may divide into branches R_2 through R_5. The archetypical media apparently had two main branches, of which only the posterior one persists today. It is designated M and its subsidiaries as M_1 through M_4. *Cubitus* may divide into two main branches, of which the first may split into two branches. The *anal veins*, associated with the third axillary sclerite, are termed 1A, 2A, 3A, *etc.*, owing to their frequent basal separation. Any *jugal veins* present within the jugum are designated as 1J and 2J.

One or more major *furrows* along which the wing always folds, fan-like, course along the surface of the generalized wing and serve as stable morphological landmarks. Among them are the cubital and jugal furrows separating the cubital and anal veins and the anal and jugal veins, respectively.

Cross veins complete the wing's circulatory pathways. These secondary veins which are missing in some insect groups, generally take their name (listed with a lower case initial letter) from the combination of veins that they connect. When two or more of them connect the same two veins, they are given the same name with an increasingly higher number from wing base outward. However, important cross veins may be given special names, as with the proximal cross vein between costa and subcosta, called the *humeral cross vein* or h; that between R_1 and the first branch of radial sector, called the radial cross vein or r; that between the forks of radial sector, called the sectorial cross vein or s; and that between M_2 and M_3, termed the medial cross vein or m.

Chapter 2 / EXTERNAL BODY STRUCTURE

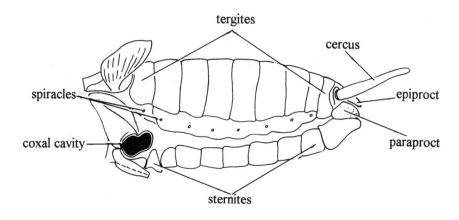

Fig. 2-17. Lateral view of a cricket metathorax and abdomen (Modified and redrawn from R. E. Snodgrass, 1935).

Most fossil insects had two pairs of wings, a large number of veins and cross veins, and a reticulate, or net, pattern of venation. These primitive insects were unable to fold their wings over the abdomen in repose because of their lack of the axillary sclerites and other structures necessary for wing folding. Insects of this type are relegated to infraclass Paleoptera of subclass Pterygota. They include most orders of past geological times as well as the living orders Ephemeroptera and Odonata. Most modern insects are placed within infraclass Neoptera of Pterygota. When at rest, they rotate their fore wings backward to conceal the hind wings and fold the hind wings at specific articulations (*Fig. 1-4 c*). Neoptera, like Paleoptera, are ancient, though presumably they evolved more recently.

Abdomen

The third and final body region, the *abdomen*, is notable for its structural simplicity. Internally it contains most of the viscera including the heart portion of the dorsal blood vessel, and caudally it bears the reproductive aperture and anus. The primitive number of abdominal segments is 12, though lesser numbers, often 10 or 11, are common in adult insects. These segments are annular in form, more or less similar, and distinctly separated one another. They lack well-developed sclerotized pleura and walking legs but usually bear appendages related to reproduction.

Each abdominal segment is an annulus consisting of a tergum above, a sternum below, and laterally usually a pair of spiracles (*Fig. 2-17*). The adult *tergum* is generally a simple plate whose anterior border is, in reality, the posterior border of the preceding segment. In many larvae, however, dorsal sclerotization involves development of a medial tergite and one or more paired laterotergites from whose fusion the adult tergum forms. The adult *sternum*, though superficially similar to the tergum, is a composite structure formed by fusion of the primary sternum with the paired, primitive leg bases. A true pleural region is lacking in the abdomen, except in certain larvae.

Abdomen

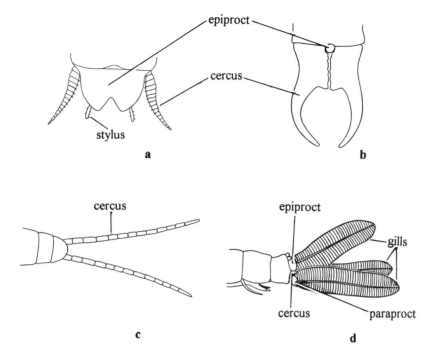

Fig. 2-18. Variations in the cerci of selected insects including (*a*) cockroach, (*b*) earwig, (*c*) stonefly naiad, and (*d*) damselfly naiad (*a*, *b*, and *c* are dorsal views, *d* a lateral view) (Redrawn from R. E. Snodgrass and other sources).

That the abdominal sternum is composite is attested to by embryogenetic and comparative morphologic evidence, especially the attachment of vestigial abdominal appendages, or *styli*, in certain apterygotes. A coxosternal derivative presumably became the definitive sternum as the styli disappeared during the course of evolution.

Embryos of certain generalized insects and adults of Protura (telsontails) have 12 abdominal segments including a distinct periproct bearing the anus. The *periproct*, which serves as the body's posterior closure, must be non-segmental, for it lacks appendages, segmental ganglia, and coelomic cavities. However, it is clearly represented in the above arthropods. It is also present, in modified form, in the dorsal *epiproct* and the ventral, paired *paraprocts* of adults of many generalized insects (*Fig. 2-17*). The eleventh abdominal segment is often reduced, and sometimes the tenth also, so the twelfth segment may come to lie behind the tenth or even the ninth or eighth, which is the source of much confusion.

The abdomen is most strongly altered caudally. Any loss of segments generally occurs here where the *anus* discharges to the outside. The anus opens from an exposed periproct in embryos and in Protura but tends to be covered in generalized insects. This covering consists of the dorsal *epiproct* and of paired, lateral *paraprocts*. The epiproct is derived from the twelfth tergite and the paraprocts from the twelfth sternite. The appendages of the eleventh segment persist as paired feeler-like *cerci*.

Chapter 2 / EXTERNAL BODY STRUCTURE

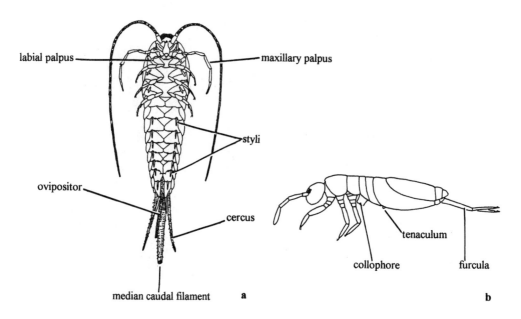

Fig. 2-19. (*a*) Ventral view of a bristletail showing its antennae, elongate maxillary palpi, short labial palpi, three pairs of legs, many paired abdominal styli, cerci, and median caudal filament; (*b*) lateral view of a springtail (*a* is modified and redrawn from Comstock, 1948; *b* is modified and redrawn from Belkin, 1972).

The medial *genital aperture* usually opens from the eighth or ninth abdominal segment in female insects and from the ninth in males. It lies in close association with the organs of oviposition and copulation.

Cerci

Cerci (sing., *cercus*) are the paired, modified appendages of the primitive eleventh abdominal segment, but they appear to arise from the tenth or ninth segments in insects in which, respectively, the eleventh or the eleventh and tenth segments are reduced or lost. They are present in most insects but vary widely in form (*Fig. 2-18*).

Styli

Multiple pairs of vestigial abdominal appendages, or *styli* (sing., *stylus*), are present in the apterygotes. Each of the first three abdominal segments in telsontails (Protura) bears a pair of styli in the conjunctival membrane between the tergal and sternal plates. There are varied numbers of styli in bristletails (Microcoryphia), in which they appear to be vestiges of the shaft of a typical arthropod walking limb (*Fig. 2-19 a*). There are three appendages on the abdominal underside in springtails (Collembola): collophore, tenaculum, furcula (*Fig. 2-19 b*). They are medial and unpaired in adults but suggest a paired origin. The collophore is a vesicle arising from beneath the first abdominal segment. The tenaculum, or catch, which

Abdomen

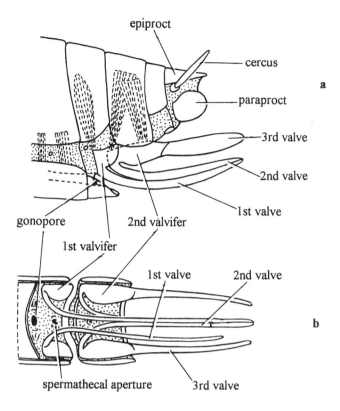

Fig. 2-20. (*a*) Lateral and (*b*) ventral views of the terminal abdominal segments and ovipositor of a generalized female pterygote (Modified and redrawn from R. E. Snodgrass, 1935).

arises from beneath the third segment bears apically divergent prongs that grasp and hold the flexed furcula. The furcula, or spring, is a leaping organ arising from the fourth or perhaps the fifth abdominal segment. It is brought into repose by its paired flexor muscles and held there by the tenaculum until released, whereupon its extensor muscles contract, violently forcing it against the ground, propelling the animal upward and forward.

Reproductive Appendages

Insects' external genitalia are derived from the eighth and ninth abdominal segments in females and from the ninth in males. They constitute the copulatory organ of both sexes and, in females, are also egg-laying organs.

The ovipositor of generalized pterygotes consists of a pair of *valvifers* from the eighth and ninth segments each and of three pairs of *valves* (*Fig. 2-20*). All three pairs of valves interlock to form an ovipositor in some Orthoptera, but in others the second pair is reduced to an egg guide between the first and third pairs. The third is the one that is lost in insects having only two pairs of valves. Sawflies and certain other Hymenoptera use the first and second pairs of valves for cutting and sawing and the

Chapter 2 / EXTERNAL BODY STRUCTURE

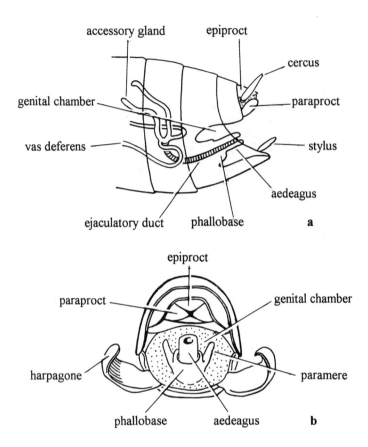

Fig. 2-21. (*a*) Lateral and (*b*) caudal views of the terminal abdominal segments and external genitalia of a generalized male pterygote (Modified and redrawn from R. E. Snodgrass, 1935).

third as a sheath to cover the ovipositor during retraction. A few insects either dispense with the ovipositor altogether or form a replacement device by modification of the posterior abdominal segments. Springtails, which lack an ovipositor, merely discharge eggs directly through the gonopore. House flies and certain other Diptera take an alternative approach. They use their elongated, protractile abdomen as a functional ovipositor.

The external genitalia of male pterygotes are exceedingly varied which makes them a useful tool for taxonomic separation but complicates direct comparison among the orders. Past specialists tended to develop their own terminology for the different insect groups, sometimes without much basis in homology, so the names used often conflict.

The external genitalia of male insects include two principle organ groups, *viz.*, phallic and periphallic. *Phallic organs* are those directly concerned with coition. They include *phallomeres*, or lobular outgrowths from the ninth sternum, the intromittent *phallus*, and various *accessory structures* surrounding the genital opening. *Periphallic organs*, in contrast, are movable or immovable processes involved in clasping.

Abdomen

The male external genitalia of generalized pterygote insects are situated within a *genital chamber* located below the anus (*Fig. 2-21 a, b*). This complex is shielded above by the medial *epiproct*, on the sides by paired *paraprocts*, and below by an unpaired *subgenital plate*. The phallus itself is a composite structure differentiated into a proximal *phallobase* and a distal *aedeagus*, or inflexion, bearing the genital aperture, and associated with them *accessory structures* (*parameres, etc.*).

These phallic organs are either medial or closely adjacent to the midline, as opposed to the laterally located periphallic organs. Among them are the paired harpagones, derived from the ninth abdominal appendages, which may take the form of movable lobes, hooks, or claspers.

The preceding description of the male genitalia of generalized insects cannot be used to describe the genitalia of specialized insects, for the latter of which one must consult detailed descriptions by Snodgrass, Tuxen, and others.

Suggested Additional Reading

Borror, Triplehorn, & Johnson (1989); Chapman (1991, 1998); Davies (1988); Gullan & Cranston (2000); Lawrence, Nielsen, & Mackerras (1991); Matsuda (1965, 1970, 1976); Richards & Davies, vol. 1 (1977); Scudder (1961, 1971); Snodgrass (1935, 1952); Steinmann & Zambori (1985); Tuxen (1970).

Chapter 3
Skeletointegumentary System

My skeleton is so protected
It can hardly be detected
But there is no slightest doubt
Most creatures wear it inside out.
— J. W. Fox

Integument

The insect body wall is a combined skin, skeleton, and food reserve composed of (from inside to outside) basement membrane, epidermis, and cuticle. Its invaginations, or sac-like inpocketings, form apodemes for support and internal muscle attachment and the linings of body orifices and apertures including the mouth, anus, tracheae, salivary ducts, and genital passages. Its evaginations, or outpocketings, form legs, wings, gills, and other appendages as well as setae and other micro- and macrostructures. The body wall thus constitutes a distinct *skeletointegumentary system*, a classification that satisfies the realities of structure and function but violates the traditional relegation of skeleton and skin to separate organ systems.

The body wall supports the insect, maintains its shape, protects it from physical forces such as pressure and evaporation and from attack by disease-bearing organisms; it bears the sensory receptors that apprize the animal of environmental stimuli; and it provides the levers, fulcra, and muscle attachments that make body movement possible. However, it necessitates periodic shedding to accommodate growth and increase in size.

The insect integument consists of three principal layers: a basement membrane facing the body cavity, a cuticle facing the environment, and, between them, a single layer of cells called the epidermis.

Epidermis

The *epidermis* is a cellular, living layer consisting of cuboidal or columnar cells aggregated into a one-cell thick epithelium (*Fig. 3-1 a*) with endoplasmic and Golgi complexes. It secretes and remains in contact with the cuticle and, along with the plasmatocytes, may secrete the *basement membrane*. The latter is a simple, flat sheet beneath the epidermis, being a layer without apparent internal structures or pores but with elastic fibers, collagen, and glycoproteins.

The epidermis is not a simple epithelium. It includes specialized *trichogen cells* from which take root articulated bristles, *tormogen cells* forming a socket for the bristles, *gland cells* that open to the outside, and several kinds of *specialized cells* (oenocytes, sensory cells, vacuoles, and hemocytes) that may adhere to the basement membrane. *Oenocytes* synthesize the hydrocarbons that contribute to the epicuticle, so they show cycles of development correlated with the molt cycle.

Chapter 3 / SKELETOINTEGUMENTARY SYSTEM

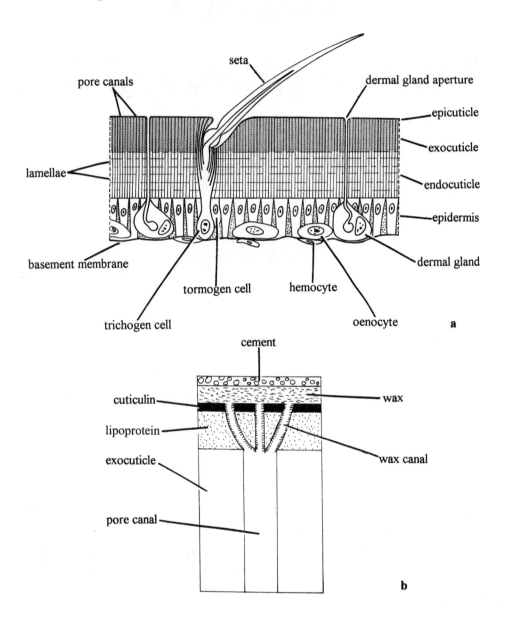

Fig. 3-1. (*a*) Idealized sectional view of the insect integument based on light microscopy; (*b*) highly enlarged, diagrammatic view of part of the exocuticle (clear area) and the epicuticle (lipoprotein + cuticulin + wax + cement layers) (*a* is modified and redrawn from V. B. Wigglesworth, 1950; *b* is after various sources).

Cuticle

Cuticle resembles plywood in that it is a laminate. It is secreted in three separate layers (from inside to outside): a flexible endocuticle, a rigid exocuticle, and a thin, usually impermeable epicuticle facing the environment.

Skeletointegumentary System: Integument

The *endocuticle* is a soft, colorless cuticular layer of chitin, protein, and resilin. The nitrogenous polysaccharide *chitin* takes the form of microfibrils arranged in horizontal lamellae within a protein matrix accounting for perhaps half the total weight of dry cuticle, and it is commonly arranged in a sheet in which the microfibrils are in parallel. *Resilin* is a colorless, insoluble, rubber-like, three-dimensional protein network resistant to deformation. It occurs where the cuticle is subject to spring-like action as, for example, at the wing base.

Numerous vertical *pore canals* (*Fig. 3-1 b*) perforate the endocuticle. These are minute, often helically coiled, duct-like spaces that provide a channel for wax secretions discussed later. They arise from the epidermis and extend to, but do not penetrate, the epicuticle. *Dermal gland ducts* also penetrate the endocuticle, but they extend to the surface. They are responsible for cement secretion atop the wax and undergo cycles of development synchronized with the molt cycle.

The *exocuticle*, the layer immediately peripheral to the endocuticle, varies in development from a thin, indistinct layer that merges with the epicuticle, as in soft-bodied insect larvae, to a thick armor that makes up much of the cuticle, as in some adult beetles, but it is intermediate in development in most insects. Like the endocuticle, it consists of chitin and protein deposited in horizontal sheet-like *lamellae* and perforated by vertical *pore canals* and *dermal gland ducts*. Each successive lamella lies in the same horizontal plane but at a slight angle relative to the previous one, such that, when viewed from above, a thickness of many sheets takes on a helicoid arrangement appearing as alternating light and dark regions.

Exocuticle is essentially altered, stabilized outer endocuticle. Its protein components are bound together and converted into rigid, horny, insoluble *sclerotin* by a tanning process under control of the hormone *bursicon*. This process involves quinones derived ultimately from the phenolic alpha amino acid tyrosine, a precursor of various hormones and of melanin. It was long supposed incorrectly that the insect exoskeleton consists largely of chitin; that it alone is responsible for the skin's rigidity; and that sclerites are simply strongly chitinized areas (the rigid exocuticle actually contains less chitin than does the flexible endocuticle).

The combined endocuticle-exocuticle, or *procuticle* (a name for newly secreted cuticle before differentiation into endo- and exocuticle), ranges from perhaps 10–200 μm (microns) in thickness but is thinner where exocuticle is reduced or absent. Such places range from the distensible abdomen of queen ants (*Fig. 3-2 b & c*), queen termites, or blood-sucking bugs, to more limited areas such as insects' *conjunctivae*, or intersegmental membranes, which are lines of flexibility that either lack exocuticle altogether or have it in the form of wedge-shaped sclerites imbedded within the folded, flexible endocuticle.

The *epicuticle* is the thin, refractile, outermost layer of hardened cuticle that lacks chitin. Though multilayered, it is less than 4 μm in thickness and is thrown into deep folds where the integument stretches or bends. It is composed of (from exocuticle outward) tanned lipoprotein, cuticulin (polyphenols), wax, and cement zones (*Fig. 3-1 b*). *Lipoprotein* arises from epidermal secretions; the *cuticulin envelope* is laid down at the plasma membrane surface; the *wax layer* (important in waterproofing) is either synthesized by the oenocytes or, in some insects, by the fat body; and the protective cement layer by the *dermal glands*.

Chapter 3 / SKELETOINTEGUMENTARY SYSTEM

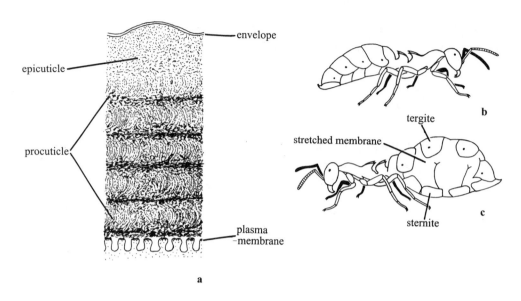

Fig. 3-2. (*a*) Sectional view of the insect cuticle based on recent electron microscopy; (*b*) a queen army ant with abdomen in the non-distended condition, as opposed to (*c*) a queen bloated with eggs. Note in *c* the separation of the abdominal tergites and sternites along the elastic intersegmental conjunctivae (*a* is modified and redrawn from M. Locke, 2001; *b* and *c* are modified and redrawn from H. R. Topoff, 1977).

The preceding interpretation of insect cuticle is based on the light microscope observations of Wigglesworth, Richards, and those who followed them over subsequent decades. However, recent studies by Locke and others using electron microscopy provide information allowing for possible simplification of the conventional terminology. Electron microscopy indicates that the cuticle consists of three basic layers, *viz.*, envelope, epicuticle, and procuticle (*Fig. 3-2 a*). The *envelope* is a cuticulin membranous structure that forms at the epidermal cells' plasma membrane surface and becomes a base for lipid layers and a template for epicuticular patterns laid down against it. The *epicuticle* arises from and supports the inner face of the envelope. The *procuticle* then assembles at the cell surface before enlarging hundreds of times beyond the narrow envelope and epicuticle that confine it. *Wax* is transported through the procuticle to the cuticulin envelope, and the *polyphenols* that tan the epicuticle and procuticle into exocuticle are secreted through it. *Procuticle* consists mainly of laid down stacks of laminae within a protein matrix.

Molting

Cuticle is capable of the limited stretching allowed by the conjunctivae and other places lacking exocuticle (*Fig. 3-2 c*), but it is tanned and virtually inextensible over the remainder of the sclerotized body. Consequently, most insects find themselves locked within a rigid suit of armor that restricts major expansion in body size. They accommodate significant size increase only when the cuticle is soft and extensible, as during the hormonally mediated changes that begin with *apolysis*, or separation of the old

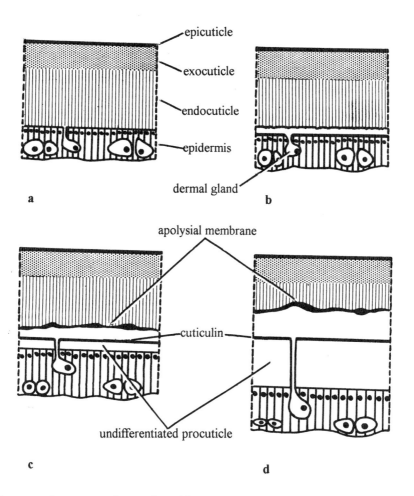

Fig. 3-3. Diagrammatic summary of events in molting and cuticle formation, with (*a*) prior to the onset of molting, (*b*) production of new epicuticle (cuticulin) atop the epidermis and withdrawal (apolysis) of the old endocuticle from the epidermis, (*c*) continued digestion of the old endocuticle, and (*d*) proliferation of new undifferentiated procuticle as digestion of the old endocuticle continues (Modified and redrawn from R. F. Chapman, 1982).

cuticle from the underlying epidermal cells. A new cuticle is secreted under the old cuticle, and then the old one is shed, resulting in creation of a new, larger skin.

Ecdysis is the term applied to the physical shedding of the old cuticle, as opposed to *molting*, the more comprehensive term involving overall cyclical growth and cuticle secretion. The two processes are often confused.

As molting begins, the epidermal cells divide mitotically and change in form (*Fig. 3-3 a*). At about the same time, possibly owing to the increase in cell number, apolysis begins under influence of the molting hormone *ecdysone*. Apolysis involves separation of the epidermis from the overlying old endocuticle. Simultaneous with apolysis occurs deposition of new epicuticle in the form of cuticulin within the withdrawal space (*Fig. 3-3 b*). Within this space, the cells secrete a gel that becomes

Chapter 3 / SKELETOINTEGUMENTARY SYSTEM

activated into *molting fluid*. This fluid, containing proteases and a chitinase, begins digesting away the old proteinaceous-chitinous endocuticle above (except for a thin layer that persists as the apolysial membrane) but is ineffective against the new cuticulin that develops below (*Fig. 3-3 c*). Then, as the old endocuticle liquifies and disappears (most of it being resorbed), a new procuticle is deposited beneath the protective new epicuticle (*Fig. 3-3 d*). Finally, the undigested old exocuticle and epicuticle left intact at the periphery are discarded as a cast skin, or *exuviae* (pl., *exuviae*). The new epicuticle can be bent or stretched only where it is folded, so it is secreted in deep pleats that accommodate whatever limited expansion the insect may experience during the following *stadium* (the time period between successive molts). No later expansion or folding is possible, except during subsequent molts, so the size attained by the animal during a given stadium is limited by the dimensions of epicuticle laid down. This explains the pleats and other surface patterns that unfold during a stadium as an insect moves about, feeds, *etc.*

Ecdysis

The new skin is now complete. All the insect need do is rupture and shed the remnants of the old skin, or exuviae, and harden the new one. Rupture takes place along the *ecdysial cleavage line*, a predetermined line of weakness located mid-dorsally along the head and thorax (*Fig. 2-4 a*). There being no exocuticle along this line, the insect's endocuticle extends directly to the thin, fragile epicuticle. The animal simply contracts its abdominal musculature forcing blood into the head and thorax when the new skin is ready and the old endocuticle has been dissolved. It may also swallow quantities of water or take in air. The head and thorax bulge increasingly under this pressure until the old skin gives way along the cleavage line. Once parts of the head and thorax are exposed (*Fig. 3-4 a*), the animal undergoes continued peristaltic movement, squirms actively, draws itself out, usually head and thorax first, followed by the abdomen and appendages until the entire body slips free of the exuviae.

Sclerotization and Melanization

Ecdysis is now complete, but molting is not. The soft, pliable skin of the newly freed insect must now be inflated, balloon-like, until the new exoskeleton hardens. The animal may swallow air or water as it maintains its abdominal musculature in continuous peristaltic contraction, raising blood pressure and keeping the skin expanded until termination of hardening. Should the body wall be perforated at this time, the resulting loss of pressure will prevent the insect from inflating itself, and its legs and wings will harden in a permanently shriveled condition. Once the expansion is complete, the undifferentiated procuticle beneath the new cuticulin layer undergoes sclerotization to form the new exocuticle.

A newly ecdysed insect is a soft-bodied, pale creature said to be *teneral*. It is subjected to immediate hardening and darkening. Hardening, or *sclerotization*, was long thought to be a simple consequence of exposure to air but is now known to involve tanning under control of the hormone *bursicon* which initiates the process. It involves the quinone conversion of cuticular protein into sclerotin, which is not only hard but amber-colored. However, hardening does not explain all cuticular coloration. Most color results from other oxidative reactions, chiefly those concerned with the production of melanin. These reactions,

Skeletointegumentary System: Integument

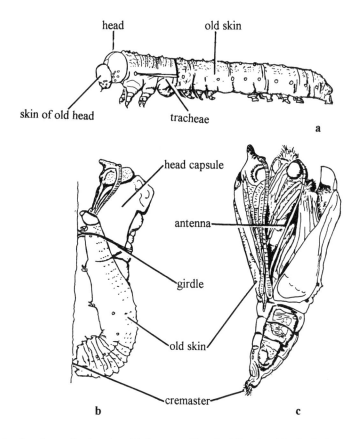

Fig. 3-4. Molting in a butterfly showing (*a*) the caterpillar shedding its skin (including that of the head capsule, detached in front) and the cuticular linings of its tracheal trunks, (*b*) the pupa emerging from the last larval skin, and (*c*) the adult emerging from the pupal skin (Modified and redrawn from E. O. Essig, 1942).

termed *melanization*, resemble hardening in that they too stem ultimately from the breakdown of tyrosine, producing blackish, brownish, or yellowish pigments. In short, new cuticle is subjected to two separate processes, sclerotization and melanization, that go forward simultaneously. Hardening occasionally proceeds in absence of darkening to produce rigid, fully sclerotized, colorless cuticle.

Molting Regulation

Factors causing insects to molt at a particular time in their development remain obscure. It is likely, however, that molting is triggered when the insect reaches a given size. This readiness is conveyed to the brain via stretch receptors causing release of *prothoracicotropic hormone*, or PTTH, to the corpora cardiaca and, at least in some Lepidoptera, to the corpora allata. These neurohemal organs then release PTTH into the hemolymph for circulation to the *prothoracic glands*. These target organs secrete *ecdysone* into the hemolymph. Many tissues, including epidermal cells, convert ecdysone into the *molting hormone*, 20-hydroxyecdysone. The latter initiates DNA synthesis and mitotic division of the

Chapter 3 / **SKELETOINTEGUMENTARY SYSTEM**

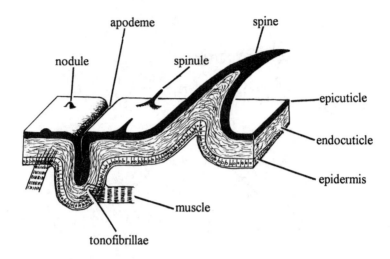

Fig. 3-5. Idealized section of integument showing an apodeme, a spine, two tonofibrillae, two nodules, and two spinules (Modified and redrawn from R. M. & J. W. Fox, 1964).

epidermal cells followed by apolysis and cuticle production. Finally, *eclosion hormone* from the central nervous system stimulates cuticle extensibility, causing ecdysis and expansion, and triggers release of the hormone *bursicon* that initiates tanning.

Wax and Cement Secretion

Wax or its precursors are secreted by the epidermal cells shortly before ecdysis and pass through the endocuticle and developing exocuticle via the pore canals and then through the epicuticle via the slender wax canals (*Fig. 3-1 b*). Eventually it coats the epicuticular surface rendering it waterproof. Finally, dermal glands secrete cement over the newly waxed surface (*Fig. 3-1 a*). *Cement* is a shellac-like protective substance that may coat the epicuticle in a near-continuous sheet or in discontinuous patches. It is not secreted, of course, in insects that lack dermal glands. Wax is also secreted during much of the intermolt period following molting.

Integumentary Processes and Outgrowths

The insect integument is seldom smooth and bare. It usually has a characteristic surface sculpturing, a variety of pits and ridges, and innumerable setae, bristles, spines, knobs, and other processes, large and small. Setae and bristles occur in patterns that tend to be similar in related taxonomic groups, so their study is sometimes useful in systematics. This technique, called *chaetotaxy*, is applicable to insects in general but is most commonly used in taxonomic separation of adult flies and fleas, larval Lepidoptera, and other "bristly" insects.

Integumentary processes are either cuticular or cellular (*Fig. 3-5*). *Cuticular processes* include nodules, ridges, spinules, and similar outgrowths of solid cuticle. Though secreted by epidermal cells, they lack an internal epidermal core. *Cellular* processes, in contrast, always contain an epidermal core. They are either unicellular or multicellular but are alike in that they never consist of solid cuticle.

Setae are among the unicellular processes that invest the body. A seta (*Fig. 3-1 a*) is a hair-like articulated process stemming from a trichogen cell enclosed within a tormogen cell socket. There are several kinds. There are, for example, clothing setae (a hairy subtype well-developed in caddisflies, especially on the wings). They are simple and unbranched. However, bees' setae are branched and adorn the body so thickly that it looks fuzzy. Some setae are not hair-like at all but stout, rigid bristles. This type is characteristic of Tachinidae and certain other "bristly" flies. Other setae take the form of flattened scales, as on the body and/or wings of butterflies and moths, mosquitoes, and certain other insects. Still other setae are classified, not by form, but by function, as with sensory setae and glandular setae. Sensory setae, which are common on the antennae, mouthparts, cerci, *etc.*, connect with a sensory neuron by which mechanical, chemical, or other stimuli may be transmitted to the central nervous system (*Fig. 4-6*). Glandular setae are hollow bristles associated with poison glands that may endow their possessor with a degree of protection.

Among insects' multicellular processes are projections that attach to the body wall by a membranous articulation permitting movement and those that are rigidly fixed to the body wall. Spurs and claws, like setae, are flexible (*Fig. 2-15 a*). *Spurs* are movable projections found at the apex of the tibia. *Claws* belong in the same morphological category as spurs but are given a different name because they are associated with the pretarsus. *Spines* are non-articulated, rigidly fixed multicellular processes. They include the stout, barb-like processes of legs which resemble spurs, except for their lack of a flexible articulation.

Pigmentary Coloration

Pigmentary colors may be classified according to their deposition site. The yellow, brown, or black pigments commonly deposited within exocuticle are *cuticular*; the red, orange, yellow, green, or white granules that control color are *epidermal*; and the blood, fat body, or gut pigments that, from deep within the body, may be visible through the semitransparent integument are *subintegumentary*. Cuticular colors are relatively permanent, as opposed to the epidermal and subintegumentary kinds. Consequently, many gaudily colored insects transform into drab, darkish specimens soon after death.

Pigmentary colors also may be classified according to the chemical composition of the responsible pigments. Among them are melanins, tetrapyroles, carotenoids, pteridines, ommochromes, and xanthoquinones whose consideration is beyond our present concern.

Physical Coloration

Coloration in certain beetles, butterflies, moths, bugs, and a few other insects does not depend on pigmentation. Theirs is purely physical, being subject to alteration or elimination by pressure, distortion, swelling, shriveling, or immersion in a medium of the same refractive index. The color produced depends on the surface elements' structural conformation. Most of these physical colors are so-called *interference colors* produced by multiple thin films or lamellae separated by a substance with a different refractive index. Interference produces the scale iridescence of some butterflies which changes with inclination of the light source. It is also responsible for the color of many beetles, flies, and other insects with a metallic luster.

Combination Coloration

Combination coloration, as seen in the hue of many beetles, butterflies, and other elaborately colored insects, is more common in insects than is purely physical coloration. Combination colors arise from pigments within the skin that act together with physical color. For example, red integumentary pigments may combine with a violet physical color to produce magenta, and yellow pigments with a physical blue may produce a green.

Skeleton

To this point, I have treated the skeletointegumentary system as if it were entirely skin, which it is not. It is a combined skeleton/skin. The two functions are inseparable, though it is convenient to discuss them separately, as I do here. I now address the purely skeletal function.

Articulation

To review, the insect body consists of a linear series of sclerotized, annular segments composed of a number of plates, or *sclerites*, each made rigid by a layer of exocuticle. Successive segments operate as a unit and interconnect lengthwise by flexible *conjunctivae*, or intersegmental membrane. The sclerites serve for protection and internal muscle attachment, and the conjunctivae connecting them permit movement of the otherwise inflexible body wall. The legs, antennae, cerci, and other appendages that attach to the body likewise consist of segment-like units, in this case lesser ones termed *articles*, or subsegments, and they, in turn, are joined by flexible membrane, in this case *arthrodial membrane*. Aside from the conjunctivae and arthrodial membranes, exocuticle is also lacking along the ecdysial cleavage line where the cuticle splits during ecdysis.

Sclerotization hardens the cuticle but does not necessarily guarantee rigidity. Most rigidity is obtained by integumentary inflections visible superficially as *sulci*, or grooves. In the head, for example, inflections of this type have fused into the *tentorium* used for rigidity and surface for muscle attachment (*Fig. 2-5 b*). Likewise, the thoracic furcae and pleurademata have fused into a strong bridge (*Fig. 2-14 b*) that prevents thoracic distortion upon contraction of the powerful flight muscles.

Several types of articulation may be distinguished. *Universal articulation* is illustrated, for example, by the abdomen's attachment to the thorax. The abdomen's lack of definite hinge points allows for relatively unrestricted movement. To a degree, it can be lifted, depressed, flexed to either side, protracted (thrust outward), retracted, or even slightly rotated. Some appendages are articulated in the same way, being telescopic and movable in various directions. However, the movements of most appendages are restricted by *condyles*, or specialized points of articulation, between successive articles. Condyles serve as a pivot, restricting movement to particular planes. There is sometimes but

Skeletointegumentary System: Skeleton

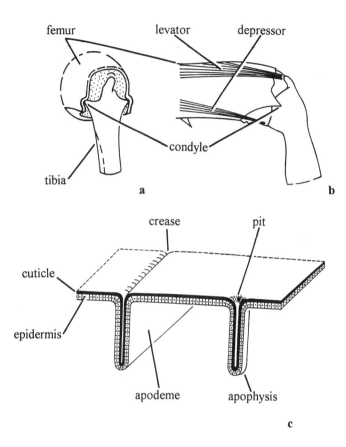

Fig. 3-6. (*a*) A three/quarter view of the dicondylic femorotibial joint, (*b*) frontal dissection of same showing the associated musculature, and (c) comparison of a plate-like apodeme (left) with an arm-like apophysis (right) (a and b are modified and redrawn from R. E. Snodgrass, 1935; c is modified and redrawn from Borror et al., 1989).

one condyle. For example, antennae and the mandibles of apterygotes pivot on a single condyle (*Fig. 2-9 a*). Such *unicondylic articulation* permits limited movement, yet assures better support than does the membranous, universal connection. However, most jointed appendages are *dicondylic*, the two condyles of which assure an efficient hinge-like motion restricted to a single plane. The attachment of the mandibles to the head capsule of biting-chewing pterygotes is an example (*Fig. 2-9 b*). The two mandibles can only abduct (pull away from the main axis of the body and, therefore, from one another) or adduct (pull toward the axis and one another). The femorotibial joint is another such hinge joint (*Fig. 3-6 a, b*).

Muscle Attachments

Muscles attach to the internal body surface by: (1) apodemes or apophyses, (2) tendons, (3) tonofibrillae, or (4) without intervening structures.

Chapter 3 / SKELETOINTEGUMENTARY SYSTEM

Apodemes are integumentary invaginations that collectively form the endoskeleton. They may be slender, arm-like structures more properly called *apophyses*, whose location is indicated by a pit visible on the external cuticular surface (*Fig. 3-6 c, right*). This type functions solely in muscle attachment. They may also take the form of extensive internal ridges or plates (*Fig. 3-6 c, left*) whose position is indicated by a crease along the external body wall. The tentorium (*Fig. 2-5 b*), phragmata (*Fig. 2-14 c*), and pleurademata (*Fig. 2-14 b*) are examples of this kind of apodeme serving both as an internal brace and for muscle attachment. Whatever their form, apodemes are hollow. Therefore, their epidermis separates from the old cuticle during molting and forms a new cuticle internal to the old one. Then, during ecdysis, the old apodeme is shed together with the overall exuviae.

Those apodemes termed *tendons* are solid, thread-like cuticular extensions that arise from given epidermal cells and often flare out at the apex for muscle attachment (*Fig. 5-7*), as opposed to *tonofibrillae* which are epidermally derived, non-striated fibrils that penetrate into the cuticle (*Fig. 3-5*).

Suggested Additional Reading

Chapman (1991, 1998); Hepburn (1976, 1985), Locke (1974, 2001); Nation (2002); Patton (1963); A. G. Richards (1951, 1953); Richards & Davies, vol. 1 (1977); Snodgrass (1935); Wigglesworth (1972, 1984).

Chapter 4

Nervous System

*(An animal is) An ingenious machine to which nature has given senses
so that it may renew itself, and so that, up to a certain point, it may be safe from
everything tending to destroy or disable itself.*
— Jean Jacques Rousseau

Nerve Structure

Neurons are the basic structural and functional units of the insect nervous system. They are ectodermally derived, generally long, branched *nerve cells* enveloped within a cell membrane (*Fig. 4-1*). Each consists of one or two *cell processes* associated with a *cell body* containing the nucleus and abundant mitochondria, Golgi complexes, and rough endoplasmic reticulum. One process, the *axon*, is an elongate fiber responsible for most of the neuron's length. It may bear a collateral branch, and both it and its collateral divide apically into fine nerve twigs. The other process, the *dendrite*, is composed of nerve twigs that sometimes arise directly from the cell body. Golgi complexes and rough endoplasmic reticulum are lacking in axons and dendrites inasmuch as protein synthesis does not occur there, only in the cell body.

Insect neurons are invested within a lipoproteinaceous sheath called *neuroglia* but lack the fatty myelin sheath typical of most vertebrate neurons. Thus, insect neurons aggregated into nerve tissue resemble vertebrate gray matter rather than white matter.

The insect nervous system consists of three separate, interconnected divisions to be discussed in the order listed: (1) *peripheral*, (2) *central*, and (3) *sympathetic*.

Peripheral Nervous Division

The peripheral nervous division is composed of many paired free nerves that radiate between the centrally located brain and nerve cord ganglia and the peripheral body wall and associated musculature. Some peripheral nerves contain axons that transmit information from sensory receptors; others innervate effector muscles or glands; and still others contain both sensory and motor axons. All of them consist of tightly packed, parallel axons bound together by an enveloping, non-cellular, supportive collagen-like *neural lamella* and a more internal *perineurium* concerned with the passage of organic substances and salts into the internal nerve tissues.

Neurons may be sensory, motor, or interneuronal in function. *Sensory neurons* (*Fig. 4-2 a*) transmit environmental impulses from receptors to the central nervous division, and *motor neurons* (*Fig.*

Chapter 4 / NERVOUS SYSTEM

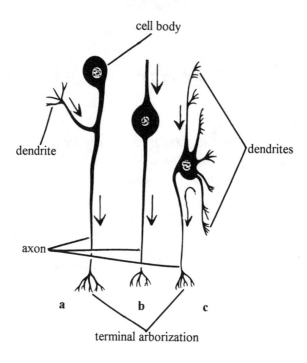

Fig. 4-1. Comparison of (*a*) monopolar, (*b*) bipolar, and (*c*) multipolar insect neurons (Modified and redrawn from R. F. Chapman, 1982).

4-2 c) receive impulses from the central nervous division via *interneurons* (*Fig. 4-2 b*) and pass them on to effector muscles or glands (*Fig. 4-2 d*).

Motor neurons (*Fig. 4-1 a*) are monopolar, having only one process. Their cell bodies are located within a central nervous division ganglion (*Fig. 4-2 c*) from which each gives rise to an axon and a collateral. The axon proceeds, always within a nerve, to an *effector* muscle or gland before breaking into its terminal arborization. The highly branched collateral dendrite ends within the ganglion where it receives connections either from interneurons or from sensory neurons.

One or more sensory neurons form a *sense receptor* consisting of a dendrite, cell body, and axon. These neurons are either bipolar (*Fig. 4-1 b*), having two processes, or multipolar (*Fig. 4-1 c*), having several processes. The cell body is always outside of, and separate from, the central nervous division, being located within or just beneath the epidermis or on a muscle and are classified accordingly. The axon of the first, bipolar type proceeds to a central nervous division ganglion before ramifying into its terminal arborization. The axon of the second, usually multipolar type, arises from receptors along the inner surface of the body wall or viscera and runs to a central nervous division ganglion.

Central Nervous Division

The central nervous division is composed of the "brain," or *supraesophageal ganglion*, and the *ventral nerve cord*. The brain (*Fig. 4-3*) is not a single ganglion but several fused pairs of ganglia located

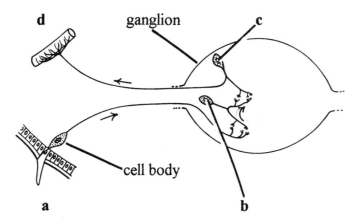

Fig. 4-2. The path of nerve conduction from (*a*) an external sense receptor and sensory neuron, to (*b*) an interneuron within a segmental ganglion (ovoid outline), and to (*c*) a motor neuron that exits to innervate (*d*) a muscle (Modified and redrawn from V. B. Wigglesworth, 1972).

above the pharynx or esophagus between the supportive arms of the tentorium. Its bulk consists mostly of neuropile (interneuronal arborizations). It receives paired *circumesophageal connectives* from the subesophageal ganglion located below the gut and, hence, has indirect nerve connections with all other parts of the body. It is an important sensory center in that it receives neurons from the simple and compound eyes, antennae, and preoral cavity. It gives off motor neurons to the antennae and labrum, so, in small part, it is also a motor center.

The bilobed fore brain, or *protocerebrum*, is the largest, most dorsal part of the brain (*Fig. 4-3*). It apparently stems from coalescence of the first pair of neuromeres during embryogenesis and innervates the simple and compound eyes. Its structures include a medial pars intercerebralis and paired protocerebral and optic lobes, as well as two smaller pairs of ocellar lobes. The ganglionic centers of the compound eyes are contained within the optic lobes; those of the paired ocelli are each contained within one of the paired ocellar lobes; and those of the median ocellus are contained within the other paired ocellar lobes, the two pedicels having united into a median ocellar trunk.

The mid brain, or *deutocerebrum*, innervates the antennae. It consists chiefly of a pair of antennal lobes located below and in front of the fore brain. Each gives rise to an antennal nerve consisting of sensory fibers associated with the antennal sense organs.

The hind brain, or *tritocerebrum*, consists of a pair of lobes attached to the underside of the fore and the back of the mid brain. Paired circumesophageal connectives arise from the floor of each tritocerebral lobe, encircle the gut below, and meet at the subesophageal ganglion in the lower head. Paired labral and frontal nerves also stem from the hind brain. The frontal nerves fuse at the frontal ganglion, and fibers lead thence into a recurrent nerve that connects with the sympathetic nervous division. The labral nerves innervate the labrum.

The ancestral nerve cord probably consisted of a pair of fused ganglia per segment. This primitive condition is lost in modern insects, in which a number of ganglia have undergone coalescence

Chapter 4 / NERVOUS SYSTEM

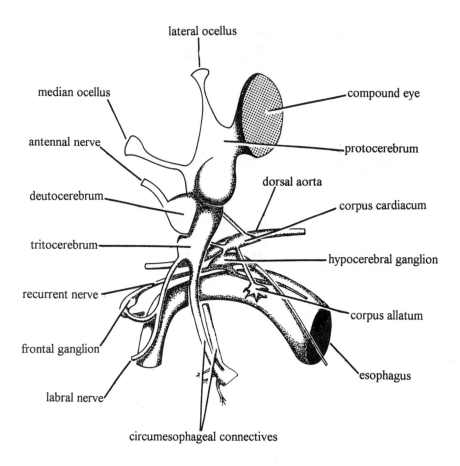

Fig. 4-3. Lateral view of a generalized insect brain and sympathetic nervous division (Modified and redrawn from F. O. Albrecht, 1953).

(*Fig. 4-4*). There are, at most, eight ganglia in the abdomen, three in the thorax, and one (the subesophageal) in the lower head, for a maximum of 12 ventral ganglia, as in certain apterygotes, mayflies, and some lepidopterous caterpillars. Certain of these ganglia are composite. For example, the eighth abdominal ganglion innervates the segments eight to 11, and the subesophageal ganglion represents three fused ganglia (mandibular, maxillary, and labial). The ten ganglia of termites are a common pattern among generalized insects, but the total may be further reduced in more advanced insects. For example, the three thoracic ganglia may fuse into a single one in certain flies; some beetles have fused their thoracic and abdominal ganglia; and certain higher Diptera and some bugs have reduced all the ventral ganglia (but not usually the subesophageal one) into a single ganglionic mass from which the entire body is innervated.

A *ventral nerve cord ganglion* is generally ovoid, as seen from above (*Fig. 4-5*). It receives paired connectives from, and gives them off to, the preceding and succeeding ganglia, respectively. It is provided with tracheae and is enclosed within an inner cellular perineurium and an outer fibrous

Nerve Structure

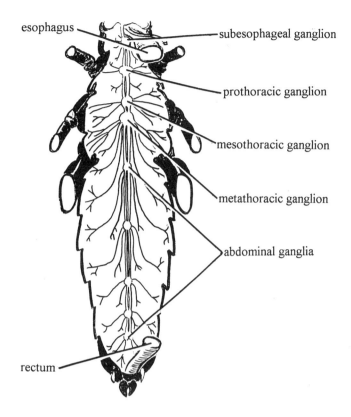

Fig. 4-4. Dorsal view of the grasshopper ventral nerve cord and associated peripheral nerves (Redrawn from various sources).

neural lamella. It gives rise to two or three pairs of *lateral nerves* belonging to the peripheral nervous division and sometimes to a short *median nerve* that arises between the connectives (*Fig. 6-2*). Each median nerve usually bifurcates into transverse nerves after issuing from its ganglion but may continue backward to enter the fore part of the next ganglion. Each consists of motor fibers associated with the spiracles and tracheae and so may constitute a possible ventral branch of the sympathetic nervous division.

Certain dragonflies, cockroaches, and other insects capable of rapid, evasive movements possess a number of *giant fibers* within the ventral nerve cord ganglia and connectives. These fibers consist of interneurons that run considerable distances without undergoing synapses. Their exceptional length and large diameter are suggestive of rapid conduction.

Sympathetic Nervous Division

The sympathetic nervous division is the center of fore gut involuntary peristaltic movement. It includes motor neurons, sensory neurons, and interneurons grouped into nerves and ganglia distributed over the fore and sometimes mid guts. These structures connect with the dorsal aorta, salivary

Chapter 4 / NERVOUS SYSTEM

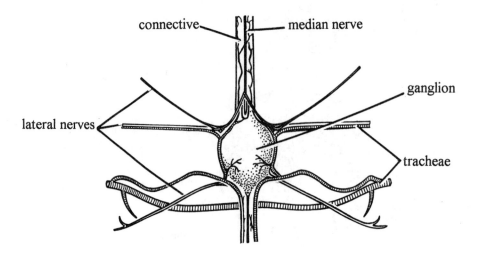

Fig. 4-5. A ventral nerve cord ganglion and its associated lateral nerves, paired longitudinal connectives, and tracheae (Modified and redrawn from R. E. Snodgrass, 1935).

ducts, and corpora allata. The overall division originates with the paired frontal connectives that emerge from the tritocerebrum and fuse into a frontal ganglion located below the brain on the roof of the esophagus (*Fig. 4-3*). This ganglion gives rise to a medial recurrent nerve that proceeds backward between the dorsal aorta and the esophagus. This nerve may terminate in a medial hypocerebral ganglion from which bifurcate paired nerves, each proceeding backward alongside the crop and ending in a stomachic ganglion. The hypocerebral ganglion, when developed, also gives rise to paired nerves leading to the corpora cardiaca. In other cases, the recurrent nerve may continue uninterrupted, giving off branches to the stomodael muscles until ending in a ventricular ganglion.

Nerve Function

Resting Potential

Each neuron is bounded by a differentially permeable cell membrane across which a *resting potential* is maintained owing to the unequal distribution of ions on either side of the membrane. However, the potential tends to run down owing to ionic leakage, so it must be maintained by energy expenditure that moves some ions against their diffusion gradient. This metabolic function is performed by the *sodium/potassium pump mechanism* which simultaneously extrudes Na^+ (sodium) from the cell and takes up K^+ (potassium).

Nerve Function

Action Potential

A nerve impulse is a complex, transient series of permeability changes that sweep along the length of the axon. Passage of this *action potential* stems from the opening and closing of ion channels affected by voltage changes. Once the action potential is initiated by inflow of Na^+, the changes in membrane potential are then sufficient to cause adjacent ion channels to open, propagating the impulse farther along the length of the axon. Action potentials of up to 140 mv may last a msec or less at a given level of the fiber. The resting potential is momentarily abolished or reversed thereby, and the nerve membrane suddenly becomes more permeable to positively charged ions than it was before. Na^+ diffuses inward, resulting temporarily in a negative charge outside the membrane and a positive one inside. After a momentary lag, K^+ moves outward, and the membrane is restored to resting condition as the pump extrudes Na^+ and takes up K^+. As with other excitable tissues, the nerve membrane requires calcium to maintain excitability.

This conceptualized impulse is localized, taking place in a given part of an axon. Its wave-like spread to other parts of the same axon involves a flow of current that proceeds from the surface inward and, in doing so, initiates an action potential at the next adjacent point; flow from there triggers an impulse at the following point; and so on until the flow reaches the axon's terminal arborization.

A neuron obeys the *All or None Law*. Subjected to weak stimuli, it does not respond; subjected to stimuli at or above threshold, it responds but always in the same magnitude, regardless of strength of stimulation. Once the response has been set off in a given part of the neuron, it passes from there through the remainder of the cell much like the burn of a trail of gunpowder; *i. e.*, it is propagated at a uniform velocity and intensity without decrement.

Starting at the dendrite and sweeping through the axon, the neuron experiences a brief refractory period following transmission of the action potential. No stimulus, regardless of strength, can evoke response in the depolarized cell during this period. However, the refractory phase is succeeded momentarily by recovery, whereupon the neuron again becomes excitable.

As noted, an impulse is transmitted at a constant velocity requiring a few msec to run the length of the neuron, and it proceeds at a constant magnitude. How can such a fiber, transmitting one kind of impulse, carry signals representative of different degrees of stimulation? Each impulse is like a wave that sweeps through the neuron. These waves succeed one another at a rate varying with the intensity of excitation by sensory receptors. Any increase in stimulation causes a proportionate increase in the number of nerve impulses. Consequently, motor neurons elicit a proportionately stronger contraction of the muscles they innervate when they conduct more impulses, assuming no potentiation or inhibition by the central nervous division. Eventually, however, the impulses result in a sustained muscle contraction called *tetanus* as the frequency of stimulation reaches a critical level.

Synapses and Neuromuscular Junctions

Not all nervous system properties are accounted for by neurons because they are not physically continuous with one another. Interpretation is required of events that occur where neurons come together and where they meet the muscle fibers that they innervate. The terminal arborization of each axon is sepa-

Chapter 4 / NERVOUS SYSTEM

rated by a microscopic gap from the dendrite of the next, between which there is no neurilemma. This space, called the *synapse*, acts as a kind of physiological valve, permitting some impulses to pass to the next neuron to set off the same impulse. A similar relationship exists between a motor neuron and its associated muscle fibers. They are separated by a *neuromuscular junction*.

Transmission of the nerve impulse through a nerve fiber is, as noted, electrochemical, but passage across the synapse is chemical. Accordingly, upon arrival of the impulse at the terminal arborization, there is a delay of a few msec before onset of the impulse in the next neuron or, in the case of a neuromuscular junction, the beginning of muscle contraction. This delay stems from the time it takes to liberate a chemical *neurotransmitter* stored within the axon's synaptic vesicles and for that substance to bridge the synapse and initiate the impulse in the next neuron or in the adjacent muscle fiber.

The chemical neurotransmitter that bridges the synapse in insects seems to be *acetylcholine* (Ach), the same neurotransmitter used by other animals. Ach is concentrated in the central nervous division of insects, and cholinesterase (the enzyme that deactivates Ach) is likewise found here, particularly in postsynaptic membranes. This suggests that Ach is the mediator in insects' excitatory synaptic transmission. However, another neurotransmitter, *glutamic acid*, seems involved at insects' neuromuscular junctions. It is liberated from *in vitro* preparations by presynaptic stimulation and initiates impulses in the adjacent muscle fiber when at physiological concentration. Still another substance, *GABA*, or γaminobutyric acid, has been implicated as the inhibitory neurotransmitter at the neuromuscular junction. GABA is known to function in this manner in crustaceans, and its action has been extended to several insects.

Summation and Inhibition

To this point, I have discussed a simple situation involving a single train of impulses crossing a given synapse via a given neurotransmitter. This single synapse is either excitatory or inhibitory, but a dendrite can receive input from more than one neuron at different synapses. What happens when several postsynaptic impulses arrive at the same synapse at approximately the same time? They can either supplement one another, so-called *summation*, or interfere with one another, so-called *inhibition*. The interplay of these phenomena is believed responsible for many characteristics of insect nerve physiology, and the balance between them determines whether or not a particular synapse will be bridged.

Nerve Pathways and Behavior

The nervous and muscular systems operate together to bring about appropriate adjustment of the insect to its dynamically changing environment. This complex nerve-muscle interplay is through nerve pathways (*Fig. 4-2*). A true reflex arc involves a sensory neuron that contacts a motor neuron directly without intervention of an interneuron. Stimulation of the sensory receptor always produces the same response in that there is no interneuron. However, most pathways involve: (*a*) a *sensory receptor* that, in response to environmental change, is stimulated to transmit the impulse to a *ventral nerve cord ganglion*; (*b*) an *interneuron* that relays the impulse to (*c*) a *motor neuron* that stimulates (*d*) an *effector*, either a muscle fiber that contracts or a gland that secretes.

Nerve Function

These pathways are nothing more than physiological abstractions. Even simple pathways of this type involve countless sensory neurons, interneurons, and motor neurons whose sequential interplay is complex and subject to modification by fatigue, summation, inhibition, and sometimes overflow into adjacent arcs. Moreover, the pathways used may be altered by the intervention of specialized ganglia that operate as *coordinating centers* for particular functions.

Given pathways may involve a single ganglion that serves as its center. However, more remote ganglia may also exert stimulative or inhibitory influence over the local ganglia. The supra-esophageal, subesophageal, thoracic, and abdominal ganglia are among such coordinating centers. Their use is characteristic of complex activities involving an orderly sequence of nerve pathways extending over a number of body segments.

Abdominal ganglia, for example, exhibit autonomy, as demonstrated by experiments in which the ventral nerve cord is severed both in front of and behind a particular ganglion, leaving it unaffected by outside coordinating centers. The ganglion, thus isolated, still functions as the center controlling local ventilation by the tracheal system, and it remains the center for other responses such as might be appropriate. Thus, the last abdominal ganglion of various moths continues to control oviposition, defecation, *etc.*, even after the ventral nerve cord in front of it is severed.

Each thoracic ganglion controls local motor functions including walking, clasping, and wing vibration, as shown by severing the nerve cord in front of and behind it. For example, a single leg of a cockroach with its ganglion isolated is able to take a step when its tarsus is stimulated.

The preceding involves experimentally isolated thoracic ganglia. What happens in intact animals? One example suffices. Pressure on the tarsus of intact, normal insects inhibits wing vibration, and loss of tarsal contact elicits wing beat. This is demonstrated when a living, mounted insect has a card acting as its footing pulled out from under it and then replaced. The animal's wings first vibrate and then cease vibrating.

As noted, each abdominal ganglion is a center in control of its own spiracular movements, and it continues ventilatory activities independently of other ganglia. Obviously, however, the uncoordinated breathing of individual segments does not provide a means by which increased or decreased oxygen needs may be met; more is needed. The answer lies in the prothoracic ganglion or some other coordinating center that regulates the rate of breathing in accordance with the oxygen - carbon dioxide tension of blood and also regulates the overall action of the spiracles, providing an efficient bellows system.

The subesophageal ganglion is the most important locomotor center of the body, as shown by decapitating an insect, which renders it sluggish and unresponsive. Decapitation removes both the brain and the subesophageal ganglion, but simple decerebration, or brain removal, prompts heightened walking for long periods. These facts suggest that the subesophageal ganglion facilitates conduction of the nerve pathways of walking in accordance with "instructions" from the brain.

The brain is the sensory and association center that integrates the body as a whole by its initiatory and inhibitory reflexes. It receives incoming environmental stimuli through the sense organs, and its extensive neuropile coordinates the body's reflexes accordingly. The brain's importance in initiatory responses is demonstrated by decerebration. The animal is rendered sluggish and unable to perform instinctive acts such as eating or seeking shelter. The obvious explanation is that the motor

Chapter 4 / **NERVOUS SYSTEM**

centers of decerebrated insects remain intact and motor coordination is unimpaired, but the local centers are not brought to the proper level of excitation in absence of brain stimulation. The inhibitory function of the brain may be demonstrated by its destruction. Brain destruction may lead to exaggerated reflexes readily induced by stimuli that are not usually effective. Once initiated, these inappropriate reflexes may continue for long periods without interruption, as observed in decerebrated insects that groom ceaselessly. Insects' so-called *circus movements* are another example. They result from destruction of one side of the brain, leaving the other side intact. The experimental animals walk continuously toward the intact side, sometimes for hours or days. The apparent explanation involves the release of cerebral inhibition on the destroyed side causing uncontrolled leg movement toward the intact side.

It follows, therefore, that the brain has several major functions. It stimulates insects to respond to environmental change by exerting control over local centers in other parts of the body which, in turn, carry out their respective activities in appropriate sequence. The brain also inhibits excessively rigid execution by lower centers. An insect may be visualized, therefore, as a complex of operating nervous circuits that, when stimulated by environmental change, is triggered into response. Stimulated one way, the insect responds with an appropriate sequence of nerve circuits; stimulated another way, it responds in another fashion. These responses generally harmonize the animal with its environment in a way that involves comparatively little centralization of nervous function.

Sense Organs and Sensitivity

The neuromuscular complex of nerves and muscles is functionally inseparable. However, nerve fibers usually cannot detect the environmental stimuli that constantly impinge on insects. Only receptors are so adapted. *Receptors* consist of specialized *sensilla*, or sense organs, that perceive energy emanating from internal and external sources, sometimes over distances, and pass them on to nerves that change this energy into nerve impulses.

Sense organs may be classified according to the stimuli they receive, as follows: (1) *mechanoreceptors* including touch receptors that detect movement, tension, and pressure and sound receptors; (2) *chemoreceptors* including taste receptors that detect chemical substances in solution and smell receptors that perceive volatile chemicals in air; and (3) *visual receptors*, chiefly eyes, that perceive light radiations. Insects also respond to certain external environmental agents for which they lack specialized receptors, and they have receptors that detect environmental change within the body itself.

Mechanoreception

Tactile impressions, the sense of touch, are the province of sensory setae, campaniform organs, and scolopophorous organs to be discussed in the order presented. *Sensory setae* are widely distributed over the insect body, especially on the antennae, palpi, tibiae, tarsi, and cerci (*Figs. 3-1 a, 4-6 a*).

Sense Organs and Sensitivity

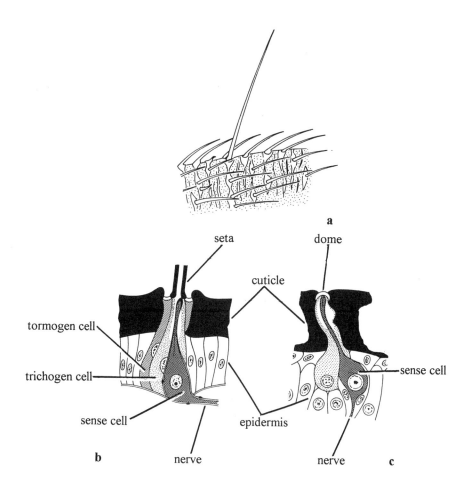

Fig. 4-6. (*a*) Part of an antennal article showing numerous short setae and a single long seta; (*b*) a sectional view of a generalized seta; and (*c*) a sectional view of a campaniform organ (*a* is redrawn from V. B. Wigglesworth & J. D. Gillette, 1934; *b* and *c* are modified and redrawn from R. E. Snodgrass, 1935).

Each sensory seta consists of a cuticular process having a movable articulation with the body and a direct connection with a sense cell. The latter is stimulated by movement of the process which often consists of little more than a peg, cone, seta, spine, or scale, either exposed or sunken into a pit. The setal sensillum, whatever its conformation, consists of the following elements (*Fig. 4-6 b*): (1) an outer tormogen cell forming a socket, (2) a middle trichogen cell that constitutes the process, and (3) an inner sense cell that innervates the process. These three enclose one another in the manner indicated or take on a serial, more elongated arrangement with partial overlap. All insects also have groups of small hairs, so-called *hair plates*, at joints in the cuticle.

Campaniform organs (*Fig. 4-6 c*) are dome-like, mechanoreceptive papillae sometimes sunken into cuticle. They are widely distributed over the insect body in cuticular areas subject to stress, especially on or near articular and arthrodial membranes. Their deformation by exoskeletal stress triggers response. The dome takes the form of an inverted cup in the center of which attaches a sense cell.

Chapter 4 / **NERVOUS SYSTEM**

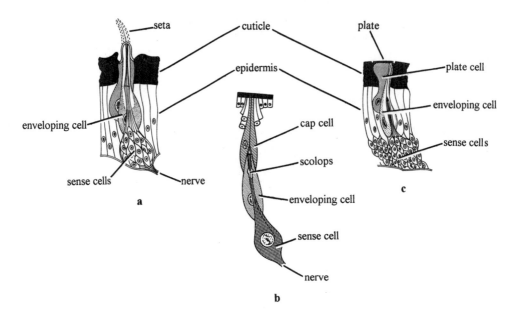

Fig. 4-7. Sectional views of (*a*) a chemoreceptive seta, (*b*) a scolopophorous sensillum, and (*c*) a plate organ (Modified and redrawn from R. E. Snodgrass, 1935).

Scolopophorous (chordotonal) *organs* are stretch receptors responsive to vibration and tension change. They consist of clusters of simple sensilla with a common attachment to the body wall. They vary from band-like bundles to elongate, thick cords, but they lack the setae, papillae, or other cuticular specializations of other mechanoreceptors. Each individual sensillum (*Fig. 4-7 b*) consists of three cells in serial sequence, with only partial overlap: an external cap cell that attaches to the cuticle, an enveloping cell, and an internal sense cell associated with a sense rod called the *scolops*.

Sensory setae, scolopophorous organs, and (on antennae) a modified scolopophorous organ called Johnston's organ are among the sensilla commonly involved in hearing.

Chemoreception

Taste and *smell* are overlapping senses. The former generally involves direct contact with chemical substances in solution and the latter either short- or long-distance perception of volatile substances in air. Insects respond by moving toward or away from the stimulus or, in the case of food, by accepting or rejecting it.

Insects' sense of taste is about 200 times more acute than that of man. It resides within the palpi, labella, antennae, tarsi, tibiae, and especially the preoral cavity, mouth, and mouthparts. Its location varies with the species, but usually there are several sites. The chemoreceptive setae of the antennae and mouthparts are taste receptors that respond either to dry chemicals on a leaf surface or to wet ones. They are characterized by a delicate cuticle and multiple sense cells. The stimulating chemicals apparently pass through the sensilla's terminal pore to trigger response. Insects' sense of smell is even

Sense Organs and Sensitivity

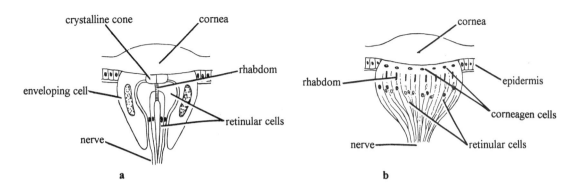

Fig. 4-8. (*a*) Sectional view of a stemma compared with that of (*b*) an ocellus (Redrawn from various sources).

more acute than is their taste. Smell is localized in the antennae but may also occur on the palpi, labella, and other body parts. *Smell receptors* may take the form of a *porous peg* or seta (*Fig. 4-7 a*) or of a *plate organ* (*Fig. 4-7 c*). Plate organs are thin, porous, ovoid or elongate cuticular plates associated with three partly overlapping cellular elements: a plate cell, an enveloping cell, and multiple sense cells. They resemble campaniform organs in their lack of a seta or other process but differ in their numerous small cuticular pores that allow passage of odors to the underlying sense cells.

Visual Reception

The term "eye" embraces any organ sensitive to light and capable of transmitting that impulse to the central nervous division. It includes both photoreceptors that receive light but do not form an image and complex "true eyes" capable of registering impressions of form, motion, and sometimes color. Internal structure varies with the kind of eye, of which insects have three: ocelli, stemmata, and compound eyes.

The *ocelli* (*Figs. 2-6 b, c, 4-8 b*) of nymphs, naiads, and adult insects are photoreceptive stimulative organs that do not form an image. Each ocellus usually has a biconvex, transparent cornea, or lens. Beneath the cornea is a thin, transparent layer of corneagen cells involved in corneal secretion and, beneath them, an extensive light-sensitive retina. The retina consists of many elongate sensory cells disposed parallel to one another. They are organized into clusters called retinulae, each cell of which bears a rhabdomere (inner microvillous face). Several such faces come together to form a common light-sensitive core called the *rhabdom*. Below the latter is usually a pigmented middle layer and below it a zone of retinular nuclei. Deeper still, the retinular cells merge into an ocellar nerve that penetrates the basement membrane and ends in the appropriate ocellar lobe of the protocerebrum.

Stemmata (sometimes called lateral ocelli) (*Figs. 2-6 a, 4-8 a*) are photoreceptors capable of perceiving limited images and motion. They occur in adults of certain bristletails, springtails, and fleas but are best known for their occurrence in caterpillars and other larvae, in which they are the only eyes present. They do not persist in adults but are replaced by the compound eyes, with which they are homologous. Stemmata differ from ocelli in number and location. There may be as few as one

Chapter 4 / NERVOUS SYSTEM

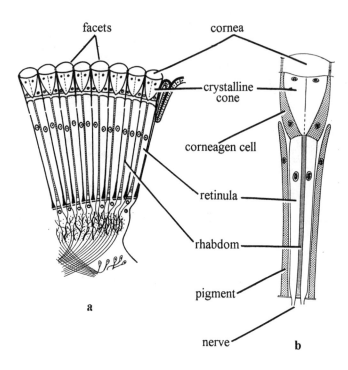

Fig. 4-9. (*a*) Section of an insect compound eye showing a number of its ommatidia and (*b*) a single idealized ommatidium (*a* is modified and redrawn from R. E. Snodgrass, 1935; *b* is modified and redrawn from R. F. Chapman, 1982).

stemma on each side of the head or as many as seven or eight in that location. They also differ in innervation and structure. They are innervated from the optic lobes of the protocerebrum, and they interpose a transparent crystalline cone reminiscent of that of ommatidia between their cornea and retina. Therefore, stemmata are essentially ommatidia-like structures that occur singly rather than massed as in compound eyes. However, they are more like ocelli with multiple rhabdoms in larval Lepidoptera, Trichoptera, and most Coleoptera and Hymenoptera.

Compound eyes differ from ocelli in their innervation and lens mechanism. They are innervated from the optic lobes of the protocerebrum, and they consist of numerous individual groups of sensory cells, or *ommatidia*, each with its own lens system. Compound eyes consist of a few to thousands of these massed ommatidia, each superficially visible on the eye as a single hexagonal facet.

The structure of an individual ommatidium is as follows (from surface inwards) (*Fig. 4-9 a, b*): cornea, a transparent, biconvex outer lens of cuticular composition (hence, cast during ecdysis); crystalline cone or inner lens consisting of four cells; corneagen or primary pigment cells that sometimes extend beneath the cornea (which they secrete) but are usually peripheral to the crystalline cone; retinulae or inner, light-receptive units surrounded by secondary pigment cells that partly overlap the primary pigment cells; and retinal cells whose nerve fibers penetrate the basement membrane to join similar fibers from other ommatidia to become the optic nerve.

Sense Organs and Sensitivity

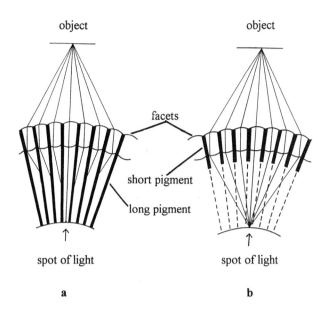

Fig. 4-10. Pigmentary distribution of (*a*) a day-adapted apposition eye compared to that of (*b*) a night-adapted superposition eye (Modified and redrawn from H. Oldroyd, 1962).

Each retinula is composed of about eight elongate retinal cells that, in cross section, are arranged like segments of an orange. From beneath, the axons exit through the basement membrane. Where the cells of each retinula meet, they form a central light-sensitive core, or *rhabdom*, containing visual pigments that absorb light and produce the photochemical effect.

Compound eyes can be likened to a massed series of minute, parallel tubes (ommatidia), each with a transparent, light-sensitive core. Each of these optically isolated units perceives a small portion of the field of vision (*viz.*, those light rays traveling in the direction of the main axis and focused on it by the lens). Therefore, according to a prevalent theory, the impression received by the eye as a whole is a mosaic of as many points of light as there are ommatidia, each varying in color and intensity according to the part of the object that it perceives.

Movement of an object within the visual field registers immediately on one or more components of the mosaic, producing alteration in visual pattern according to the number of ommatidia involved, the size and distance of the object, and its velocity. Movement of a small, nearby object is readily detected but that of a large, nearby object stimulating many ommatidia simultaneously is less readily detected, especially when it moves slowly. There is no variable focusing mechanism. The usual image formed by an insect's compound eye is a slightly blurred, upright, coarse mosaic of nearby objects.

Insects have two types of compound eye. The *apposition* eye of diurnal insects (*Fig. 4-10 a*) features ommatidia that focus at the base and retinulae entirely shielded from each other by long secondary pigment cells. This adaptation assures that only light rays directly normal to the lens affect the rhabdoms, oblique rays being absorbed. The *superposition eye* of nocturnal insects (*Fig. 4-10 b*),

77

Chapter 4 / NERVOUS SYSTEM

in contrast, features ommatidia that focus at the mid point and retinulae that are not completely shielded by the secondary pigment cells. Hence, oblique light rays entering an ommatidium may pass into adjacent ommatidia where, along with perpendicular rays, they become effective stimuli, and relatively few rays are dissipated by absorption. Accordingly, the apposition eye produces a comparatively acute, mosaic image in bright light and the superposition eye a continuous, blurred image in dim light.

The compound eye is capable of color perception; *i. e.*, it perceives varying wave lengths of light independently of their brightness. The cabbage butterfly and a few other insects that see into the red range of the color spectrum are attracted to red flowers or red models, but most insects are less sensitive than vertebrates to these longer wave lengths and are blind to deep reds. However, insects are more sensitive than vertebrates to greens, blues, and ultraviolet.

Suggested Additional Reading

Bullock & Horridge (1965); Chapman (1991, 1998); Davies (1988); Gullan & Cranston (2000); Nation (2002); Patton (1963); Richards & Davies, vol. 1 (1977); Snodgrass (1935); Wigglesworth (1972, 1984).

Chapter 5

Muscular System

Whatever moves is moved by another.
— Thomas Aquinas

There are two main types of insect muscle: skeletal and visceral. The first kind, *skeletal muscle*, is associated with the skeletointegumentary system. Skeletal muscles originate on one sclerite and insert on another, crossing and compressing one or more flexible conjunctivae and bending the body there. They are involved in maintenance of posture, internal movement, and gross locomotion. The second kind of muscle, *visceral muscle*, invests the internal organs. Visceral muscles often insert on themselves, directly or indirectly, "squeezing" structures associated with them.

Muscle Structure

Visceral Muscle

Insects' visceral muscle is striated, distinguishing it from vertebrates' smooth muscle. These muscles invest the alimentary canal, reproductive ducts, and sometimes the Malpighian tubules, and they form the dorsal blood vessel, dorsal and ventral diaphragms, and certain other internal organs. They lack strict origins and insertions. They take two general forms: (1) an irregular, branched, intercommunicating network of muscle fibers (*Fig. 5-1 a*) with a superficial resemblance to vertebrates' cardiac muscle, and (2) regularly disposed layers of circular and longitudinal fibers (*Fig. 5-1 b*). The first, branched type is characteristic of the lining of the crop, and the second, layered type is typical of most of the alimentary canal, except at localized points where the circular fibers become accentuated into occlusive *sphincters*.

Skeletal Muscle

Insects' skeletal muscles, like those of vertebrates, are paired and metameric in conformity with the body's bilateral symmetry and segmentation but differ from their vertebrate counterparts in being more numerous and more complex. Certain caterpillars, for example, have over 2,000 separate muscles, as opposed to the less than 700 found in the human body.

Insects' skeletal muscle usually consists of long, slender bands. One of their extremities, the *origin*, attaches to a relatively stationary sclerite and the other, the *insertion*, to a sclerite that is relatively movable upon contraction of the muscle. Each individual muscle is composed of numerous strongly

Chapter 5 / MUSCULAR SYSTEM

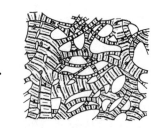

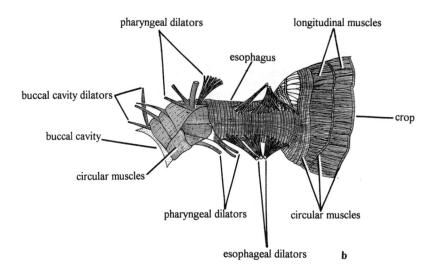

Fig. 5-1. (*a*) Irregular, branched, interconnecting muscle fibers in the crop of a blow fly; (*b*) the layered musculature of a caterpillar stomodaeum (*a* is modified and redrawn from G. S. Graham-Smith, 1934; *b* is modified and redrawn from R. E. Snodgrass, 1935).

striated muscle fibers (cells) bound together by a connective tissue *perimysium*. Each fiber is a syncytium, having numerous nuclei per cell, and consists of a number of closely parallel *myofibrils* differentiated from the sarcoplasm, or general muscle protoplasm.

The histology of insect skeletal muscle varies with the taxonomic group and even within the same individual animal. Among the several recognized skeletal muscle types are: (1) a type with a thick peripheral layer of sarcoplasm (*Fig. 5-2 a*); (2) a type with myofibrils uniformly distributed throughout the cross-section (*Fig. 5-2 b*); (3) a type with centrally distributed nuclei (*Fig. 5-2 c*); and (4) "fibrillar muscle," a kind with numerous refractile sarcosomes (*Fig. 5-2 d*). Sarcosomes are large mitochondria that supply the biochemical energy for contraction.

A muscle fiber or cell, whatever its histology, is characterized by lengthwise alternating regions of strong and weak refringence to light. These optically different areas coincide at a given level in adjacent myofibrils, hence the entire fiber takes on a striated appearance, even in the living, unfixed state (*Fig. 5-2 f*). These cross-striations reflect an underlying structural arrangement caused by the alignment of alternating anisotropic and isotropic bands associated with two different submicroscopic

Muscle Structure

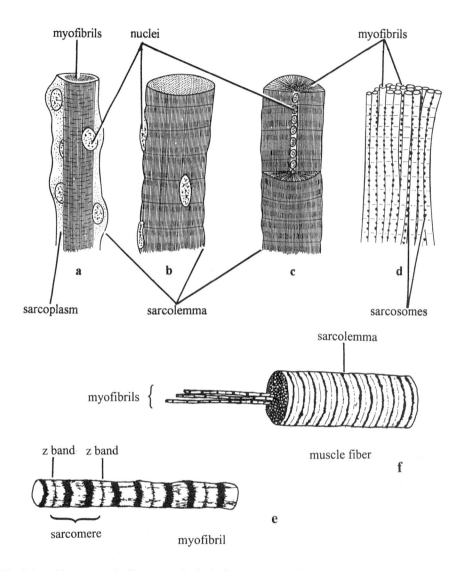

Fig. 5-2. Selected insect muscle fibers, or cells, including (*a*) a type with a thick peripheral sarcoplasm, (*b*) a type with uniformly distributed myofibrils, (*c*) a type with centrally distributed nuclei, and (*d*) fibrillar muscle, a type with numerous refractile sarcosomes. (Note that *d* is drawn at a more enlarged scale than are *a*, *b*, and *c*). Fig. 5-2 *f* is a three-quarter section of an enlarged muscle fiber showing the many myofibrils of which it is composed including three that are partly extracted from the bundle. Fig. 5-2 *e* is an even more enlarged three-quarter section of a single myofibril showing the z bands that constitute one of the myofibril's sarcomeres (*a – d* are modified and redrawn from R. E. Snodgrass, 1935; *f* and *e* are modified and redrawn from Romoser, 1981, after Herried).

protein filaments. Electron microscopy reveals that, in addition to these longitudinal bands, there are thin transverse Z lines delineating individual contractile segments called *sarcomeres* (*Fig. 5-2 e*) on each fiber. Thus, a succession of sarcomeres (not single myofibrils) runs the length of each fiber.

Chapter 5 / MUSCULAR SYSTEM

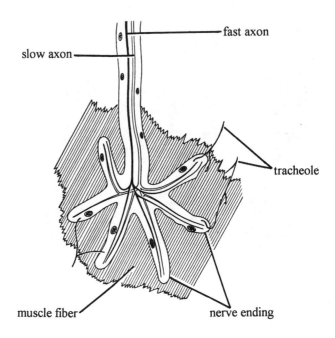

Fig. 5-3. A claw-like motor end plate (Modified and redrawn from G. Hoyle, 1965).

Innervation

Transmission of a nerve impulse to a muscle fiber is through a motor end plate (*Fig. 5-3*) consisting of nerve branches resting in a shallow groove on the muscle fiber's surface. This structure commonly takes on the appearance of a "claw" spreading over the muscle surface, often accompanied by tracheole (respiratory) branches. Here, the neural lamella fuses with the basement membrane of the muscle sarcolemma. The individual nerve twigs, or terminals, may be simple or branch one or more times.

I have said little so far to characterize insect muscle innervation as different from that of vertebrates. There is, however, one fundamental difference: insect muscle fibers lack vertebrate's single motor end plate per fiber, usually having a series of end plates distributed at regular intervals over the length of the muscle fiber. This is because insect muscle membrane (sarcolemma) does not conduct impulses nearly as well as does vertebrate sarcolemma. Moreover, a pair of axons usually leaves each nerve to become distributed, side-by-side, throughout the muscle. This *double innervation* involves one member of the pair, the *fast axon*, that is responsible for rapid muscle contraction, and the other, the *slow axon*, for tonic and slow contractions. Some fibers even receive a third, inhibitory axon. Such *polyneural innervation* may involve more than two excitatory and two or three inhibitory neurons.

Muscle Function

The physiological response of insect muscle to stimulation by a fast axon is much like that of vertebrate muscle. A single stimulus delivered to the muscle may be too weak to induce contraction, or it may be at or above threshold and sufficiently strong to cause contraction. The well-known "All or None Principle" of physiology demands either contraction with fixed, maximal force or no contraction according to the level of stimulation. However, insect skeletal muscle does not behave this way. It is capable of graded responses, contracting either forcibly or weakly depending on the number of axons activated and their discharge frequency. The force of the overall contraction elicited depends on graded contractions within individual fibers.

Single Muscle Twitch

When a single electrical impulse at or above threshold is delivered either through a neuron or, experimentally, directly onto the muscle surface of an insect, there follows a momentary *latent period* without apparent change though internally the muscle is coupling excitation to contraction. The *contraction phase* follows, indicated by a rising curve on the electronic record, and then the *relaxation phase* occurs, indicated by a falling curve that eventually reaches base level. There is also a brief *refractory period* within the relaxation phase during which the muscle cannot contract unless the strength of stimulation is increased. The resulting tracing of a single muscle twitch on the electronic record is a skewed, bell-shaped curve. The curve's height under load indicates twitch force, and its rate of rise is a function of intrinsic muscle speed. The latter increases at elevated temperatures and decreases at reduced temperatures.

The surface events that occur during the twitch are of interest. A relaxed muscle maintains an electrical potential across its external membrane. This *resting potential*, like that of nerve fibers, is upset by nerve stimulation resulting in local depolarization near the synapse. If of sufficient magnitude, this excitation elicits an *action potential*, and the associated current spreads down fine T-tubules ramifying within the muscle fibers and releases internal calcium ions that stimulate shortening of the contractile elements. The T-tubules, which are invaginations of cell membrane, transmit excitation through the thickness of the muscle fiber so that deep-lying myofibrils are excited simultaneously with more peripheral ones. This surface excitation is coupled to the *sarcoplasmic (endoplasmic) reticulum* which releases calcium ions within the muscle fiber, activating the contractile proteins. These ions must be pumped back into storage to allow for relaxation.

Summation and Tetanus

A second stimulus administered during a muscle's latent period or early during contraction results in a heightened contraction called *summation*. It may be produced by a series of stimuli delivered at a frequency greater than the muscle's recovery period. In this event, a step-wise curve is traced on the electronic record reflecting an overall magnitude of contraction greater than that of a single muscle

Chapter 5 / MUSCULAR SYSTEM

twitch. Then, should the frequency of stimulation be increased, there occurs no relaxation at all as the individual contractions summate and become no longer discernible on the curve. A steady state of contraction called *tetanus* now persists at a level of magnitude greater than that of a single muscle twitch and continues for a time before fatigue sets in. Then the muscle cannot resume contraction until physiological recovery has been achieved.

Slow, deliberate movements of appendages result from slow-axon stimulation and vigorous, rapid movements from short, high-frequency impulses via the fast axon. The fast axon innervates all fibers of a given muscle, so it elicits a brief, powerful contraction not subject to graded control, except by frequency (*Fig. 5-4 a*). The slow axon innervates only a portion of the fibers of a muscle (sharing some end plates with the fast axon), so it elicits no significant contraction upon receipt of a single shock. However, repetitive stimulation causes slow, smooth contractions starting at a frequency of about 10 per sec (*Fig. 5-4 b*) and increasing progressively with heightened frequency up to about 150 contractions per sec. Combinations of slow-axon activity with occasional bursts of fast-axon activity occur during execution of normal movement. Stimulation via the inhibitory axon causes the muscle to relax more quickly.

Most muscles have a direct 1:1 relationship between their stimulation by an excitatory axon and their contraction. This includes the flight muscles of butterflies, grasshoppers, and other insects with a relatively slow wing beat. Their wings beat in direct response to individual contractions of the *synchronous muscles* that drive them (*Fig. 15-9 a*). However, the fibrillar-type flight muscles of bugs, beetles, flies, and bees and other Hymenoptera have a high wing-beat frequency, and one genus of biting midge has a wing-beat frequency well over 1,000 cycles per sec. Most muscles cannot undergo nerve-stimulated contraction, followed by relaxation, at frequencies exceeding 100 beats per sec. Muscles such as these, their rate of contraction far exceeding the rate of nerve stimulation, are termed *asynchronous* (*Fig. 15-9 b*). The wing motion of asynchronous muscles stems only partly from muscle contraction. The remaining energy is derived from oscillation of the elastic thoracic system including cuticle, muscles, wings, and resilin pads, as discussed in Ch. 15.

Muscles can be specialized for fast activation and relaxation or for slow contraction and recovery. The jumping muscles of orthopteroids and the flight muscles of orthopteroids and of Lepidoptera are fast muscles capable of twitch responses to single nerve impulses. They have an extensive sarcoplasmic reticulum which allows for a large quantity of calcium ions to be released and recaptured rapidly. Much of the energy expenditure of such muscles is used to pump sodium, potassium, and other ions across the cell and T-tubule membranes to restore excitability and put calcium back into storage allowing for relaxation. The time requirements for these processes set upper limits on the frequency of repetitive movements such as flying or stridulation when these movements are controlled by individual nerve impulses.

Paradoxically, the flight muscles of some of the most efficient fliers with high wing-beat frequencies are slow-type muscles with relatively little sarcoplasmic reticulum. It takes several nerve impulses for them to release enough calcium to allow for contraction, and they are slow to relax after cessation of nerve impulses. An occasional subsequent nerve impulse maintains them in an excited, contracted state. The antagonistic indirect (dorsal longitudinal and tergosternal) flight muscles simultaneously go into a state of nearly isometric contraction. For their wings to move between the up and

Muscle Function

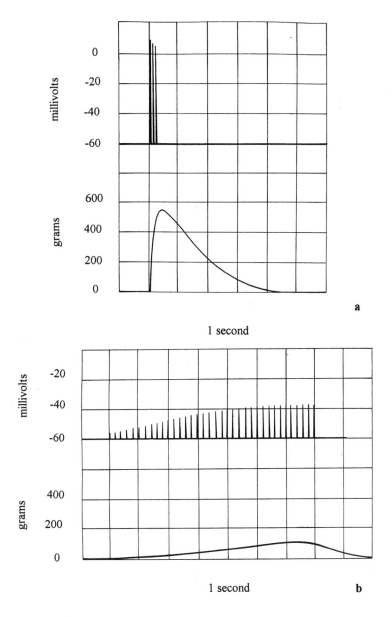

Fig. 5-4. Comparison of muscle reaction to stimulation by (*a*) a fast nerve fiber and by (*b*) a slow nerve fiber, with the top of each graph indicating muscle electrical charge and the bottom the resulting muscle tension (Modified and redrawn from G. Hoyle, 1958).

down positions, the wing articulation must force the pleural walls of the thorax laterally against resistance which can be increased by contraction of the pleurosternal muscles. As the moving wing articulation passes the metastable position of greatest resistance, thoracic elasticity snaps the wing toward the other extreme position, suddenly stretching the antagonistic muscle. Sudden stretch of an excited

Chapter 5 / MUSCULAR SYSTEM

muscle produces a coordinated increase in tension, returning the muscle through the metastable position, and the cycle is repeated as long as both muscles remain sufficiently excited.

Wing-beat frequency is determined by the mechanical properties of the system, the thoracic dimensions, and wing loading. The advantage of the system is that the amount of energy spent on pumping ions is greatly reduced. Moreover, the muscle fibers themselves change little in length throughout the cycle, so relatively little energy is lost in the fibers' viscoelastic deformation.

Inhibitory neurons can speed relaxation. Often they innervate only a portion of a muscle, allowing it to be used in different ways depending on whether all or only some of the fibers contract.

Flying honey bees burn mainly glucose. Migratory locusts and some other migratory insects start flight with carbohydrates but switch to burning fat during prolonged flights. The flight muscles of many insects capable of prolonged flight are completely aerobic but do not incur a significant oxygen debt because extensive tracheation supplies all the oxygen needed. These muscles appear pinkish owing to their numerous mitochondria, but muscles such as the jumping muscles of Orthoptera have few mitochondria and appear whiter. They tend to be anaerobic, to accumulate lactic acid, and to fatigue rapidly.

Tonus

Active, living muscle is in a constant state of *tonus*, or partial contraction, responsible for posture. It involves spontaneous contraction of a few muscle fibers resulting from low-frequency discharge of the slow axon. A steady discharge of these slow-axon impulses takes place in a standing insect, causing tonic contraction of the depressor muscles of the legs, lifting the animal off the ground. It is against this steady, spontaneous tonic background that the slow and fast movements of the body are superimposed. Tonus produces a slight heat as it exerts tension on the body parts with which it is associated.

Biochemistry of Contraction

The principal components involved in the contractile process are the fibrous proteins *actin* and *myosin* which, in the presence of salts that normally occur in muscle, interact and slide past one another. The resting whole muscle of insects contains the sugar glycogen, phosphogen, and ATP. Glycogen and ATP decrease during contraction, and pyruvic acid and lactic acid increase. Therefore, lactic acid accumulates after long periods of vigorous activity, except in indirect flight muscle in which LDH (lactic dehydrogenase) is deleted, and glycogen and ATP become depleted. This physiological condition is known as *fatigue*.

The energy used for the sliding process that causes shortening, or contraction, is derived from the breakdown of ATP which is continually replenished from phosphogen as long as it is available. The energy used for the recovery that follows contraction stems from the breakdown of glycogen into lactic acid, part of which is used to replenish the depleted glycogen. Thus, contraction can be anaerobic and recovery aerobic.

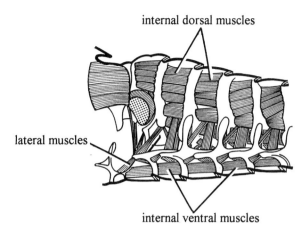

Fig. 5-5. Internal view of the musculature of abdominal segments one to five (Modified and redrawn from R. E. Snodgrass, 1935).

Muscle Pattern

Muscles are not actively extensible. They cannot lengthen; they only contract. Once contracted, they must be returned to their original resting position by some antagonistic force. The force used by visceral muscle is pressure either of blood or of food. The force used by skeletal muscle is provided either by skeletal muscles called *antagonists* that contract in opposition to opposing skeletal muscles or by cuticular elasticity. For example, the femur has a ventral depressor that pulls the tibia downward and a dorsal levator that lifts it upward (*Figs. 3-6 b, 5-7*). This simple example involves only two antagonists. In other instances, several muscles may function together, always in opposing groups, to produce more complex movements.

Actions

As noted, a muscle's function is to contract, thereby bringing its more movable end, the *insertion*, closer to its less movable end, the *origin*. In doing so, it pulls on any membrane, appendage, or sclerite to which the insertion attaches, causing movement of the limb or other part with respect to the body as a whole. These movements are called *actions*.

Myology

Myology is the study of the pattern of muscles in an animal or an animal group. The visceral muscles of insects are disposed either in layers or in an irregular network. Skeletal muscles, in contrast, are paired, individual bands that conform with the body's bilateral symmetry and metamerism. They are numerous, complex in pattern, and present a myology summarized below.

Chapter 5 / **MUSCULAR SYSTEM**

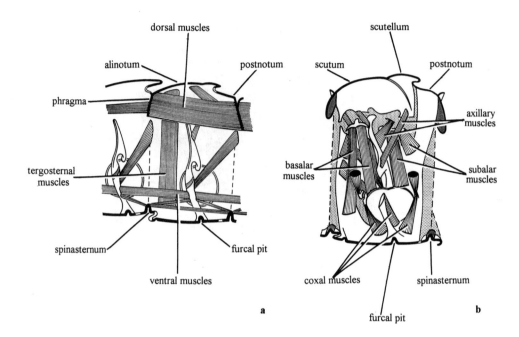

Fig. 5-6. Internal view of a pterygote mesothorax chiefly showing (*a*) the dorsal, ventral, tergosternal, and oblique muscles and (*b*) the axillary, alar, and coxal muscles (Modified and redrawn from R. E. Snodgrass, 1935).

The abdomen's comparatively simple musculature is important in many activities including breathing. Its chief muscle groups (*Fig. 5-5*) include dorsal muscles that attach to successive terga, ventral muscles that attach to successive sterna, lateral muscles involved in abdominal compression, spiracular muscles, and genitalic muscles. The dorsal and ventral muscles act as *retractors*, telescoping the abdomen back into the body.

The thorax, adapted as it is for locomotion, has a more specialized musculature than does the abdomen. Its chief muscle series (*Fig. 5-6 a*) include dorsal muscles, especially the paired, longitudinal bands running from pre- to postphragmata that warp the notum upward, raising the wing base and depressing the wings, and tergosternal muscles that depress the notum, thereby levating (lifting) the wings against the pleural fulcra. (The antagonistic dorsals and tergosternals are often called *indirect wing muscles* in that they move the wings without actually inserting on them). There are also (*Fig. 5-6 b*) axillary muscles from the pleuron that insert on the third axillary sclerite and flex the wing, and alar (basalar and subalar) muscles that insert on the alar sclerites and rotate and extend the wing. (The antagonistic axillaries and alars are so-called *direct wing muscles* whose contractions directly elicit wing depression). Finally, the thorax has ventral muscles and various leg muscles (*Fig. 5-7*), including *extrinsic ones* that arise from the tergum, pleuron, or sternum and insert on or near the coxa causing movement of the leg as a whole, and *intrinsic ones* within the leg itself where they cause flexion and extension of the articles of the leg.

Muscle Pattern

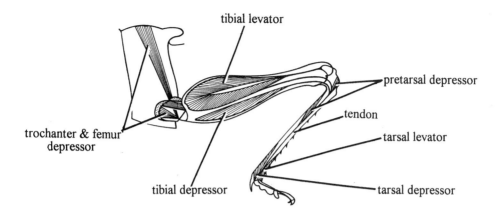

Fig. 5-7. Musculature of a grasshopper's hind leg (Modified and redrawn from R. E. Snodgrass, 1935).

The head, the most highly modified region of the body, has a correspondingly complex myology that includes *cervical muscles* that arise from the prothorax and cervix and insert on the tentorium and epicranium of the head, causing levation, depression, retraction, protraction, and rotation of the head; and *intrinsic* and *extrinsic maxillary* and *labial muscles* (*Fig. 2-10 a, b*) that originate on the head capsule or tentorium and insert on the maxilla or labium, respectively; *extrinsic mandibular muscles* (the mandibles have no intrinsic muscles) (*Fig. 2-9*); and *antennal muscles* (*Fig. 2-7*), adapted appropriately either to segmented or to annulated structural types.

Suggested Additional Reading

Chapman (1991, 1998); Davies (1988); Gullan & Cranston (2000); Miller (1975); Nation (2002); Patton (1963); Richards & Davies, vol. 1 (1977); Smyth (1985); Snodgrass (1935); Wigglesworth (1972, 1984).

Chapter 6
Endocrine System

Listening to the inner flow of things.
— James Russell Lowell

Two organ systems whose primary function is body coordination control activities of many of the same organs and tissues. One, discussed earlier, is the nervous system whose rapidity of conduction by nerves provides an instantaneous, "pin point" response. The other, the endocrine system, provides a slower, longer-lasting, more diffuse response through secretions disseminated directly into the bloodstream. These secretions, called *hormones*, are specific excitatory or inhibitory substances liberated into the tissues or blood for transport to other parts of the body where, in minute quantity, they modify growth, development, or other physiological processes. They consist of a wide variety of chemical substances such as steroids, lipophilic sesquiterpenoids, biogenic amines, and simple and complex peptides. Steroids penetrate the cell membrane and react with intracellular molecules and nuclear receptors. Peptide hormones do not penetrate the cell membrane but bind to receptor molecules within the membrane and act as secondary intracellular messengers. They affect specific tissues only and always in reduced concentration.

Internal organs without an obvious function are invariably suspected of being endocrine. The classical method of proving this role, as outlined in any general textbook, seems simple but is difficult to carry out on insects because the endocrine tissues of most of them are small and highly ramified, limiting possibilities of successful excision and collection of material for chemical analysis; because some of insects' endocrine organs secrete several different hormones and others have non-endocrine as well as endocrine functions; and because much of insects' overall endocrine activity is by *neurosecretion*, which is a function of neurons with gland-like characteristics. Neurosecretion involves specialized cells at certain parts of the brain and nerve cord ganglia that produce so-called *neurohormones* that activate or inhibit other endocrine organs or act on other targets. These cells periodically become packed with secretory granules that either pass directly into the bloodstream or move down the axon for release elsewhere in neurohemal organs. Neurosecretory cells are often diffuse and intricately interwoven within strictly nervous tissue, making isolation virtually impossible. A strict separation of the endocrine system from the nervous system is invalidated on this basis leading some authorities to go so far as to recognize a combined neuroendocrine system.

Chapter 6 / ENDOCRINE SYSTEM

Hormone Classes

Three classes of hormone, *viz.*, ecdysteroids, juvenile hormones, and neurohormones, are involved in insects' growth and reproduction. *Ecdysteroids*, or molting hormones, are steroid-derived substances necessarily stemming from a dietary source. One of them, *ecdysone*, is a known prohormone, being secreted by the prothoracic glands before conversion in other tissues into the active molting hormone, 20-hydroxyecdysone. Ecdysteroids are also secreted by the ovaries of adult insects for use in ovarian maturation and/or other purposes. The singular symbol *MH* is used in the text that follows to refer to these ecdysteroid compounds.

The so-called *juvenile hormones* are, in reality, a number of related sesquiterpenoid compounds involved in the control of metamorphosis and reproduction. Several slightly different forms have been isolated. The singular symbol *JH* is used to refer herein to any of these juvenile hormones whose function is to maintain larval characteristics and inhibit metamorphosis.

Neurohormones are the final, largest group of hormones. They are chemical messengers of peptide or small protein composition that regulate aspects of reproduction, development, and metabolism. *Prothoracicotropic hormone*, or PTTH, activates the prothoracic glands to secrete molting hormone. Other neurohormones regulate the activity of the corpora allata to secrete juvenile hormone, control and release *eclosion hormone* (itself a neurohormone) to initiate ecdysis, and release *bursicon* to control the hardening and darkening of the newly molted cuticle. *Adipokinetic hormones* secreted by the corpora cardiaca promote carbohydrate and lipid metabolism, cardiac acceleration, and affect the flight musculature. *Diuretic* and *antidiuretic hormones* secreted by the brain and/or ventral nerve cord ganglia are other neurosecretions that might be mentioned, as well as those neurosecretions involved in the contractile activity of muscles, control of coloration (pigmentotropins or chromatropins), activation of glands that excrete sex pheromone, induction of egg diapause, *etc.*

Structure and Function

Among the important endocrine organs of insects are the brain neurosecretory cells, corpora cardiaca, corpora allata, the prothoracic glands, and the subesophageal ganglion, to be discussed in the order given (*Table 6-1*). There is also the ring gland which, in higher Diptera, encircles the aorta. It is a composite organ formed by the fusion of glands functionally equivalent to the prothoracic glands, corpora allata, and corpora cardiaca.

Brain Neurosecretory cells

The insect protocerebrum bears dorsomedial clusters of large, often lobulated neurosecretory cells (*Fig. 6-1*) in the region of the pars intercerebralis. These cells are generally arranged in two medial and two lateral clusters flanked by the paired protocerebral and optic lobes and separated by the medial protocerebral furrow. They stain deeply with acid fuchsin and certain other dyes and include

Structure and Function

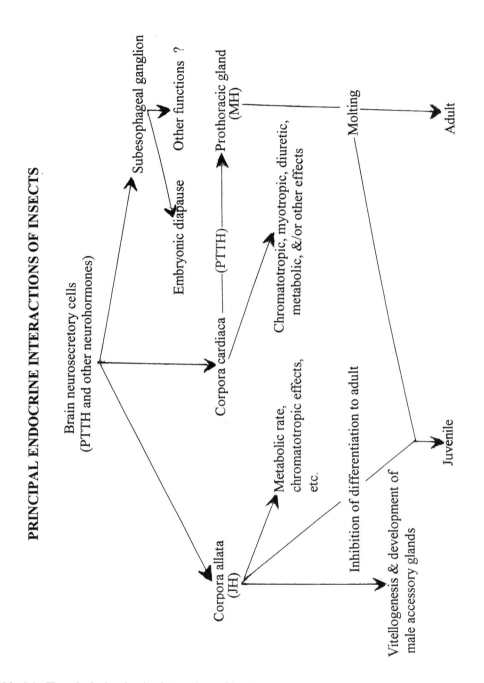

Table 6-1. The principal endocrine interactions of insects.

several histological types. Axons from each medial cluster usually cross inside the brain and exit as a nerve to the corpus cardiacum on the opposite side of the body. Axons from each lateral cluster exit without crossing as a separate nerve to the corpus cardiacum on their side of the body. They terminate

Chapter 6 / ENDOCRINE SYSTEM

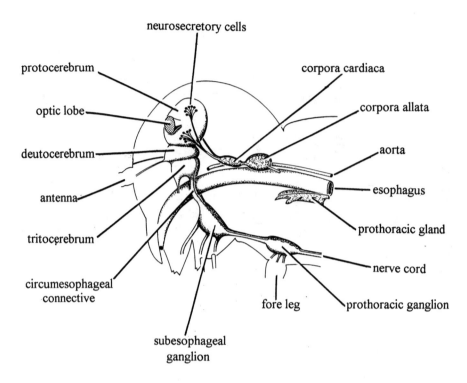

Fig. 6-1. Location of selected endocrine organs with respect to the insect brain and subesophageal and prothoracic ganglia (Modified and redrawn from various sources).

in the corpus cardiacum near the surface where, immediately or after storage, they release neurosecretory granules into the blood or affect intrinsic cells. Other axons from the pars intercerebralis traverse each corpus cardiacum and exit as a nerve to the corpus allatum.

The brain neurosecretory cells produce *prothoracicotropic hormone*, or PTTH, which initiates molting by activating the prothoracic glands. This neurohormone and others produced in the perikarya of the brain neurosecretory cells pass along the axons in neurosecretory granules. Different neurohormones originating in the brain control many physiological processes including regulation of the activity of the corpora allata that secrete the juvenile hormone involved in metamorphosis and reproduction.

The brain neurosecretory cells undergo cyclic changes correlated with secretory activity and development, as shown during pupal diapause (temporary arrest in growth and development). These cells give no evidence of neurosecretion until diapause breaks, whereupon granules appear within the cells of the pars and pass to the corpora cardiaca, which swell accordingly. Some granules pass thence to the corpora allata through their respective nerves. Not surprisingly, severance of either or both of the nerves of the corpora cardiaca causes accumulation of granules forward of the cut and lack of it in the corpora cardiaca and the corpora allata.

Structure and Function

Corpora Cardiaca

The corpora cardiaca are neurohemal organs closely associated with the dorsal aorta just behind the brain (*Figs. 4-3, 6-1*). They are round or elongate in outline and generally paired though sometimes fused into a medial corpus cardiacum. They have nerve connections with the hypocerebral ganglia along the midline, the brain neurosecretory cells forward, and the corpora allata behind. Their milky white tissue differs from the translucent bluish tissue of nerve ganglia and of the corpora allata.

Each corpus cardiacum includes nervous and glandular tissues. The nervous elements constitute a neurohemal organ consisting of neurosecretory axons that terminate in the corpus cardiacum or extend through it into the corpus allatum. The glandular elements include deeply staining, neuron-like cells with "pseudopod" extensions and non-staining interstitial cells. The former, scattered among the latter, give evidence of being modified neurons specialized for secretion. Their "pseudopods" proceed to the surface of the corpus cardiacum where they contact, and presumably pass secretions into, the dorsal aorta.

The apparent functions of the neurohemal part of the corpora cardiaca are to store and release into the bloodstream PTTH and other hormones received from the brain neurosecretory cells. The glandular part of the corpora cardiaca secretes hormones that affect metabolic processes such as blood sugar level and lipid mobilization (release of lipids from the fat body into the blood to be used as fuel during long migratory flights). Other hormones produced and/or released by the corpora cardiaca affect muscle contraction, heart beat, coloration, and other physiological processes.

Corpora Allata

One or two glandular bodies, the corpora allata, lie immediately behind or beneath the corpora cardiaca in pterygotes as well as in apterygotes, except springtails. These glands are ovoid or elongate in outline (*Figs. 4-3, 6-1*). Each connects by a nerve to a corpus cardiacum and often, by another nerve, to the subesophageal ganglion. The corpora allata are usually paired but during development may fuse into a medial corpus allatum. They may also fuse with the corpora cardiaca or, together with the latter, contribute to an unpaired retrocerebral gland.

The appearance of the corpora allata correlates with their cells' glandular activities under control of the brain neurosecretory cells-corpora cardiaca axis. A new phase of activity follows each molt. The cells swell cyclically during their period of heightened secretory activity and then shrink and become vacuolated during their inactive period.

The corpora allata secrete JH, or *juvenile hormone*, which is a sesquiterpenoid that is not stored but released as it is produced. It is lipophilic and associated with a carrier protein in the blood. It is destroyed by JH-esterases. JH levels are reduced just before molting but increase shortly afterward, except during the final molt (during which JH may be completely absent). In some insects, juvenile hormone is a mixture of terpenoids released at different times, so it is properly referred to, not as JH, but as JHs. In juveniles, it is responsible for inhibiting maturation. In adults, it stimulates vitellogenesis in ovaries and development of the male accessory glands; it increases metabolic rate; and it often modifies other physiological processes.

Chapter 6 / ENDOCRINE SYSTEM

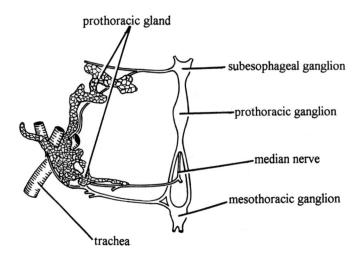

Fig. 6-2. Dorsal view of the subesophageal, prothoracic, and mesothoracic ganglia with respect to a dissected-free median nerve, trachea, and prothoracic gland (Modified and redrawn from V. B. Wigglesworth, 1972).

Prothoracic Glands

These paired, richly tracheated, lobular or bead-like strings of secretory cells innervated by the subesophageal or prothoracic ganglia are closely associated with the main tracheal trunks (*Figs. 6-1, 6-2*). They occur in the thorax of all immature pterygotes, except juveniles with the functionally equivalent part of the ring gland or the functionally equivalent ventral gland. They degenerate upon the final molt though their humoral factor has been detected in the ovaries of adults. They persist in adult apterygotes, which consequently continue to molt throughout life.

Prothoracic gland tissue is semitransparent in the living condition but stains deeply in histological preparations. These lobulated cells, characterized by polymorphonuclear nuclei, undergo periodic secretory activity correlated with the molt cycle.

MH, the ecdysteroid hormone of the prothoracic glands, is a cholesterol-like steroid. Cholesterol or phytosterols are apparently obtained from the diet because insects cannot synthesize steroids. The term MH includes a number of such substances properly referred to as MHs. It is convenient, however, to refer to it in the singular, as I do below. The substance usually secreted by the prothoracic glands, *ecdysone*, is a prohormone, so the actual MH is usually 20-hydroxyecdysone. MH acts on the epidermis to cause molting. It induces the epidermal cells to synthesize nucleic acids and proteins, to detach from the old cuticle, and to lay down a new cuticle, and it simultaneously influences other body parts. Thus, MH is responsible for directing growth and morphogenesis. Its efficacy may be demonstrated by implanting active prothoracic glands or by injecting or slowly infusing the hormone itself into insects in which it does not normally function, as, for example, in pupae, inducing them to molt. PTTH plays only an indirect role in the process, activating the prothoracic glands to secrete MH.

The ventral glands are paired organs located in the ventrolateral, caudal part of the head in Orthoptera and other relatively unspecialized pterygotes. Their function is the same as that of the prothoracic glands and, like them, they degenerate after the final molt, except in termites, the silverfish, and certain other apterygotes. Molting alternates with reproduction throughout adult life in the latter.

Ring Gland

This composite structure, the ring gland, is found behind the brain in higher Diptera. It combines the several retrocerebral glands into a ring encircling the dorsal aorta. In its maximal development, this ring includes the corpora cardiaca and corpora allata, the hypocerebral ganglion, and the pericardial gland (whose structure and function are similar to those of the prothoracic and ventral glands). The ring gland is capable of producing/releasing various neurohormones including PTTH as well as MH and JH.

Subesophageal and Other Ventral Nerve Cord Ganglia

Neurosecretory cells occur in the subesophageal ganglion in a number of insects (*Fig. 6-1*), and nerves sometimes connect this important coordinating center of the nervous system with the corpora allata or the prothoracic glands. The secretions of these cells are usually peptides but, on occasion, may be biogenic amines. These facts suggest an endocrine as well as a nervous function. These subesophageal neurosecretions control embryonic diapause in the silkworm moth.

Neurosecretory cells are also present in other ganglia of the ventral nerve cord in association with small neurohemal organs located laterad of the nerve cord.

Growth and Metamorphosis

Metamorphosis refers to possession of a number of postembryonic body forms in individuals of the same species. It is realized as a step-wise pattern of development involving several immature stages and a terminal adult stage, each more or less different from the preceding and separated from it by a molt. Not only does metamorphosis involve change, it is a directional change toward the adult. Control of these developmental events is mediated by the metamorphosis hormones secreted in predetermined sequence to assure perfect overall timing.

Molting is initiated by sensory input that activates certain of the brain's neurosecretory cells. In the assassin bug, for example, this input involves stimulation of stretch receptors following a blood meal. In other insects, it may be photoperiod, temperature, diet, crowding, or other factors. Whatever the immediate cause, the brain neurosecretory cells are somehow induced to secrete PTTH, triggering molting. Inasmuch as these cells are incapable of inducing molting in absence of the prothoracic glands or their homologues, the role of PTTH is to stimulate the prothoracic glands to secrete MH which, in turn, initiates molting. Then JH, operating simultaneously with MH, modifies expression of the molt by inhibiting differentiation of adult characteristics.

Chapter 6 / **ENDOCRINE SYSTEM**

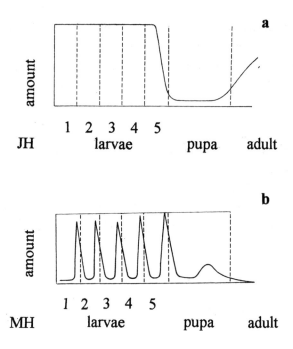

Fig. 6-3. Schematic comparison of (*a*) JH and (*b*) MH levels during insect development.

The same three hormones are involved in all molts, except the final one. The sequence is initiated by PTTH, but postembryonic development, once underway, depends on the relative concentrations of MH and JH. MH, whenever secreted, stimulates the synthetic activities that cause molting, but the kind of cuticle produced depends on the JH that is secreted simultaneously. The presence of JH prevents metamorphosis to the adult. Then the final secretion of MH causes a terminal molt that, in absence of the adult-inhibiting influence of JH, produces adult cuticle (*Fig. 6-3*).

Postembryonic growth and maturation are, therefore, products of two opposing forces, MH and JH. MH imposes on the insect a series of molts resulting in growth and differentiation toward the adult stage. JH delivered in relatively large doses inhibits differentiation of adult characteristics; in small doses or in large doses delivered late, the insect pupates; and, in absence of JH, the animal transforms into an adult.

An important aspect of the MH-JH interplay involves the latter's indirect role in production of the former. In the presence of JH, the prothoracic glands enter a resting phase at the end of each cycle of activity. Following an appropriate interval, they resume activity and secrete MH, inducing the next molt. However, prothoracic glands that go through a cycle of activity in absence of JH, as happens during the final molt, break down and are resorbed in the resulting adult. The adult, now lacking prothoracic glands, cannot molt again, whereupon the corpora allata resume activity and begin influencing reproduction.

The results of several experiments illustrate these relationships. Individuals of the assassin bug, for example, go through five nymphal stages after hatching. Each nymph takes a single, massive blood meal inducing the next molt, which occurs some weeks after gut and body distension. The molt

Growth and Metamorphosis

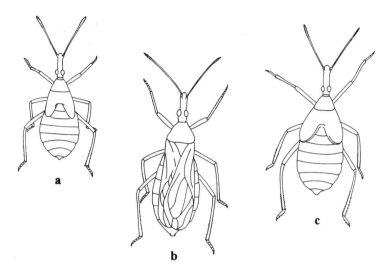

Fig. 6-4. Arrested development in a reduviid bug showing (*a*) a 5th-stage nymph that normally molts into (*b*) a winged 6th-stage adult but instead becomes (*c*) a large, wingless 6th-stage nymph under the influence of implanted corpora allata (Modified and redrawn from V. B. Wigglesworth, 1974).

is triggered following a critical period of some days following the meal. If a 5th-stage assassin bug does not feed or if fed and decapitated *before* the critical period has elapsed, the animal fails to molt though it may continue living for months afterward. However, if the insect is decapitated *after* the critical period or if the decapitated insect receives a blood injection from another insect decapitated *after* the critical period, the animal molts into a headless adult. This suggests the involvement of blood-borne factors. Once appropriately stimulated, the insect's brain neurosecretory cells secrete PTTH inducing the prothoracic glands to secrete MH. This also explains the head critical period. Inasmuch as several post-feeding days are required for production of the PTTH needed to activate the prothoracic glands, decapitation prior to this time stops molting, but decapitation subsequent to it has no effect on the molt that follows.

Larvae of the cecropia moth and of certain other holometabolous insects can be ligated, yet continue to live, divided, for long periods of time. The ligature may be tied at different levels of the body, separating the larva into fore and hind sections between which blood flow is interrupted. If the ligature is tied behind the thorax *after* the thoracic critical period (*Fig. 6-5 c*), both the fore and hind sections pupate. If tied at this level *before* the thoracic critical period has elapsed (*Fig. 6-5 b*), the fore section containing the brain and prothoracic glands pupates, but the hind section remains larval.

These experiments relate only to PTTH, MH, and molting. Other tests demonstrate the complex MH-JH interplay that eventually results in maturation. A 5th-stage assassin bug (*Fig. 6-4 a*) generally molts once more into a winged 6th-stage adult (*Fig. 6-4 b*). However, if an early 5th-stage juvenile is injected with JH-containing blood from a 4th-stage nymph, it molts into a large 6th-stage nymph (*Fig. 6-4 c*) rather than into the expected adult. Likewise, implantation of active corpora allata into

Chapter 6 / **ENDOCRINE SYSTEM**

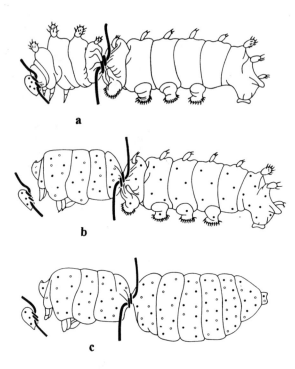

Fig. 6-5. In *a*, a caterpillar ligated *before* release of PTTH fails to pupate because the head-thorax ligature prevents the hormone from spreading to the prothoracic glands; in *b*, after release of PTTH (*dots*), the thorax pupates, but the thorax-abdomen ligature prevents MH (*circles*) from the prothoracic glands from spreading into the abdomen which remains larval; and in *c*, the entire body of a caterpillar ligated *after* release of PTTH and MH pupates. In *b* and *c*, the head-thorax ligature is irrelevant, and the head may be removed altogether (Modified and redrawn from C. M. Williams, 1950).

early last-stage immatures prolongs their juvenile life, and surgical removal of these glands from immatures results in their precocious attainment of adult-like attributes in a diminutive body.

These experiments suggest that, in the presence of JH, the next molt can only produce larval cuticle. Then, when JH is no longer secreted, the juvenile molts into an adult. In short, growth and differentiation result from a balance struck between MH, inducing molting, and JH, inhibiting adult differentiation.

Control

Some modern insecticides used to control noxious insects have become dangerously concentrated in the environment. Their side effects have affected, not only the pests at which they are targeted, but other animals including humans. Moreover, some insects are becoming physiologically resistant to these chemical pesticides. In casting about for an answer to this dilemma, entomologists have turned to the use of insect hormone analogues.

Growth and Metamorphosis

We have seen that JH is necessary for normal development of juveniles, and its continuing presence prevents metamorphosis to adults. The resulting creatures are unable to reproduce and eventually die. It follows that certain hormone analogues may have potential use as insecticides during these critical life stages. For example, a certain European species of red bug proved puzzling to entomologists during early attempts to rear it in the United States. This insect was reared successfully in cages lined with paper of European or Japanese origin but not in cages lined with American paper. The bugs either remained permanent nymphs or molted into sterile adults when reared on American paper though other insect species were unaffected. The responsible "paper factor," *juvabione*, has been traced to balsam fir, the principal source of pulp in USA but not elsewhere in the world. Here, then, in juvabione is an extractable JH analogue with selective action against this red bug but not other species. Although this particular bug is not a pest, these findings suggest that some of the thousands of known terpenoids and related compounds may likewise prove to be JH analogues with potential use in control.

Today several potent JH analogues available as commercial preparations are being used against pest insects including, for example, methoprene (trade name *Altosid*) against mosquitoes and pyriproxyfen (trade name *Sumilarv*) against destructive moth larvae. An important feature of these analogues is that they penetrate the insect cuticle, so they are effective by topical application and can be sprayed.

It is also likely that the future will see increasing attempts to use MH-based substances or agents that block MH synthesis as means of control. MH is released at precise rates and times as normal development and molting progress. Certain developmental events are initiated by low concentrations of MH and require days to go to completion. Research shows that, if insects are exposed to abnormal concentrations of MH, or, if exposed to the hormone at an improper time, certain developmental events may be by-passed, whereupon a new, often incomplete cuticle is secreted prematurely and internal morphogenesis is interrupted. The affected insects may die after having been induced to molt by injecting or topically applying MH or after having been fed the substance.

Suggested Additional Reading

Bursell (1970); Chapman (1991, 1998); Davies (1988); Nation (2002); Nijhout (1994); Richards & Davies, vol. I (1977); Wigglesworth (1954, 1959, 1970, 1972, 1974, 1984); Williams (1950, 1958, 1967).

Chapter 7
Digestivoexcretory System

*Tell me what you eat, and I
will tell you what you are.*
— Brillat-Savarin

Insects have a combined digestivoexcretory system. This designation satisfies the realities of structure and function but violates zoologists' traditional relegation of the digestive and excretory processes into separate systems. This combined system processes food stuffs to yield energy for metabolism and then eliminates wastes resulting therefrom. The system's *digestive division* ingests food, transports it along the length of the gut, digests it into simple food molecules, absorbs these molecules into the bloodstream, and egests (defecates) the undigested food residues. The system's *excretory division* is concerned with excretion, defined as maintenance of the body's internal environment within physiological limits. Excretion itself is a complex process involving removal or storage in insoluble form of nitrogenous and other metabolic wastes and restoration of balance among the body salts, water, and other substances.

Food and Its Treatment by the Digestive Tract

Included among materials that insects consume are such non-nutritive, albeit valuable substances as water, vitamins, salts, and minerals. Also included are so-called "true foods," *viz.*, carbohydrates, fats, and proteins. These complex organic materials of plant or animal origin needed in quantity for metabolism and growth usually require degradation for absorption into the bloodstream but subsequently may be reconstituted into more complex molecules for storage or construction of body tissues.

Carbohydrates

The sugar molecules that insects use for most of their energy for muscular activity are termed *carbohydrates*. They occur singly in simple molecules such as glucose or fructose, or combined into larger molecules such as the double sugars maltose, sucrose, cellobiose, and trehalose, or in complex molecules such as starch, glycogen, or chitin.

Fats

Glycerol and fatty acids combine to form *fats* which serve as insects' common storage food and also provide energy during flight. These compact, readily oxidized food molecules yield much energy per

Chapter 7 / DIGESTIVOEXCRETORY SYSTEM

unit of weight. They contain the same elements as carbohydrates but have a reduced oxygen content. They also may contain nitrogen, with or without phosphorus.

Proteins

Proteins are long-chained polymers of amino acids, about ten different ones of which are essential for normal insect nutrition. They may be used in construction of new tissue or enzymes, or they may be deaminated and then metabolized within the fat body to yield the energy of their phosphate linkages. Amino acids other than proline are seldom used as energy sources, except during starvation.

Enzyme Adaptation to Food

Insects' digestive processes are handled by a battery of enzymes specialized for particular foods. For example, cockroaches, many other insect omnivores, and plant feeders such as lepidopterous caterpillars have an appropriately broad complement of enzymes including protease, lipase, maltase, invertase, and especially amylase. There is a corresponding enzymatic reduction in insects of more restricted food habit. Thus, nectar-feeding adult butterflies secrete only invertase and sometimes maltase; blow fly adults (which are also sugar-eaters) secrete amylase, invertase, and maltase but have weak proteases; and the blood-sucking adults of the tsetse fly have strong proteases but weak amylase. Moreover, a few insects are adapted to digest such seemingly resistant substances as chitin, keratin, beeswax, and even cellulose. Most so-called "wood-eating" insects do not themselves digest the cellulose that composes the bulk of their diet but rely upon gut symbionts (often in special gut pouches) to digest it for them.

Digestive Tract Structure

The insect digestive tract (*Fig. 7-1*) extends between the ends of the body and opens outside through an anterior *mouth* and *preoral cavity* and a posterior *anus*. It is a tube that is either long or short and either straight, gently curved, convoluted, or tightly coiled. It may have one or more *caeca*, or it may lack these blind pouches. It is fixed to the body wall at the pharyngeal and rectal musculature but elsewhere is only loosely attached by tracheae and connective tissue.

Insects' alimentary canal consists of three principal regions: fore gut, mid gut, and hind gut. The *fore gut*, or stomodaeum, and the *hind gut*, or proctodaeum, arise embryonically from ectodermal invaginations from the anterior and posterior ends of the body, respectively. The *mid gut*, or mesenteron, develops as an entodermal sac that, during embryogenesis, connects to, and then becomes confluent with, the invaginating fore and hind guts.

Fore Gut

The fore gut of most insects is a long tube specialized lengthwise into organs of varied structure and diameter innervated by a dorsal branch of the sympathetic nervous division. It includes the following tissues arranged concentrically around the *lumen*, or gut cavity (*Fig. 7-2 a*): intima, epidermis, base-

Digestive Tract Structure

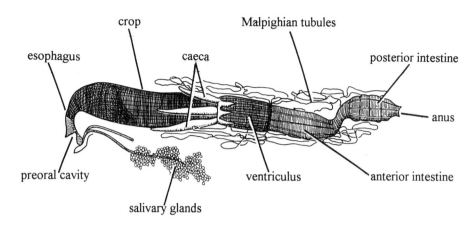

Fig. 7-1. Lateral view of a generalized insect's major digestivoexcretory organs (Modified and redrawn from R. E. Snodgrass, 1935).

ment membrane, longitudinal muscle (visible as numerous cross-sectioned fibers), and circular muscle (visible as concentric, overlapping fibers). The intima is a cuticular membrane confluent with, and homologous to, the external body cuticle, so it is cast during ecdysis. It is often thrown into irregular longitudinal ridges, and its internal surface may bear short "hairs," spines, or spicules that commonly point backward toward the anus. The epidermis is a single layer of often flattened, indistinctly separated epithelial cells continuous with, and homologous to, the epidermis of the outer body wall. Closely applied to the epidermis is the basement membrane. Peripheral to the basement membrane are longitudinal and then circular muscle layers, both consisting of striated fibers. The longitudinal fibers either insert on the intima or form a knot-like plexus without epidermal connections. The circular fibers insert on one another and form a continuous, often well-developed muscular sheath. There is also an ill-defined connective tissue membrane (not shown in *Fig. 7-2*) lining the outer surface of the digestive tract, bearing nerves and tracheae and separating the tract from the surrounding body cavity.

The several fore gut organs of a generalized insect (*Fig. 7-1*) include (lengthwise from mouth backward): buccal cavity and pharynx, esophagus, crop, and proventriculus. The *buccal cavity* is the anterior part of the pharynx. It is an ingestive chamber readily identified by fan-like extrinsic muscles that anchor and dilate it (*Fig. 5-1 b*). It bears the mouth and is not to be confused with the preoral cavity. The latter is not actually part of the digestive tract but merely the space enclosed by the mouthparts.

Paired *salivary glands* are closely associated with the preoral cavity and mouth. Strictly speaking, they too are not part of the digestive tract, but they discharge into it and are vital in its function. They are relatively small and restricted to the head in some species; they lie within the thorax in others; and, in still others, they are so extensive that they reach into the abdomen. Whatever their size and location, they consist either of simple or of convoluted tubes, lobes, or highly branched structures. Each paired salivary duct into which they flow may be provided with a reservoir, and the com-

Chapter 7 / DIGESTIVOEXCRETORY SYSTEM

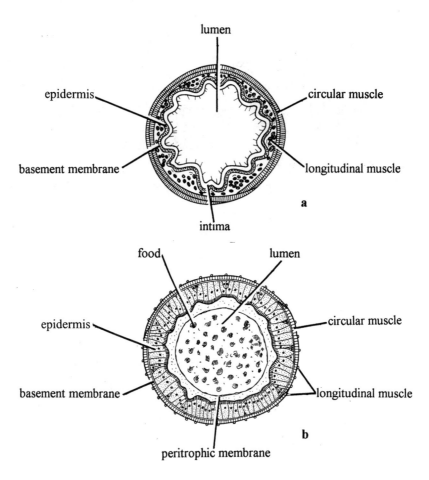

Fig. 7-2. Cross-sections of (*a*) generalized crop and (*b*) ventriculus, the latter showing within its lumen the food-enclosing peritrophic membrane (Modified and redrawn from R. E. Snodgrass, 1935).

mon duct formed by their confluence discharges into the preoral cavity at the hypopharynx. These glands secrete *saliva*, a mixture of digestive enzymes and mucus. In blood-sucking parasitic species, saliva contains an anticoagulant that keeps the host blood flowing; in lepidopterous caterpillars, it produces silk for constructing the cocoon; and in many ants and wasps the enlarged salivary glands produce fluids that adults and larvae exchange during trophallaxis (see Chapter 17).

The *esophagus* is an unspecialized food passageway that expands gradually as it joins the crop. However, it takes the form of a long, slender tube extending directly to the mid gut in insects in which the crop and proventriculus are lacking (*Fig. 7-3 a*).

The *crop* (*Figs. 7-1, 7-3 b, 7-4*) is often a capacious, symmetrical dilatation of the alimentary canal that may constitute much of the fore gut. Its intima tends to be thick and its epithelium flat; its muscle layers are weak; and its wall may be thrown into longitudinal folds allowing for distension. However, in most butterflies and moths (*Fig. 7-3 c*), it consists only of a ventral esophageal diverticulum.

Digestive Tract Structure

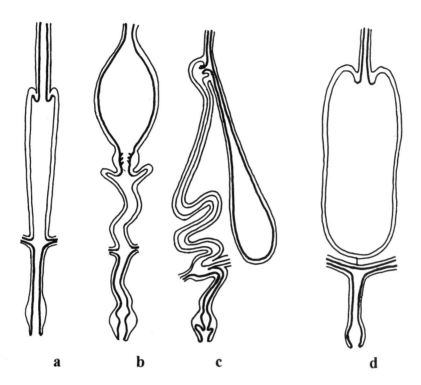

Fig. 7-3. Diagrammatic views of selected insect alimentary canals including those of (*a*) primitive insects and many larvae, (*b*) many generalized insects, (*c*) "higher" Diptera, and (d) larval Hymenoptera, with the fore- and hind-gut structures indicated by a *heavy* internal line and mid gut structures by a *light* line. The point of attachment of the Malpighian tubules is indicated by paired broken tubes (Modified and redrawn from V. B. Wigglesworth, 1972).

The *proventriculus* is often the most specialized part of the alimentary canal. In many larvae, it is a flexible, constricted, posterior segment of the fore gut; in the honey bee, it is a muscular valve with recurved spines; and, in crickets and cockroaches (*Figs. 7-3 b, 7-4*), certain beetles, and other insects that eat hard foods, it is a specialized gastric mill with powerful musculature, a deeply folded internal surface, and a hardened, complexly armed intima. This armature is often useful in identification of genera and species.

The *stomodael valve* (*Fig. 7-4*) marks the posterior end of the fore gut. Its sphincter, when contracted, inhibits regurgitation but, upon relaxation, allows for backward passage of food from the crop into the ventriculus.

Mid Gut

The mid gut, or mesenteron, has an entodermal epithelial layer unlike the ectodermal epithelia of the fore and hind guts. In cross-section, it differs from the fore gut in its lack of an intima and in its possession of a *peritrophic membrane*, its thick digestive epithelium of tall columnar cells, and its

Chapter 7 / DIGESTIVOEXCRETORY SYSTEM

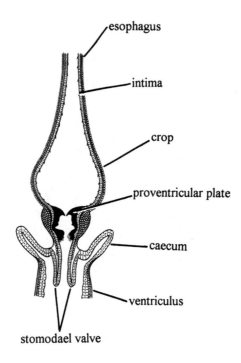

Fig. 7-4. Longisection of an orthopteran crop, proventriculus, and anterior ventriculus (Modified and redrawn from R. E. Snodgrass, 1935).

weakly developed, reversed muscle layers (longitudinal muscles peripheral to circular muscles) (*Fig. 7-2 b*).

The ventriculus is the main organ of the mid gut (*Figs. 7-1, 7-4*). It is usually a uniform tube, either straight or coiled, but it may be divided into two or more regions of different diameter, structure, and function. It may lack *caeca* or bear numerous such blind pouches that increase the effective secretory and/or absorptive surfaces. In certain larval Hymenoptera and Neuroptera (*Fig. 7-3 d*), the ventriculus is a blind sac in which food residues accumulate until, at the end of metamorphosis, the alimentary canal completes itself, ending a classical case of constipation!

The fore and hind guts are protected internally by a cuticular intima that is missing in the mid gut, apparently replaced there by the peritrophic membrane (*Fig. 7-2 b*). This thin, colorless mucopolysaccharide sheath encloses food within the mid gut. It is produced either by secretion from a ring of cells adjacent to the proventriculus or by delamination of successive, concentric layers from the walls of the entire midgut. It resembles the fore gut and hind gut intima in being chitinous but differs from them in that it is entodermally derived. It apparently protects the soft, vulnerable mid gut cells from mechanical damage and prevents the food mass from bulging into the caeca, except in instances in which it is also secreted within the caeca. Through its selective permeability, it may also render the feeder resistant to tannins, other chemicals, or even harmful bacteria accidentally ingested with the diet.

Hind Gut

The hind gut, or proctodaeum, shares many structural features with the fore gut and, like it, develops from an ectodermal invagination. However, its epidermal cells are flat or columnar, usually larger than those of the fore gut, and give little evidence of a glandular function. Moreover, the hind gut intima is thinner than that of the fore gut and is freely permeable to water. The muscle layers of the hind gut are poorly developed, except around the rectum, and they are more varied than those of either the fore or the mid guts. The circular fibers are usually located internal to the longitudinal fibers though additional circular bands may develop peripherally, along with prominent extrinsic muscles. The latter are fan-like bundles arising from the body wall that support and dilate the hind gut. Hind-gut innervation is from the last abdominal ganglion.

The hind gut is commonly divided into *anterior* and *posterior intestines* separated externally by a constriction (*Fig. 7-1*) and internally by a rectal valve. An additional organ, the pylorus, may be developed between the mid gut and the anterior intestine. When present, as in lepidopterous caterpillars, the pylorus is constricted and bears an internal pyloric valve near the origin of the Malpighian tubules.

The anterior intestine receives the *Malpighian tubules* whose presence makes it recognizable when there is no pylorus. Any tract forward of the tubules is arbitrarily mid gut and any behind it is anterior intestine. The anterior intestine may be straight, pear-shaped, or convoluted, and it may be simple or divided into an anterior ileum and a posterior colon, often of different diameter and histology and separated by a visible external constriction. The anterior intestine usually opens into the posterior intestine through a rectal valve.

The posterior intestine is often differentiated into an expanded rectal sac anterior to the rectum. The *rectum* voids caudally through the *anus*, which is closed by an anal sphincter. It is usually thin-walled except at the rectal pads, which are regions of columnar epithelial cells lined internally by cuticle and provided with tracheae. There are usually six rectal pads arranged radially within the rectum.

Digestive Tract Function

Fore Gut

The *pharynx* has dilator muscles for *ingestion* (food intake) and, to a lesser extent, *digestion* because, as food is taken in, it is mixed with saliva containing enzymes responsible for limited digestion and mucoid substances for lubrication. Indeed, some insects extrude saliva directly onto the food mass initiating limited digestion before entry into the gut. The salivary enzymes usually include amylase, invertase, and sometimes also protease and/or lipase, the occurrence of which accord roughly with diet. Relatively larger amounts of saliva are swallowed with dry than with moist food.

The esophageal intima's thickness suggests that it functions purely in *conduction*, conveying food morsels from the pharynx to the crop by active peristaltic movement.

Chapter 7 / DIGESTIVOEXCRETORY SYSTEM

The *crop* acts as a temporary storage organ for ingested food in, among other insects, grasshoppers and their allies, bees, wasps, and some beetles and true flies, in which it occupies the thorax and may extend into the abdomen. It is so capacious in cockroaches, for example, that individuals fed to repletion on oil and sugar can survive for weeks on the stored food. Arguably, the crop is also a site of digestion from swallowed salivary enzymes and from enzymes regurgitated forward from the ventriculus. The crop's recorded pH values vary, being partly a function of the ingested food's pH. There is no appreciable fore gut buffering.

Food is driven backward through the alimentary canal by active propulsive contractions of the gut wall controlled by the sympathetic nervous division. These contractions, known as *peristalsis*, have been widely studied in the crop, in which they begin and sweep backward from the anterior end of the organ. There are also antiperistaltic contractions that progress forward, as well as gross contractions that churn the food rather than move it definite directions. These contractions are initiated upon ingestion, being under nervous control, and are independent of peristalsis elsewhere in the gut.

The *proventriculus* (*Figs. 7-3 b, 7-4*) is structurally varied suggesting that it performs a number of tasks. It pulverizes food in live katydids in which wax "pencils" inserted into it become indented and in cockroaches whose strong anterior denticles break larger food morsels into smaller ones. A valvular function is also indicated for the cockroach proventriculus because it restrains food until the pieces have been partly digested. Finally, the specialized proventriculus of the honey bee seems to have a propulsive function, funneling pollen grains into the mid gut without crushing them.

The *stomodael valve* (*Fig. 7-4*) is intimately associated with the valvular function of the proventriculus. It consists of folds that project backward into the ventriculus. They ensure enclosure of the food mass by the peritrophic membrane secreted by the proventriculus, and, together with the associated sphincter, inhibit forward regurgitation of food particles from the ventriculus into the crop.

Mid Gut

The mid gut is divided into functionally different zones, some of which may secrete, some absorb, some function both ways, and some take part in intermediary metabolism (cellular reactions not immediately concerned with energy release). The largely columnar mid-gut digestive cells seem capable of carrying out both secretion and absorption, whatever their zone, perhaps through alternating periods of secretion and absorption. The absorption rate seems controlled by the crop's rate of release of food into the mid gut.

The physiologically active mid gut cells give histologic evidence both of merocrine secretion in which the cell membranes rupture to discharge enzymes, leaving an intact nucleus capable of resumption of secretion at a later time, and of holocrine secretion in which the cell membranes rupture and discharge the entire content necessitating replacement by regenerative cells.

The mid gut is usually buffered, so its pH remains relatively constant within the species. These values, which condition the medium and influence enzymatic activity, usually range between pH 6.0 to 8.0.

Most *digestion* and *absorption* of food take place within the mid gut cells. Uptake varies according to food type, but generally only small molecules are taken in. Glucose absorption is faster than that of amino acids. The difference seems correlated with the transformation of glucose into trehalose (which is insects' main blood sugar) during passage through the gut wall. This conversion ensures a low glucose concentration within the blood which, in turn, favors rapid absorption. In contrast, the relatively high amino acid concentration of insect blood ensures a shallow diffusion gradient between gut and blood and a correspondingly reduced rate of absorption. Fats differ from carbohydrates and proteins in that they may be absorbed either unchanged or digested into their constituent fatty acids and glycerol.

Digestive enzyme control is poorly understood in insects which range from occasional to frequent feeders. There are no known continuous feeders. This is despite the fact that laboratory-confined aphids seem to excrete continuously and perhaps feed semi-continuously. Perhaps enzymes are secreted continuously in some species, obviating need for a triggering mechanism. This adaptation is appropriate both to insects with a spacious crop and to those living on, within, or near a food that can be ingested relatively continuously, assuring an uninterrupted supply for enzymatic attack. Intermittent-feeding predators and blood-suckers probably have a discontinuous enzyme flow involving neurosecretory control.

The food mass is held within the ventriculus by a pyloric valve that opens from time to time to admit a quantity of food to the anterior intestine.

Hind Gut

From the early days of insect physiology onward, the mid gut has been regarded as the principal site of food digestion and absorption. This idea is based on experimentation and the organ's lack of a chitinous intima. If true, this leaves water absorption and the voiding of undigested food residues to the hind gut. The hind gut must also degrade/absorb excess digestive enzymes and inorganic materials, perform an excretory function by restoring water, salts, amino acids, and other essential substances to the blood, and be active in intermediary metabolism. There is an exception, however, to the generalization that digestion does not occur in the hind gut. This exception occurs in the case of termites and other insects with symbiotic hind-gut microorganisms.

The hind gut's pH is less varied than that of the mid gut, and it is often more acidic, partly due to secretions from the Malpighian tubules.

Passage of the gut contents through the anterior intestine is comparatively rapid, and the residues remain moist. Passage through the posterior intestine is slower, and food may remain there for an appreciable time, being subject to much excretory and metabolic activity.

Egestion, or elimination of food residues through the anus, has been studied in grasshoppers in which it is elicited by electrical stimulation of the last abdominal ganglion. The digestive tract tapers to a smaller diameter at the colon, indicating occurrence of compression and water absorption there. The colon constricts, when full, separating a short piece from the semicontinuous food column, and breaking the food-enclosing peritrophic membrane. Finally, the newly formed fecula is extruded through the anus by peristalsis as the colon returns to its characteristic S-shape.

Fecal pellets, or *feculae*, vary greatly in form and water content. They range from a dry powder in mealworms, to sculptured, obliquely aligned deposits in larval sawflies (*Fig. 7-5 a*), to a formless, paste-like deposit in many butterflies (*Fig. 7-5 b*), to lobulated masses in certain larval caterpillars (*Fig. 7-5 c*), to elongate, spindle-like pellets in grass-feeding grasshoppers (*Fig. 7-5 d*), and to a clear fluid extrusion called honeydew in aphids.

Chapter 7 / **DIGESTIVOEXCRETORY SYSTEM**

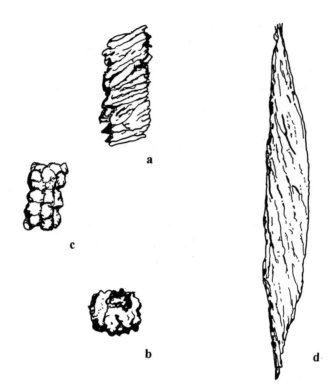

Fig. 7-5. Fecal droppings, or feculae, of (*a*) larval sawfly, (*b*) butterfly, (*c*) larval tent caterpillar, and (*d*) grass-feeding grasshopper (*a*, *b*, and *c* are modified and redrawn from H. B. Weiss & W. M. Boyd, 1950 & 1952; *d* after S. K. Gangwere).

Malpighian Tubule Structure

*All the vital mechanisms, varied as they are,
have only one subject, that of preserving constant
the conditions of life in the internal environment.*
— Claude Bernard

Malpighian tubules are present in most insects. They are slender excretory vessels closed at one end and open into the digestive tract at the other. Lying within the body cavity, they are everywhere bathed directly by blood, except at their apices in certain insects in which they fuse with the rectum. They tend to occur in multiples of two and are generally fewer in number than 100 though certain grasshoppers and their allies may have up to 250 of them. They are sometimes slender, highly convoluted, and relatively few in number, but in other cases they are short and numerous. They may discharge directly into the digestive tract (*Fig. 7-1*) or indirectly via individual swellings called *ampullae* that communicate with the gut (*Fig. 7-6*), or several tubules may fuse basally to form a *ureter*, or combined duct, in contact with the

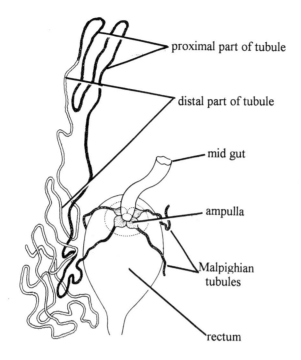

Fig. 7-6. The excretory division of a bug showing only one of its four Malpighian tubules in its entirety (Modified and redrawn from V. B. Wigglesworth, 1972).

gut. They usually discharge into the gut at the confluence of the mid and hind guts, but they may also empty in front of the pyloric sphincter. They are traditionally ascribed an ectodermal (*i. e.*, proctodael) origin rather than the entodermal origin that a mid-gut derivation requires. Malpighian tubules are lacking in springtails and aphids, and they are reduced to papillae in telsontails, Diplura, and stylopids.

Four principal types of Malpighian tubules have been described: an orthopteran type (*Fig. 7-7 a*) in which they lack lengthwise differentiation and lie free within the body cavity; a coleopteran type similar to the orthopteran except for the cryptonephridial tubule apices imbedded within the rectum; a hemipteran type in which the tubules are differentiated into opaque proximal and clear distal sections and lie free within the body cavity (*Fig. 7-6*); and a lepidopteran type (*Fig. 7-7 b*) which is a cryptonephridial version of the opaque proximal/clear distal hemipteran type.

A Malpighian tubule (*Fig. 7-8*) is a single-layered duct. It consists, in cross-section, of a layer of epithelial cells whose internal surface faces the tubule cavity and whose external surface is supported by basement membrane. Each Malpighian cell has a central zone with a brush border composed of vertically arranged protoplasmic processes, an intermediate zone containing prominent nuclei, and a peripheral zone filled with vacuoles and mitochondria. Evidence from electron microscopy suggests that the constantly replaced mitochondria characteristic of the periphery migrate from there into the central zone processes. This transfer could be related to *secretion* (movement of substances against their diffusion gradient by expenditure of energy).

Chapter 7 / DIGESTIVOEXCRETORY SYSTEM

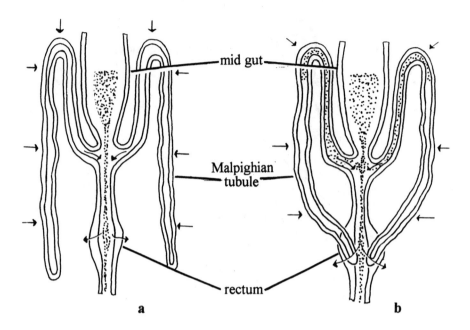

Fig. 7-7. The comparative Malpighian systems of (*a*) an orthopteran and (*b*) a lepidopteran, with arrows indicating the direction of movement of water, ions, and other substances. Note the longitudinal tubule differentiation and cryptonephridial condition of the latter and the lack of these adaptations in the former (Modified and redrawn from R. L. Patton, 1953).

Hemipteran- and lepidopteran-type Malpighian tubules exhibit functional differences correlated with their lengthwise histologic differentiation. Their proximal sections have cilia-like extrusions and an opaque, granular content. Their distal sections have a honeycomb structure, numerous mitochondria and cellular inclusions, and a clear fluid content. These features suggest that the proximal sections are absorptive and the distal ones secretory. However, the entirety of the tubules seems secretory in orthopteran- and coleopteran-type systems.

Associated Structures

The outer surface of a Malpighian tubule is richly supplied with tracheoles that anchor it and facilitate gas exchange. It may also be invested with striated intrinsic muscles that, when present, actively contort it, presumably mixing the content and moving it toward the gut.

Malpighian Tubule Function

Removal of the amino radical, termed *deamination*, allows the protein residue to be metabolized but produces toxic ammonia as well as water and carbon dioxide. Ammonia can be excreted only in a strong dilution not feasible in an insect practicing water conservation. Hence, terrestrial insects usu-

Malpighian Tubule Function

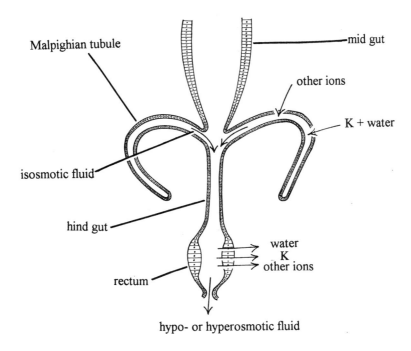

Fig. 7-8. Diagrammatic view of water and salt circulation in the Malpighian tubules and hind gut of a generalized insect, with arrows indicating the direction of movement of water, ions, and other substances (Modified and redrawn from R. H. Stobbart & J. Shaw, 1974).

ally do not excrete ammonia. They excrete non-toxic uric acid and related end-products that tend to crystallize out of solution and can be retained safely for long periods as a solid waste.

Insect blood is often nearly saturated with insoluble *uric acid*, but many insects have fat body or Malpighian tubule enzymes that enable them to degrade the breakdown products of uric acid in addition to uric acid itself. Some insects are able to oxidize uric acid into allantoin; some convert allantoin into allantoic acid; and some split urea from arginine. Trace amounts of soluble urea may also be present. Ammonia is excreted by some aquatic insects, carried in bound form to the excretory organs and then released by deamination. Inorganic calcium carbonate, calcium oxalate, and food-derived salts are likewise common insect excretions to the extent they are not required for metabolism.

Excretory Mechanism

Filtration in vertebrate kidneys uses blood pressure as the driving force. This cannot happen in insects with their negligible, erratic blood pressure. What, then, is the driving force? The answer seems to lie in some combination of *diffusion* and *secretion*, as outlined below.

Insoluble uric acid is found in insect blood, yet somehow it enters the Malpighian tubules and is converted into crystalline form and eliminated (*Fig. 7-8*). How? Organic molecules and inorganic ions apparently enter the Malpighian tubules from the blood by passive diffusion except for uric acid

Chapter 7 / DIGESTIVOEXCRETORY SYSTEM

and potassium, both of which are secreted into the tubule lumen against a concentration gradient. Uric acid secretion is presumably in association with potassium. The potassium ions move into the tubules against a steep gradient, thereby carrying water with them, inducing a flow of liquid urine. This potassium movement and, hence, that of water, proceeds at a rate proportionate to the potassium level of the blood. The resulting filtrate contains a high concentration of potassium ions, moderate concentrations of chloride and phosphate ions, and a reduced concentration of sodium ions, as well as many low molecular-weight substances. It passes rapidly down the tubules, somehow flushing out uric acid, probably as soluble sodium or potassium salts. Most of the water and salts are then resorbed as uric acid or urate precipitate out. This resorption may occur in proximal portions of the Malpighian tubules and in the rectum of insects with longitudinal tubule differentiation, but it occurs only in the rectum in insects lacking tubule differentiation. Urine production and resorption are apparently independent of one another. A diuretic hormone secreted by the brain or ventral nerve cord neurosecretory cells stimulates the Malpighian tubules to increase urine production, and one or more other hormones regulate resorption in the lower part of the Malpighian system and rectum.

Osmoregulation

Insects, like other animals, must maintain a proper concentration of water within the body. This balance is a function of water loss through evaporation, excretion, and egestion as offset by water gain through eating, drinking, and cuticular absorption, and either uptake or removal of inorganic salts with potential for altering ionic, hence, osmotic balance. Interplay of these influences is regulated chiefly by the digestivoexcretory system.

Fresh-water insects take in water with the food and also absorb it through permeable regions of their cuticle, their blood being hypertonic to the aquatic environment around them. They offset this tendency to gain water by eliminating a copious urine, and they minimize salt loss by selective resorption of salts within the rectum. Terrestrial insects, in contrast, are confronted more with problems of water conservation than with difficulties of salt retention. They may conserve water by drinking, by practicing nocturnal behavior, by possessing an impermeable cuticle and valved apertures, by reclaiming water from their excreta, by retaining the water of metabolism, and/or by absorbing water through the cuticle.

Other Excretory Organs

Fat Body

The insect *fat body* consists of a network of closely adherent mesodermal cells derived from the coelomic sac walls. This tissue is present in all stages of all insects, especially Holometabola. It usually takes the form of opaque yellowish or whitish strands, lobes, sheets, or masses closely associated with the digestive tract, situated next to the epidermis, or located elsewhere within the abdominal

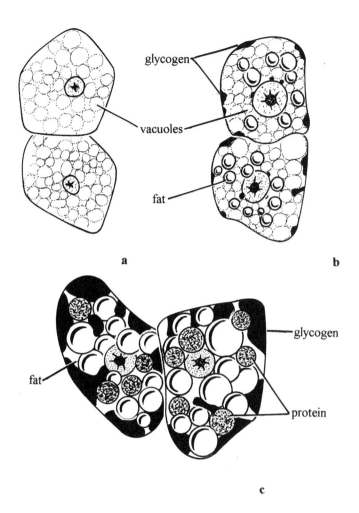

Fig. 7-9. Fat body cells of a mosquito at intervals before and after a meal, with *a* showing the cells' appearance under starved conditions *before* feeding, *b* showing their appearance 2 days *after* feeding, and *c* showing them a week *after* feeding (Modified and redrawn from V. B. Wigglesworth, 1972).

cavity (*Fig. 8-2*). Fat body is freely exposed to blood everywhere within the body cavity in relation to which it carries out its several functions. It is essentially the insect's "liver" and, hence, the site of intermediary metabolism; it converts food stuffs released into the blood by absorption (converting carbohydrates, for example, into trehalose, insects' main blood sugar); it functions in excretion by incorporating free ammonia into less toxic uric acid; and it is the body's main storage organ.

Fat cells are characterized by cytoplasmic vacuoles containing fat and other food reserves. They are relatively free of inclusions in young or unfed insects (*Fig. 7-9 a*), but they increase in size, number, and vacuolization with feeding and growth (*Fig. 7-9 b*). Finally, in well-fed animals, they may become so laden with fat globules, protein spheres, and glycogen rosettes that the cell boundaries become indistinct (*Fig. 7-9 c*). The fat body yields these reserves should the insect be deprived of

Chapter 7 / DIGESTIVOEXCRETORY SYSTEM

food at any time, and it restocks them upon resumption of feeding. It also liberates the contained food into the blood at the onset of molting and metamorphosis, sometimes severely reducing its overall size. The organ then regenerates from surviving cells toward the end of the pupal period, whereupon it again masses food in the adult for use in egg production, hibernation, and starvation.

The fat body is composed mostly of *fat cells*, but *urate cells* and/or *mycetocytes* are also scattered throughout the fat tissue of some species, especially cockroaches. Cockroaches depart from most other insects in that they store their nitrogenous wastes internally within urate cells and do not void urate-coated feculae to the exterior, even during nitrogen stress. Mycetocytes are cells containing symbiotic microorganisms, the functions of which remain obscure, but it appears that both urate cells and mycetocytes are involved in urate storage, mobilization, and utilization in cockroaches.

Nephrocytes

Insect *nephrocytes* are large, mesodermally derived, vacuolated cells that accumulate and store ammonia carmine and other injected dyes, as well as hemoglobin, chlorophyll, albumen, and other molecules too large to be taken up by the Malpighian tubules. They occur either singly or in small, irregular clumps or lobes in various parts of the body. They include paired strands, the so-called *pericardial cells* (*Fig. 8-2*), closely associated with the heart. Their presumed function, whatever their location, is excretion of colloidal particles, particularly protein molecules too large for removal by the Malpighian tubule/rectal complex. They either transform these wastes into assimilable substances or convert them into a form that the Malpighian tubules can eliminate. They may also be a site of hormone synthesis.

Digestive Tract

The important role of the *hind gut* in resorption of water and mineral constituents was mentioned earlier. The Malpighian tubules void into the digestive tract a filtrate containing the small-molecule constituents of blood roughly proportionate to their concentration in blood. The *rectum* and *rectal pads* then resorb those constituents required by the body and reject others, thus maintaining the proper blood volume and composition (*Fig. 7-8*). Hence, the Malpighian tubules and rectum may be said to act as a complex concerned chiefly with excretion of small toxic molecules. The Malpighian tubules ensure removal of any potentially harmful substances from the blood, and the rectum recovers those useful substances that would otherwise be lost. Uric acid granules accumulate in the mid gut in larval Hymenoptera (*Fig. 7-3 d*) though the region does not connect with the hind gut until maturity. Likewise, springtails and aphids, both of which lack Malpighian tubules, handle at least part of their excretion within the gut. This fact is made evident by the fact that the mid-gut epithelium of springtails is periodically cast, and injected fluorescein dye is excreted through their gut wall.

Integument

Uric acid and other wastes may find their way into, and become more or less permanently sequestered within, the epidermis or cuticle to which they impart color. Such *storage excretion* occurs in many insects. In butterflies and moths, for example, uric acid and pteridines lodge within the wing scales, producing the characteristic white, yellow, and other colors of these insects. It may be said, therefore, that storage excretion removes potentially toxic wastes at the same time as it pigments the body in a "useful," adaptive manner.

Suggested Additional Reading

Bursell (1970); Chapman (1991, 1998); Cochran (1985); Davies (1988); Gullan & Cranston (2000); Maddrell (1971); McFarlane (1985); Nation (2002); Patton (1963); Richards & Davies, vol. 1 (1977); Snodgrass (1935); Wigglesworth (1972, 1984).

Chapter 8
Circulatory System

When veins are replete the spirits will stir.
— Beryn 7-8

In order to live, insects must: (1) transport digested food from the alimentary canal to the cells and tissues for metabolism or storage; (2) move the resulting metabolic wastes outwardly to the point of elimination; (3) disseminate hormones which are the chemical messengers that integrate many body functions; (4) store water, food, salts, and other valuable substances; (5) provide a hemodynamic system for body movement, ecdysis, and metamorphosis; (6) buffer against metabolites that threaten harmful internal environmental change; and (7) combat invading bacteria and other injurious foreign objects. They satisfy these diverse needs by means of the circulatory system which is of an *open type* and, thus, different from the *closed system* of vertebrates and most other complex animals.

Insects' open circulatory system consists of a slender *dorsal blood vessel* that pumps blood forward into the head cavity and *hemocoele* (general body cavity) which, unlike vertebrates' true coelom, lacks an endothelial lining. The dorsal blood vessel thus perfuses all tissues including the Malpighian tubules, muscles, digestive tract, and nerves in a nutritive fluid carrying cells, food, salts, hormones, wastes, and gases. This blood, or *hemolymph*, is both a circulating medium analogous to vertebrate blood and a cell-bathing medium analogous to vertebrate lymph.

Insects' tracheal system obviates handling the gas exchanges carried out by the circulatory system of most other complex animals, so they are freed to evolve an open circulation effective in handling storage, metabolic, and other functions but too sluggish to be of much use in respiration. Inasmuch as insects' circulatory system performs negligible respiration, they do not maintain a definite blood volume but rely on water shifts to survive dry conditions. Notwithstanding their fluctuations in blood volume, they are able to regulate the osmotic pressure of the blood primarily by altering amino acid and salt concentrations. Their blood being a storage medium in constant exchange with the fat body, they survive long periods without feeding, as during the arrested growth and metabolism of diapause. They can rely on simple *diffusion* to supplement their weak circulation both because their highly ramified tracheae, Malpighian tubules, nephrocytes, and fat body are directly bathed in blood and because their body dimensions are so small.

Chapter 8 / CIRCULATORY SYSTEM

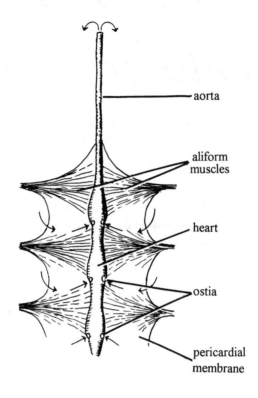

Fig. 8-1. The dorsal blood vessel showing a few anterior heart chambers and the aorta, with arrows suggestive of blood flow (Modified and redrawn from R. E. Snodgrass, 1935).

Structure

Dorsal Blood Vessel

The dorsal blood vessel (*Fig. 8-1*) is usually a straight pulsatile tube extending along the mid-dorsal line just beneath the body wall from near the posterior end into the head. It is composed of circular muscle fibers bounded inside by a delicate limiting membrane and outside by a connective tissue sheath. It includes a dilated abdominal *heart* and a slender thoracic *aorta* that empty into the head in front of and beneath the brain. The heart and aorta are both contractile, have a similar histology, and function in a coordinated manner. However, they are distinct in form, and the heart also includes valves, or *ostia*, that are lacking in the aorta. These valves assure a one-way flow of blood into the heart and direct it forward into the head.

The heart of most insects consists of segmentally arranged dilatations, or *chambers*, which beat in sequence to push the blood forward as the ostia close to prevent back flow. However, the heart is sometimes without external constrictions, its fundamental segmentation indicated only by the presence of paired ostia.

Structure

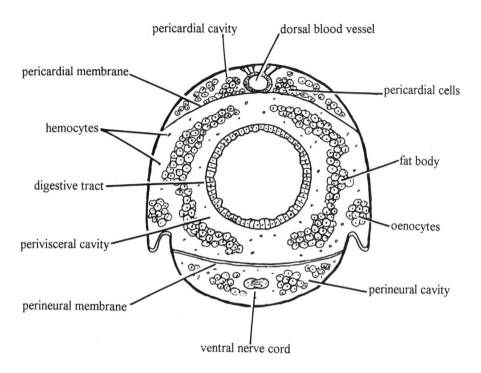

Fig. 8-2. Diagrammatic cross-section of the body wall, body cavities, dorsal aorta, digestive tract, fat bodies, ventral nerve cord, and associated structures (Modified and redrawn from R. E. Snodgrass, 1935).

One might expect insects to have one chamber for each thoracic and one for each abdominal segment except the last, but this condition does not occur in modern insects, in all of which the heart is shortened and its numbers of chambers reduced. Chambers range from many, each with a pair of ostia, as in most insects, to only a single chamber and a single pair of ostia, as in certain juvenile Odonata.

The dorsal blood vessel is suspended from the body wall by means that range from a mid-dorsal suspensory ligament, to radiating connective tissue strands (*Fig. 8-2*), to the fenestrated pericardial membrane, to the dorsal transverse muscles (*Fig. 8-4*), and to tracheae.

Pulsatile Organs

Pulsatile organs are present in the meso- and metathorax of many insects (*Fig. 8-4*). They consist of a muscular diaphragm whose pulsations pump blood into the wing veins. Similar devices may be present elsewhere to promote blood flow through the legs (*Fig. 8-5*), antennae, or other appendages.

Heart Ostia, Diaphragms, and Sinuses

Heart ostia commonly take the form of paired, slit-like, valved openings that extend inward and forward from their side of a given chamber. Modern insects depart from the presumed primitive plan consisting of one pair of ostia per segment. Certain segments may lack ostia altogether, and any ostia present may be non-functional.

Chapter 8 / CIRCULATORY SYSTEM

The insect body is divided by a horizontal septum extending across the body cavity immediately below or at the level of the heart. This *pericardial membrane*, or dorsal diaphragm (*Figs. 8-1, 8-2*) consists of paired fibromuscular sheets that meet along the midline, usually below the heart, often enclosing the aliform muscles. These sheets are sometimes relatively continuous, separating the heart and its surrounding *pericardial cavity* from the *perivisceral cavity* below. More often, they are web-like and perforated by numerous openings.

The *aliform muscles*, or dorsal transverse muscles, located within the pericardial membrane are paired, segmentally arranged, striated muscle bands that originate on the terga and insert on the lateral or ventral heart wall. They correspond in number to the number of heart chambers, so they are mostly abdominal. They are often fan-like or wing-like in form, explaining why the term *alary* is often applied to them.

There is a similar *perineural membrane*, or ventral diaphragm (*Figs. 8-2, 8-4*), in some insects. It stretches horizontally across the body cavity immediately above the nerve cord. Where present, it consists mostly of the *ventral transverse muscles*. It separates the *perineural cavity* formed thereby from the perivisceral cavity (above) except at lateral and caudal openings.

Blood

Plasma

Insects' blood, or *hemolymph*, resembles mammalian lymph more than mammalian blood. It consists of a fluid, or *plasma*, consisting of about 85% water, together with dissolved salts, uric acid, proteins, free amino acids, enzymes, glucose, trehalose, fats, pigments, gases, and suspended *hemocytes*, or blood cells, all freely bathing the exposed organs and tissues and entering the cavities of the legs, wings, and other appendages. The plasma carries nutrients from the gut to the tissues and especially to the fat body for purposes of intermediate metabolism. It transports lipids, amino acids, and carbohydrates to the tissues, nitrogenous wastes to the Malpighian tubules, and hormones from their sites of release to the target tissues. Insect blood also has an important hemodynamic function in the support and movement of soft-bodied larvae and the ecdysis and subsequent expansion of cuticle during molting.

Most investigators looking into the composition of insect hemolymph have based their analyses on whole blood. Their results reflect a wide variability with taxonomic group, life stage, sex, food, and study methods. This variability compromises attempted comparisons with mammalian plasma. Nevertheless, one may generalize that insect blood contains about the same array of chemical constituents as does mammalian plasma though in far different proportions. It has a relatively low chloride but high bicarbonate ion concentration and an amino acid concentration about 10 times that of humans; it has a 9:1 ratio of trehalose to glucose. Total blood sugar concentration is about 10 times that of humans.

Insects' plasma is transparent or lightly tinted by pigments and other dissolved substances. Its pale yellow, brown, green, or other color is a species characteristic but not a taxonomic characteristic. Its pH values usually range between 6.4 to 6.8.

Function

Scattered mesodermal cells overlying the incipient internal organs give rise to muscle, fat body, and other structures during organogenesis. Other cells, left over, are destined to become hemocytes. They circulate freely within the hemocoele and dorsal blood vessel, or they circulate but do not enter the heart, or they attach to the body tissues, or they associate with hemopoietic organs. *Hemopoietic organs* are blood cell-forming organs that occur in various locations throughout larval life but disappear in adults. They contain *phagocytic cells* that engulf cell debris and *stem cells* that give rise to hemocytes to be released into the circulation.

Hemocytes

Hemocytes are nucleate blood cells that vary from round or ovoid in outline when in the passive, free–floating phase to flattened, stellate, or drawn into long processes when in the active, adherent phase. Past attempts at classification have yielded many types according to the species examined and the technique used to prepare the cells for microscopy. Some recognized types are doubtlessly valid, but others probably resulted from failure to distinguish among different aspects of the same cell type.

There are several common types of hemocytes: (1) *Prohemocytes* (*Fig. 8-3 a*) are small, round or ellipsoidal, relatively scarce cells with a large nucleus and a scant, deeply staining basophil cytoplasm. (2) *Granular hemocytes* (*Fig. 8-3 b*) are round or discoidal, mostly non-motile cells common within the hemocyte complex. They vary in size and have a relatively small nucleus within an extensive cytoplasm containing acidophil granules. (3) *Plasmatocytes* (*Fig. 8-3 c*) are blood cells with an extensive, basophil cytoplasm and an irregular outline. They are varied in size and form, often drawn into amoeboid processes, and frequently phagocytic. They are the generalized type of blood cell from which granular and spherular hemocytes may arise. (4) *Coagulocytes* are fragile, easily rupturing cells that initiate the clotting process. (5) *Spherule cells* (*Fig. 8-3 d*) are round or ovoid, non-motile cells containing large refractile spherules. (6) *Adipohemocytes* (*Fig. 8-3 e*) are non-motile cells with an excentric nucleus and conspicuous lipid and granular inclusions. Their distinctive nucleus and relatively small size distinguish them from fat body cells. (7) *Oenocytoids* (*Fig. 8-3 f*) are distinctive non-motile hemocytes that vary in size and shape but contain canaliculi, granules, or crystals within an extensive basophil or neutrophil cytoplasm.

Hemocytes may be counted using a standard hemocytometer, yielding totals up to 100,000 cells per mm^3 of blood, but the total varies widely with species and taxonomic group and may be misleading if the technique used is not standardized. Moreover, hemocytes alternate between adherent phases, characteristic of resting insects, and free-floating phases, characteristic of actively moving or flying insects. Cell counts vary with activity.

Function

Blood percolates among the organs and tissues within the body cavity in accordance with pressure gradients established by movements of the body parts during breathing, walking, flying, and other activities which direct the flow, as modified by less vigorous movements of the pericardial membrane

Chapter 8 / CIRCULATORY SYSTEM

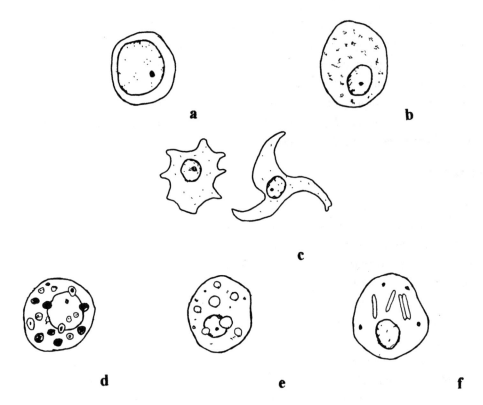

Fig. 8-3. Selected hemocyte types including (*a*) prohemocyte, (*b*) granular hemocyte, (*c*) two plasmatocytes with one (*at right*) drawn into amoeboid processes, (*d*) spherule cell, (*e*) adipohemocyte, and (*f*) oenocytoid (Modified and redrawn from W. S. Romoser, 1981).

and organs that determine the dorsal anterior flow and the perineural membrane and organs that determine the ventral posterior flow. The blood tends to pool when insects are inactive.

General Circulation

The heart itself has little to do either with blood circulation or with blood pressure. It does produce a minimal circulation during inactivity as rhythmic peristaltic waves originate at its posterior end and work forward into the aorta. These *systolic pulsations* help establish a slight pressure gradient that displaces blood already in the head backward into the perivisceral and perineural cavities (*Fig. 8-4*).

The perineural membrane, where present, channels blood backward and assures nerve cord irrigation. Its associated muscle fibers appear to undulate, facilitating circulation, helping pass blood backward, and allowing it to escape into the perivisceral cavity above through openings along the sides and at the posterior end. Blood already in the perivisceral cavity is then pushed and aspirated by the slight pressure gradient into the pericardial cavity surrounding the heart. It reaches the heart by flowing around the open posterior margin of the pericardial membrane (if the latter is continuous) or by entering along the length of the membrane (if it is web-like). Finally, blood is

Function

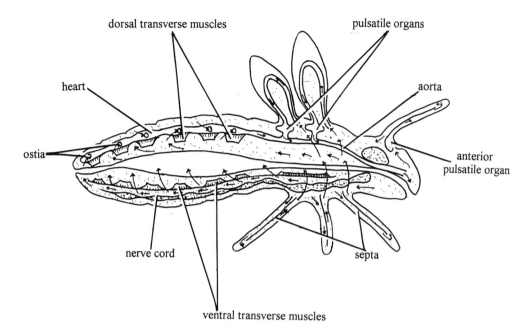

Fig. 8-4. Diagrammatic lateral view of the heart and aorta, the body cavity, digestive tract, nerve cord, and associated structures, with arrows suggestive of blood flow (Modified and redrawn from V. B. Wigglesworth, 1972).

gently sucked into the posterior heart chambers by the fall in pressure associated with *diastole*, or heart relaxation.

The lips of the ostia separate during diastole, as the heart muscles relax and the organ fills with blood drawn in by partial vacuum. This negative pressure is created by pull of the elastic suspensory fibers, usually in indirect combination with aliform muscle contraction. Then the heart muscles contract and the ostial lips close during *systole*, preventing back flow as blood is gently squeezed forward into the aorta. The hind-most chamber pulsates first, followed by the next forward chamber, and so on until the continuous, undiminished wave reaches the anterior end of the aorta, discharging blood into the head.

Systole is an obvious result of peristalsis by the heart's circular muscles, and it provides the brain with a minimal blood supply. However, as noted above, most forward blood flow is caused by body movements as directed by the pericardial membrane. Diastole was once thought to result from the aliform muscles' direct pull on the heart. Now we realize that these muscles usually attach below the heart and insert on the upwardly convex pericardial membrane (*Fig. 8-2*). These contractions probably enlarge the pericardial sinus, displacing blood toward the heart, and pull on the elastic suspensory fibers that are ultimately responsible for heart dilation. Blood flow is relatively constant in absence of body movement and increases with activity. Heartbeat itself does not increase even during such extreme activities as flying because, as noted, the heart provides only a minimal basal circulation. However, the beat generally doubles with each temperature increase of 10°C.

Chapter 8 / **CIRCULATORY SYSTEM**

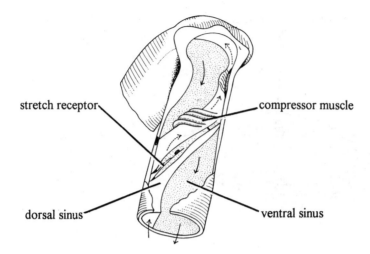

Fig. 8-5. Cut-away view of the pulsatile organ of a bug's fore leg, with arrows suggestive of blood flow (Modified and redrawn from V. B. Wigglesworth, 1972).

Pulsatile organs (*Fig. 8-4*) are auxiliary pumping devices for wing circulation. They gently pump blood into the anterior and then into the posterior wing veins before returning it to the general circulation. Similar pumping devices (*Fig. 8-5*) may be present in the limbs.

Heartbeat Innervation, Control, and Reversal

Pulse, the rate of heart contraction, varies with the species, individual, and conditions. It generally accelerates under increased temperature and heightened metabolism. A wave of contraction is sometimes so rapid that the entire dorsal vessel seems to contract as a unit, but sometimes it is so slow that separate contractions are visible at different points. The heart in certain species may beat in absence of heart-associated nerve cells and ganglia, and even *in vitro* heart fragments with severed nerve connections may beat for hours. This evidence suggests that heartbeat is *myogenic*, produced by the heart muscles themselves in absence of nerve stimulation. However, the dorsal blood vessel usually has paired lateral nerves with branches to the heart wall, aliform muscles, and lateral vessels, if any. These cardiac nerves may be joined by segmental nerves from the ventral nerve cord, but in other cases only segmental nerves are present. The existence of this innervation, the fact that the heart responds to experimental stimulation of them, the fact that acetylcholine may alter the beat, and the fact that increased temperature accelerates heartbeat roughly in accordance with the van't Hoff-Arrhenius equation suggests that the heart is also partly *neurogenic*, being under nervous control. Perhaps the heart is myogenic in some insects and neurogenic in others. Control may also vary with life stage, juveniles having a more myogenic heart and adults a more neurogenic one. Perhaps the heart is inherently contractile but has a rate and amplitude of contraction subject to nervous and/or neurosecretory modulation.

Function

The heart beats forward most of the time, pushing blood into the head, but occasionally it beats backward. Periods of such backward, antiperistaltic heart contraction may intervene between periods of normal, forward peristalsis. This deep-seated heartbeat reversal is not fully understood but possibly could result from mechanical blockage by fat or other tissues.

Hemolymph Volume and Pressure

Blood volume can be measured by bleeding the insect and noting the difference in body weight as corrected for specific gravity or by dilution methods which relate the results to body weight. Fully hydrated insects have a blood volume of 16% to 20% of body weight, but most insects can tolerate tremendous reductions in blood volume resulting from dehydration or water loss. Some insects with a blood volume of less than 5% appear to be almost dry. Blood volume apparently varies with age, developmental stage, physiological condition, *etc*.

Insects always have a low and sometimes even a slightly negative blood pressure. Even under these reduced levels, however, localized muscle contractions may force blood into particular parts of the body cavity, exerting momentary hydraulic pressure at that point, resulting in a relatively high pressure there.

Several consequences arise out of insects' possession of an open circulation. They require both the positive hydraulic pressure of propulsion and the negative pressure of suction to achieve circulation. Even so, their blood flow is sluggish at best. This is demonstrated by their complete mixing time, or *circulatory rate*, which varies from about 10 to 30+ min, as opposed to 2 to 4 min in humans. Blood flow also varies at different times and in different body parts, being more rapid from the aorta to the head than in the body sinuses to the heart. It may proceed within the sinuses either in distinct "jerks" synchronous with heart contraction or slowly and discontinuously, sometimes to the point of momentary cessation.

Hemocyte Functions

Hemocytes are derived from the embryonic mesoderm. New ones arise either from the mitotic division of existing, circulating prohemocytes (*Fig. 8-3 a*) or from previously undifferentiated cells produced by the hemopoietic organs. Hemocytes' best-established role may lie in *phagocytosis*, or engulfment of blood particulates within the size range of bacteria, cell fragments, dead hemocytes, dyes, and other small foreign objects. For example, the breakdown of larval tissue during molting and metamorphosis leaves behind in the plasma a visible cellular debris. This cloudiness disappears upon restoration of the normal blood count after ecdysis, presumably as a result of phagocytosis largely by the plasmatocytes (*Fig. 8-3 c*) but also by phagocytic cells within the pericardial membrane. Sometimes, however, the blood contains parasites or other foreign objects too large to be neutralized by a single hemocyte. In this event, the invading object is encapsulated, chiefly by plasmatocytes, which aggregate around it, adhere to it, become cemented together into a capsule, and thereby neutralize it.

Prohemocytes (*Fig. 8-3 a*) are non-phagocytic germinal cells responsible for postembryonic hemocyte multiplication. As noted above, they also may act as stem cells from which plasmatocytes

Chapter 8 / CIRCULATORY SYSTEM

(*Fig. 8-3 c*) differentiate. Spherule cells (*Fig. 8-3 d*) and granular hemocytes (*Fig. 8-3 b*) are probably normal developmental stages of plasmatocytes. Coagulocytes may also be derived from them. Oenocytoids (*Fig. 8-3 f*), like prohemocytes, are non-phagocytic. They seem involved in growth and metamorphosis and perhaps also in encapsulation.

Clotting occurs almost instantaneously in insect blood. It is apparently associated with coagulocytes which are specialized cells distinguishable under phase-contrast microscopy. They disrupt upon contact with a foreign object whereupon they presumably release a factor that promotes coagulation. Perhaps they release materials forming islets of coagulation, or they extrude cytoplasmic threads. The plug thus formed, however it arises, seals the wound and stops bleeding until healing is completed. Along with plasmatocytes, coagulocytes may help encapsulate bacteria or parasites.

Suggested Additional Reading

Bursell (1970); Chapman (1991, 1998); Davies (1988); Gullan & Cranston (2000); Jones (1977); Nation (2002); Patton (1963); Richards & Davies, vol. 1 (1977); Salt (1970); Snodgrass (1935); Wigglesworth (1972, 1984); Woodring (1985).

Chapter 9

Respiratory System

*So long as men can breathe or eyes can see,
So long lives this, and this gives life to thee.*
— Shakespeare

Structure

Gas exchange in most insects is facilitated by a system of internal air tubes, both large and small, which pipe air to within a few μm of cells deep within the body. This network supports and ties together the internal organs. The larger air conduits, called *tracheae*, and their fine branches, or *tracheoles*, ramify within the body in a manner somewhat analogous to a vertebrate's system of blood vessels and capillaries. Air enters the system through *spiracles*. These openings to the outside are usually valved to minimize water loss and to provide for a directed air flow, inward at some spiracles, outward at others.

The primitive tracheal plan of insects probably involved a pair of spiracles in each body segment (*Fig. 9-1*), tracheae from which branch repeatedly, passing upward to the dorsal blood vessel and associated dorsal musculature, downward to the ventral nerve cord and related ventral musculature, and medially to the digestive tract, other viscera, legs, and wings. Then longitudinal *tracheal trunks* interconnect the spiracles and tracheae of each segment with those of preceding and succeeding segments.

This primitive plan has been altered in living insects by migration or loss of some spiracle pairs and by modification of certain tracheae and disappearance of others. The system develops embryonically the same way in all insects, the tracheae arising as a series of simple, paired metameric, ectodermal invaginations at the sides of the body. These openings become spiracles, and the invaginations, their interconnections, and their branches become tracheae and eventually tracheoles.

Spiracles

The morphologist Snodgrass postulated that ancestral insects had a total of 17 pairs of spiracles on the head, thorax, and abdomen but noted that only the mesothoracic, the metathoracic, and eight or fewer of the abdominal pairs occur today in adult insects. However, a ninth and occasionally a tenth pair of spiracles are found in certain insect embryos. Thoracic spiracles are generally located on or near the pleura and abdominal spiracles at or near the articular membrane separating terga from sterna.

The spiracles of certain apterygotes (*Fig. 9-2 a*) are simple openings without either a closing apparatus or an atrium. This kind of structure presents little physiological handicap for animals living on or about moist soil, as do most of these primitive, wingless hexapods. However, most pterygotes

Chapter 9 / RESPIRATORY SYSTEM

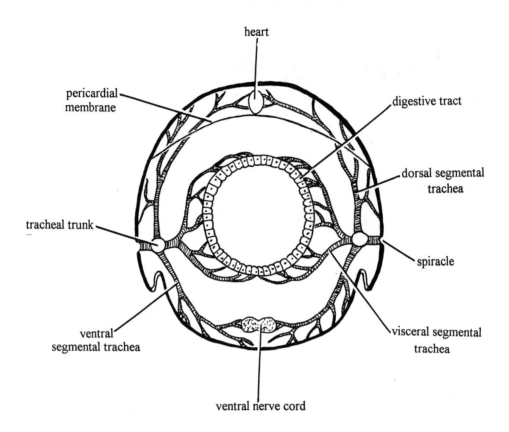

Fig. 9-1. An idealized cross-section of the insect body wall, digestive tract, heart, ventral nerve cord, and tracheal trunks (Modified and redrawn from R. J. Elzinga, 1981).

inhabit dry environments for which they have evolved a waterproof cuticle, a specialized excretory system, and valved spiracles kept closed for periods of time to conserve water. Spiracles are commonly sunken into an atrium (*Fig. 9-2 b, c*) that is often rugose and fitted with a filter and a valve (*Fig. 9-2 d*). The filter, where present, consists of interlaced setae that allow for passage of air but exclude dust and other foreign particles, and the valve consists of muscles and associated cuticular parts capable of occluding the air passageway. Valve structure varies with the species, taxonomic group, and even the life stage and location on the body.

A lip-like external valve is common on the insect thorax. A metathoracic spiracle of a grasshopper, for example, is a slit-like opening guarded by two elastic lips joined by a ventral lobe. The opening is closed by contraction of an occlusor muscle that inserts on the lobe to pull the lips downward and together. Dilation is by natural elasticity. An internal valve varying from a simple pinchcock to a true valve is common on the insect abdomen. Orthopteroids, for example, have an internal double-bar pinchcock consisting of two sclerotized processes on the opposite walls of the atrium adjacent to the tracheal mouth, with an occlusor muscle stretched between them.

Structure

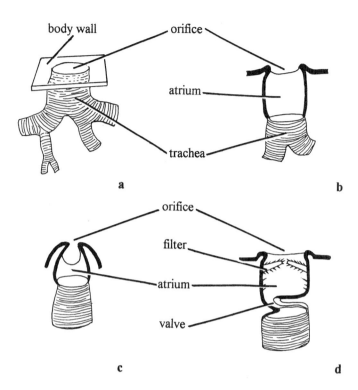

Fig. 9-2. Selected spiracle types including (a) an open spiracle without an atrium, (b) an atriate spiracle, (c) an atriate spiracle with a lip-like closing apparatus, and (d) an atriate spiracle with a filter and an internal closing valve (Modified and redrawn from R. E. Snodgrass, 1935).

Tracheae

Tracheae (*Fig. 9-3*) are characteristic of a number of arthropod taxa but reach their highest expression in insects. At the air interface within a trachea is an intima, or layer of chitin, protein, and cuticulin that is continuous with the external body cuticle and, therefore, shed when the insect molts. Peripheral to the intima is an epithelium of flat polygonal cells responsible for secreting the intima, and, peripheral to that, on the hemolymph side of the epithelial cells, is a non-living basement membrane.

Air-filled tracheae are characterized by a silvery color and a cross-striated appearance. The striations, called *taenidia* (*Fig. 9-4 a*), are spiral thickenings of the intima adapted to resist tracheal compression and collapse. Taenidia may also terminate in a line along each side of a trachea (*Fig. 9-4 b*), permitting the tube's collapse and flattening into an *air sac* along that line. Air sacs also have taenidia, but they are often widely spaced rather than arranged in a tight spiral, as in tracheae.

Tracheoles

The finest tracheae give rise to microscopic tracheoles (*Fig. 9-3*). These apical respiratory "capillaries" arise from specialized epithelial cells called tracheoblasts. The latter develop long, finger-like

Chapter 9 / RESPIRATORY SYSTEM

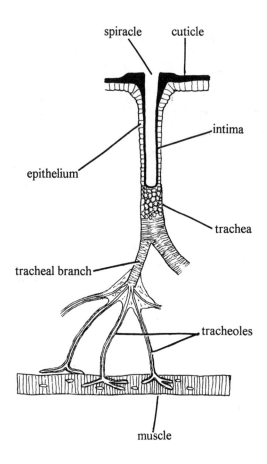

Fig. 9-3. An idealized view of relationships among a spiracle, a trachea, a tracheole, and a muscle (Modified and redrawn from Ross *et al.*, 1982).

processes that extend into the body tissues, and, within them, a cavity or tracheole develops and lengthens, eventually becoming continuous with the originating trachea. Tracheoles usually branch dichotomously into a fine network before ending blindly within the tissues and organs. This network, once formed, is not stationary but grows toward any oxygen-deficient zones that develop within the body upon injury or growth and becomes increasingly complex with each molt, apparently in relation to metabolic need.

Tracheae and tracheoles were once thought distinguishable on the basis of the presence (in tracheae) or the absence (in tracheoles) of taenidia. Now, however, electron microscopy shows that there are taenidia-like spiral thickenings within tracheoles too, indicating that the two are separable only on the basis of size, development, and lining. Arbitrarily, therefore, tracheoles are small in diameter (1.0 µm or less), terminal, intracellular, non-shedding tracheal branches.

Structure

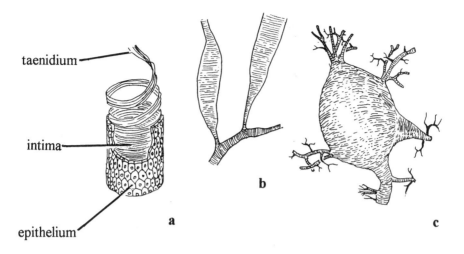

Fig. 9-4. A cut-away view of (*a*) a trachea showing one of its taenidia dissected free and superficial views of (*b*) two elongate, relatively small air sacs and of (*c*) a large, globose air sac (Modified and redrawn from R. E. Snodgrass, 1935).

Air Sacs

Tracheae expand at various points in the body to form collapsible air sacs ranging from minute vesicles, to elongate bags (*Fig. 9-4 b*), and to voluminous sacs (*Fig. 9-4 c*) that, when inflated, appear to be glistening white but, when collapsed, are hard to see. Air sacs differ from other parts of the tracheal system in the thinness of their walls and in the lack of uniform orientation of their taenidia. They are lacking altogether in apterygotes and in holometabolous larvae but occur throughout the pterygote insects, particularly in strong-flying representatives. They act like a bellows to aid breathing and are also compressible, yielding space for limited growth of the internal organs within the rigid exoskeleton.

Molting

The cuticular lining of tracheae is cast at each molt, and a new, larger intima forms in its place before the old intima is withdrawn. The longitudinal tracheal trunks break between adjacent spiracles during this process, allowing the old lining to be withdrawn through the spiracles and shed with the exuviae (*Fig. 3-4 a*). Thus, for a brief period during the molt, each trachea has a double intima.

Spiracular Pattern

Insect respiratory systems may be classified, as follows, based on numbers of spiracles and their position on the body:

A polypneustic system (*Fig. 9-5 a*) is one in which there are at least eight pairs of functional spiracles, as occurs in adults of most orders. An oligopneustic system is one in which there are reduced numbers of functional spiracles, for example, the single thoracic and the single postabdominal pair of

Chapter 9 / RESPIRATORY SYSTEM

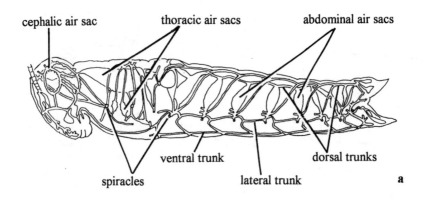

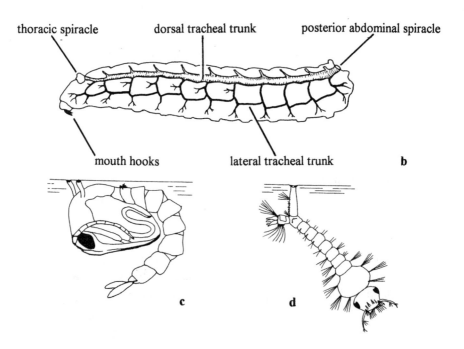

Fig. 9-5. (*a*) The polypneustic respiratory system of a grasshopper showing particularly its thoracic and abdominal air sacs; (*b*) a dipteran maggot's amphipneustic respiratory system consisting of a thoracic pair and of a postabdominal pair of spiracles; (*c*) a mosquito pupa at the water surface drawing air through its paired thoracic spiracles; and (*d*) a mosquito larva drawing air through its postabdominal spiracles at the tip of a long siphon (*a* and *b* are modified and redrawn from R. E. Snodgrass, 1935; *c* and *d* are modified and redrawn from V. A. Little, 1963).

spiracles of moth flies and certain other dipterous larvae (*Fig. 9-5 b*), the single thoracic pair of mosquito pupae (*Fig. 9-5 c*), and the single postabdominal pair of mosquito larvae (*Fig. 9-5 d*). An apneustic system is one that lacks functional spiracles altogether, as in some larval Diptera that rely on cutaneous respiration and various aquatic juveniles that use tracheal gills.

Function

The polypneustic type is probably ancestral based on its widespread occurrence among generalized insects. Moreover, larvae with oligopneustic and apneustic systems often have segmentally arranged *stigmatic cords*. These cords take the form of delicate strands from the lateral tracheal trunks to points on the cuticle with remnants of spiracles or where spiracles, if any, might be expected to occur. They are clearly vestigial tracheae and suggestive of a polypneustic derivation.

Function

Cutaneous Respiration

Some aquatic insect larvae and endoparasites with an apneustic respiratory system depend on respiration through the cuticle. However, such cutaneous respiration cannot meet the needs of aquatic insects other than a few small species having a great surface area with respect to volume. The immatures of damselflies (*Fig. 2-18 d*), mayflies (*Fig. 19-4 b*), and a few other aquatic insects have specialized tracheal gills that can be experimentally inactivated, often without adverse effect. This suggests that these juveniles do not rely on their gills to meet all their respiratory needs. Cutaneous respiration must also play a role, particularly during exposure to low oxygen tension. Some terrestrial insects likewise rely partly on cutaneous exchange, but gas diffusion through the body wall does not contribute to more than a minor percentage of the respiratory need. Those terrestrial adults that rely on cutaneous gas exchange include telsontails and certain springtails, both of which are small-bodied creatures without a tracheal system. Some larger terrestrial insects may also avail themselves of cutaneous exchange notwithstanding possession of a polypneustic tracheal system and a thick cuticle. If so, any cutaneous exchange in these insects is restricted to places where the integument is soft, thin, and permeable.

Tracheal Respiration

Tracheae branch so profusely, tracheoles insinuate themselves so deeply within the tissues, and distances at the tubule apices are so minute that the effect is of air being piped almost directly to every cell of the body. In accordance with the steep gradient established by the tissues consuming oxygen and releasing carbon dioxide, oxygen diffuses rapidly into the system and carbon dioxide out of it during those brief intervals when the spiracles open. It follows that small insects may meet their respiratory needs by cutaneous means alone when inactive, but most insects cannot do so, especially in flight.

Oxygen uptake in tracheate insects is mostly a function of the tracheal system, which is *not* true of carbon dioxide elimination. This is because carbon dioxide diffuses through animal tissues many times more rapidly than oxygen. Thus, carbon dioxide diffuses outwardly in all directions, some of it through tracheae and some through the general body surface. Small, soft-bodied insects probably exhibit maximal cutaneous carbon dioxide emission and large, heavily sclerotized insects minimal emission of this type. The carbon dioxide within insect tissues readily combines with water to form carbonic acid which dissociates into bicarbonate.

Chapter 9 / RESPIRATORY SYSTEM

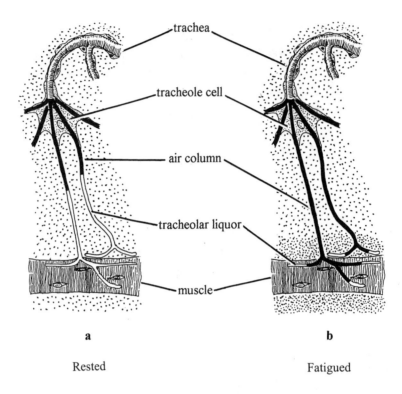

Fig. 9-6. (*a*) The rise and (*b*) fall of tracheolar liquor (shown in light) under rested and fatigued conditions, respectively (Modified and redrawn from R. E. Snodgrass, 1935).

Tracheoles contain a fluid, so-called *tracheolar liquor*, that passes through their walls into the surrounding tissues. It fluctuates in distribution, sometimes being restricted to the tracheole apices (*Fig. 9-6 b*) but other times rises far upward toward the tracheae (*Fig. 9-6 a*). The explanation lies partly in acid metabolite accumulation within the tissues surrounding the tracheoles as a result of metabolic activity, fatigue, or asphyxiation. As these metabolites become increasingly concentrated, they apparently cause an increased osmotic pressure that draws the liquor from the tracheoles, pulling the air column (shown in dark in *Fig. 9-6*) closer to the tubule apices. This movement toward the surrounding tissues increases the amount of available gas and probably compensates for the oxygen deficit there. The opposite occurs in rested tissue. Here, the osmotic pressure in tissues surrounding the tracheoles falls, and the liquor rises toward the tracheae, probably from weak capillarity.

Breathing

Simple cutaneous diffusion of oxygen may prove adequate for small, sedentary insects and for those soft-bodied ones that supplement their tracheal respiration with cutaneous exchange. However, it is insufficient for larger, more heavily sclerotized insects as well as for smaller ones during their periods of vigorous activity, especially flight. As insects become larger, the volume of their oxygen-demanding tissues increases by the cube of their linear dimensions, but the oxygen supply obtainable by diffusion alone grows only in proportion to the cross-sectional area of their tracheae. Thus, oxygen must travel farther through their tracheae, and the combined tracheal cross section does not increase nearly as fast as does body mass. They would soon experience respiratory bankruptcy if they relied solely on diffusion, but they do not. Larger, more active insects supplement whatever passive gas diffusion they have by active *breathing*. This mechanical facilitation of the gas exchange consists of an enhanced, usually directed air flow through the major tracheae and air sacs by muscles that compress and move the internal organs, compressing the air sacs. The atmosphere is effectively brought closer to the body tissues thereby, so the distances that the respiratory gases must diffuse are reduced to the length of the smaller non-ventilated tracheae.

Breathing involves a bellows mechanism that is possible only in insects with segmental sclerotic plates and associated muscles and nerves. The changes in pressure are produced by rhythmic expansion and then collapse of the air sacs alternately forcing air into, and then out of, the tracheae. The expansion involves lateral dilator muscles that separate the terga and sterna (*Fig. 9-7 b*) making the abdominal cross-section greater and/or intersegmental protractor muscles that lengthen the abdomen (*Fig. 9-7 c*). Then the collapse of the air sacs is either by the abdomen's natural elasticity and/or by muscular contraction. The muscles involved in expiration include the lateral tergosternal muscles, acting as compressors that pull the terga and sterna together (*Fig. 9-7 a, b*) and/or the intersegmental longitudinal muscles, acting as dorsal and ventral retractors that telescope the abdomen (*Fig. 9-7 c*). The muscle contractions of flight may also pump the tracheal system. The additional oxygen needed for such strenuous activity as flight is provided both by rhythmical changes in the thoracic volume and by the flight muscles' pumping during contraction and relaxation.

Most tracheae (*Fig. 9-4 a*) are air passageways that tend to be circular in cross section. They are supported by relatively uninterrupted taenidia that, like an automobile radiator hose, are resistant to compression though capable of bending and of telescopic lengthening and shortening. Other tracheae are broken by a line of flexibility along the sides and so are compressible into a flat ribbon (*Fig. 9-4 b*) that functions in breathing. Air sacs (*Fig. 9-4 c*) are thin-walled bags with weak taenidia lacking uniform orientation and, thus, are collapsible upon increased pressure exerted through the blood and muscle action. Their collapse forces air out of the tracheal system, and their subsequent expansion sucks it in again.

Not all insects or insect relatives are breathers. Selected non-breathers include apterygotes, certain insect larvae whose soft body renders them incapable of pumping, and fleas whose rigid exoskeleton likewise does not permit abdominal expansion and contraction. However, even these animals are capable of controlling the direction of air flow within their tracheae though they are unable to breathe. Fleas, for example, normally keep all of their spiracles closed except for the enlarged eighth abdominal pair, which opens and closes rhythmically at a rate varying with metabolism. Then, during active periods that require more oxygen, they open all of their spiracles. This differential use of spiracles reflects the dual role that they play. They provide for gas transfer at the same time as they ensure minimal water loss, remaining open only to the extent required by metabolic need and closed the rest of the time. Only a few insects that live in moist environments have large, open-chambered spiracles.

Chapter 9 / **RESPIRATORY SYSTEM**

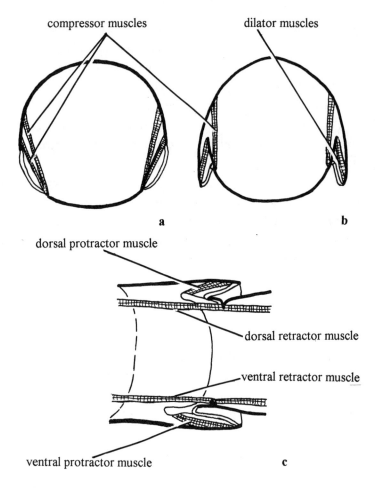

Fig. 9-7. Selected abdominal breathing mechanisms including (*a*) a cross-section of a segment with compressor muscles only, (*b*) a cross-section showing both compressor and dilator muscles, and (*c*) a longisection of successive abdominal segments with protractor and retractor muscles (Modified and redrawn from R. E. Snodgrass, 1935).

Many patterns of breathing may be found among large, active insects. A common pattern among grasshoppers (*Fig. 9-5 a*), for example, involves abdominal protraction (extension) accompanied by opening of the first four and closing of the last six pairs of spiracles, causing inspiration, followed by abdominal retraction (telescoping) accompanied by closing of the first four and opening of the last six pairs, causing expiration. However, the flow reverses at times, and sometimes air moves inwardly and outwardly from each spiracular pair independently. And, of course, the flow increases upon heightened physical activity. The massive increase in oxygen consumption occasioned by flight requires increased air flow through the pterothoracic tracheae to the flight muscles. Part of this increased flight need is answered by thoracic pumping. This pumping involves simple thoracic distortion, producing large-volume changes in the pterothoracic air sacs, enhancing ventilation of that oxygen-demanding zone. This pumping requires that spiracle pairs two and three remain open during flight, handling thoracic needs, while spiracle pairs one and four-to-ten open and close rhythmically in synchrony with the abdominal ventilation, providing for the needs of the rest of the body.

Breathing Control

Breathing resembles other complex activities in that it requires a sophisticated control, but little is known of its mechanism in insects. Decapitation (which eliminates brain and subesophageal ganglionic influences) does not stop breathing, as the headless insect remains sensitive to changes in gas concentration. Basic control must reside elsewhere, apparently in the segmental ganglia of the ventral nerve cord. Each segmental ganglion is autonomous, exerting primary control over its own pair of spiracles through nerve impulses continuously sent to the spiracular opening and closing muscles. However, each ganglion is also subject to a thoracic center that acts as a pacemaker to coordinate overall breathing. The thoracic pacemaker in certain Orthoptera, for example, resides in the metathoracic ganglion. Such secondary control is necessary because a series of autonomous segmental ganglia can only pump air independently of one another. Only secondary control can produce a directed stream of air throughout the system.

The mechanism by which the respiratory centers adjust to metabolic need is uncertain. It could be mediated by some unknown neurosecretion. The mechanism, whatever it is, is apparently activated by carbon dioxide tension. Spiracular opening is triggered by formation of lactic or pyruvic acids within the ganglia. The duration of the open period seems related to carbon dioxide concentration within the blood and oxygen concentration in air. Thus, the spiracles open and breathing accelerates when carbon dioxide tension in the form of bicarbonate rises or oxygen tension falls. This trend maximizes and eventually becomes toxic as the carbon dioxide concentration rises to approximately 15 per cent and that of oxygen falls to about 8 per cent. Normal concentration of these two gases in air is about 0.03 per cent and 21 per cent, respectively.

Aquatic Respiration

Aquatic insects are of two types: either primarily aquatic creatures that live a submerged existence and use dissolved oxygen obtained from the water, or secondarily aquatic insects that depend on atmospheric oxygen. The former rely on the pressure differential between the dissolved oxygen of their aquatic medium and the oxygen within their closed tracheal system or tissues, whereas the latter merely live in water but take their oxygen directly from air.

Tracheal Gills

Naiads, many larvae, and some pupae have thin-walled, sac-like, plate-like, or filamentous extensions of the body surface covered by a thin cuticle and supplied internally with a network of fine, air-filled tracheae and tracheoles. These *tracheal gills* occur on the head, thorax, and/or abdomen, or even within the hind gut. Their apparent function is to increase the respiratory surface with respect to volume. Being

Chapter 9 / RESPIRATORY SYSTEM

thin-walled, they readily permit gas exchange into their enclosed tracheae. The gas molecules, once within this sealed system, then move to the tissues as in all other insects. Moreover, the gills may be moved in an organized, rhythmically repeating manner that sets up water currents, preventing accumulation of oxygen-depleted water around them. This "paddling" behavior is clearly aquatic breathing. Tracheal gills necessarily share whatever respiratory function they have with cutaneous exchange, and their efficacy probably varies. They are vital in some insects' gas exchange but in others may be useful only during periods of low oxygen tension.

Inasmuch as primarily aquatic insects with tracheal gills are apneustic, the partial pressure of oxygen (consumed during metabolism) is always lower within their tracheae than in the surrounding medium and that of carbon dioxide (released during metabolism) is always higher. Therefore, the two gases move in opposite directions across the gill surface according to their respective diffusion gradients. Spiracles are present but sealed in these apneustic juveniles, so the terrestrial adults into which they transform at maturity are already provided with structures needed for breathing upon leaving the water.

Tracheal gills vary in structure and location according to the taxa. Damselfly naiads (*Figs. 2-18 d, 13-1 c*) have three leaf-like external gills at the tip of the abdomen. They provide respiratory surface and questionably may serve as rudders for swimming. Dragonfly naiads (*Fig. 19-4 d*) have internal rectal gills that are obviously respiratory, yet naiads from which these structures have been removed experimentally experience only a reduced rate of mortality, and those whose rectum is plugged show only a slight decrease in oxygen uptake in water and none in air, presumably because of the presence of small thoracic spiracles that supplement their gas exchange. Mayfly naiads (*Fig. 19-4 b*) usually have paired leaf-like or plumose tracheal gills along the sides of the abdomen. The rhythmic beating of these gills sets up water currents over the body surface and facilitates cutaneous gas exchange according to the oxygen concentration. Mayfly naiads from which the gills have been removed show a decrease in oxygen uptake at high but not at low oxygen tensions. Larval caddisflies usually have five or six pairs of abdominal gills, and stonefly naiads (*Fig. 13-1 b*) have paired gills on the thorax and near the leg bases

Blood Gills

Mosquito and midge larvae have sac-like, cuticular evaginations of the body wall filled with blood but devoid of tracheae. Such *blood gills* appear not to be important in respiration owing to their lack of tracheation but probably function in osmoregulation. Bloodworm larvae carry a hemoglobin similar to that of humans in their blood. This respiratory pigment is probably not a carrier under ordinary oxygen tensions (in which it remains saturated), but it traps and liberates oxygen when the pressure falls to about 1 per cent of an atmosphere or less, as often happens at the bottoms of the deep lakes that these juveniles inhabit.

Oligopneustic Respiration

Mosquito larvae (*Fig. 9-5 d*) and various beetle adults are secondarily aquatic, being adapted for living in water but making their gas exchange with the atmosphere oligopneustically. They regularly return to the surface to breathe air through their postabdominal spiracles. Certain dipteran larvae produce an oily water-repellent secretion around their spiracles. Soldier fly larvae have spiracles ringed by semihydrofuge setae not ordinarily wetted by water. The cohesion of water molecules to themselves is greater when in contact with the surface film than is its adhesion to these "hairs," so they are not wetted. The non-wettable side is presented to the atmosphere when the larva is at the water surface, and the wettable side is exposed to the water beneath, resulting in an umbrella-like spreading of the hairs which supports the creature against the surface film and exposes the spiracles to air. The hairs then deflect over and close the spiracles when the larva sinks, as the wettable side is again exposed to water.

Air Stores

Relatively long periods of submergence occur in various aquatic beetles and bugs with *air stores* into which spiracles open. Backswimmers, for example, are bugs that float on the water surface, venter upward, with outstretched hind legs. A series of hairs along either side of their ventral keel form an air-filled channel in contact with an air store beneath the wings. This channel is exposed to air, and the air store is replenished in the insects' normal upside-down position at the water surface. Predacious diving beetles (*Fig. 9-8 a*) may submerge for longer periods. They break the water surface periodically, exposing the enlarged hind spiracles whose semihydrofuge hairs support them against the surface film as they breathe, replenishing their subelytral air store. Water scavenger beetles (*Fig. 9-8 b*) do much the same, replenishing their ventral gas bubble by breaking the water film with their antennae.

Air stores are hydrostatic devices that contribute to buoyancy, so the insects that have them float to the surface and expose the spiracles to open air when they stop swimming or release their hold on submerged vegetation. Air stores also act as a kind of physical gill that allows their possessor to breathe while under water. The air store of certain adult diving beetles, for example, is sufficient for submergence of only 20 min duration, but the beetles can remain underwater upwards of 36 hrs. As they draw off oxygen, the partial pressure of that gas within the air bubble apparently falls below that of the surrounding water, so oxygen diffuses inward in an attempt to restore equilibrium. This tendency of oxygen to diffuse inward into the bubble is greater than that of nitrogen (which could also restore the equilibrium) to diffuse outward as a result of its increased partial pressure. Consequently, the beetle continues extracting oxygen from the medium so long as any nitrogen remains within its bubble.

Certain beetles (*Fig. 9-8 c*) and bugs use plastron respiration to remain underwater indefinitely. A *plastron* is a thin air film permanently appressed to the body surface, usually by semihydrofuge hairs, into which the tracheae open so that oxygen can pass directly to the tissues within. The plastron of certain creeping water bug nymphs is among the most efficient known. Young nymphs of these insects rely solely on cutaneous respiration, and adults of these bugs supplement their cutaneous means with a plastron that opens into the spiracles. The volume of air carried within the plastron is small but constant, so the air film operates as a gill rather than as a mere air store.

Chapter 9 / **RESPIRATORY SYSTEM**

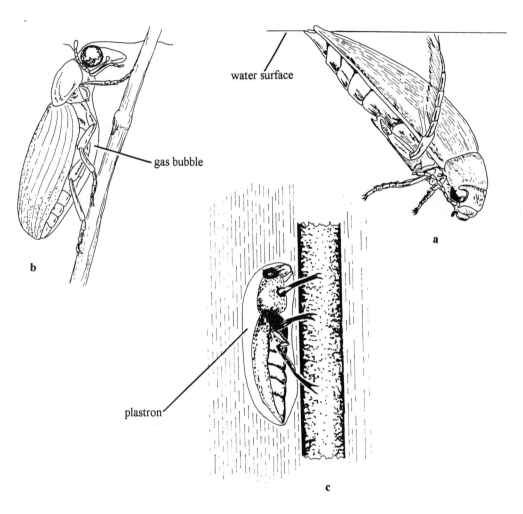

Fig. 9-8. Secondarily aquatic insects including (*a*) a predacious diving beetle replenishing its subelytral air store upon breaking the water surface with its hind spiracles, (*b*) a water scavenger beetle replenishing its ventral gas bubble upon breaking the water film with its antennae, and (*c*) a submerged riffle beetle within its overall plastron (*a* and *b* are modified and redrawn from M. D. Atkins, 1978, and *c* from S. C. Kendeigh, 1974).

Suggested Additional Reading

Bursell (1970); Chapman (1991, 1998); Davies (1988); Gullan & Cranston (2000); Nation (1985, 2002); Patton (1963); Richards & Davies, vol. 1 (1977); Snodgrass (1935); Wigglesworth (1972, 1984).

Chapter 10

Reproductive System

*The germ cells are the only living bonds not only between generations
but also between species, and they contain the physical basis
not only of heredity but also of evolution.*
— E. G. Conklin

The reproductive mechanisms evolved by insects are eminently successful. Most of them produce numerous eggs that they deposit in places appropriate to development of their offspring and with a timing that assures maximal reproductive potential. Moreover, the life cycle may involve several generations within a relatively short time period. The result is an incredibly high reproductive total which, fortunately, is never met in nature owing to attrition by negative environmental influences.

Aquatic animals can simply liberate their zygotes into the surrounding water, but terrestrial insects cannot do so. They must protect their delicate germ cells and embryos from the desiccating terrestrial environment. Most insects protect their inherently aquatic gametes through internal fertilization and their embryos by an egg shell resistant to desiccation.

Insects' *ova*, or eggs, are usually stimulated to undergo embryonic development upon fusion with *spermatozoa*, or sperm. This result is assured by sex organs, ducts, glands, and a coupling apparatus complementary in the two sexes. These structures vary with the species and taxonomic group, resulting in the pronounced differences in mating, impregnation, and oviposition now to be discussed.

Structure

Females

The reproductive structures of female insects (*Fig. 10-1 a*) consist of paired *ovaries* each having a number of *ovarioles* discharging into a lateral oviduct that converges with its mate into a *common oviduct* that exits at the gonopore. The latter communicates through the medial *copulatory pouch*, or vagina, with the secondary copulatory opening. One or more spermathecae and accessory glands as well as the female external genitalia are associated with the copulatory pouch.

The ovaries lie on either side of the digestive tract, usually enclosed within a connective tissue sheath in which are also embedded numerous tracheoles and muscles. Each ovary consists of the parallel ovarioles (*Fig. 10-1 a*), the number of which is often characteristic of the species. An individual ovariole (*Fig. 10-2 a*) consists of a *terminal filament*, a short *germarium* that gives rise to oogonia, a

Chapter 10 / REPRODUCTIVE SYSTEM

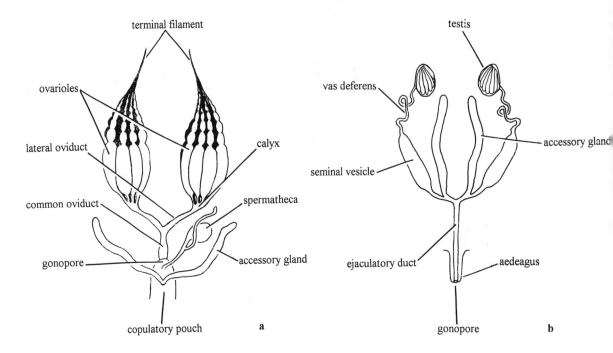

Fig. 10-1. (*a*) The female reproductive system of a generalized insect showing the ovaries, sex ducts, and associated organs; (*b*) the male reproductive system of a generalized insect showing the testes, sex ducts, and associated organs (Modified and redrawn from R. E. Snodgrass, 1935).

vitellarium consisting of a longitudinal series of developing eggs, and a stalk-like *pedicel*. The thread-like terminal filament attaches the ovariole to the body wall, dorsal diaphragm, or fat body. The several pedicels on a side usually fuse into a dilatation called the *calyx* that empties into the lateral oviduct.

Ovarioles range from panoistic (*Fig. 10-2 a*), a type without nutritive cells, as in many primitive or generalized orders, to polytrophic (*Fig. 10-2 c, d*), a type with an alternating succession of oocytes and nutritive cells, as in some specialized orders, and to acrotrophic (*Fig. 10-2 b*), a type with long cords by which the nutritive cells connect with the descending oocytes, as in bugs, neuropterans, and beetles.

The mesodermally derived lateral oviducts fuse into the common oviduct which is a muscular tube stemming from a medial ectodermal invagination from the posterior margin of the female's eighth abdominal sternum.

The term *bursa copulatrix* is sometimes applied to what is properly the copulatory pouch (*Fig. 10-3 a*). The term is more appropriately restricted to a diverticulum separate from the copulatory pouch (*Fig. 10-3 b*), in which case the bursa is the opening that receives the aedeagus and associated male parts during coitus, leaving the copulatory pouch to serve for egg discharge. Females with this genital configuration, such as the higher moths and butterflies, have two genital openings, a ventral copulatory one leading into the bursa and a more posterior vaginal one, in which case the bursa connects with the vagina through a narrow sperm duct.

Structure

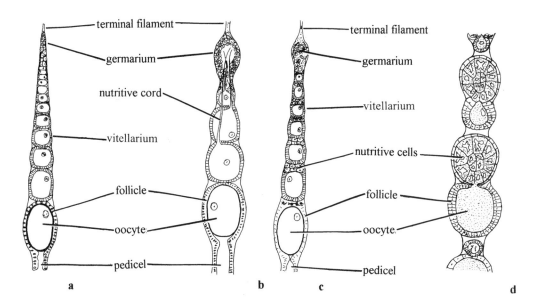

Fig. 10-2. Comparison of the structure of (*a*) panoistic, (*b*) acrotrophic, and (*c*) polytrophic ovarioles, with (*d*) an enlargement of the basal section of (*c*) (*a*, *b*, and *c* are modified and redrawn from V. B. Wigglesworth, after Weber, and *d* is modified and redrawn from R. F. Chapman, after Hopkins and King).

The *spermatheca* (*Fig. 10-1 a*) is a cuticular invagination from the copulatory pouch via a slender duct commonly provided with a gland. It is single except in certain flies.

Male Organs

The male reproductive system (*Fig. 10-1 b*), like the female system with which it is homologous, is Y-shaped, consisting of paired ovoid *testes* and paired *vasa deferentia* which converge into a medial *ejaculatory duct* that discharges to the outside at the gonopore. One or more pairs of accessory glands may be associated with this system. The testes and vasa deferentia are mesodermal derivatives and the ejaculatory duct an ectodermal derivative. The testes may be located above, alongside, or below the digestive tract and are covered by a membranous peritoneum. Each testis (*Fig. 10-4 a*) consists of one or more *testicular follicles* varying in number according to the species, and each follicle empties into the *vas deferens* on its side via a short, stalk-like *vas efferens*. The tubular vasa deferentia consist of an outer peritoneum and an inner epithelium separated by circular muscle. They commonly dilate at a point along their length into a reservoir called the *seminal vesicle* before they converge into the medial ejaculatory duct. The latter is a muscular tube that invaginates mid-ventrally from the male's ninth abdominal segment and discharges through the gonopore at or near the tip of the intromittent *aedeagus*. The sac-like or tubular, straight or coiled accessory glands may have a muscular wall.

Chapter 10 / **REPRODUCTIVE SYSTEM**

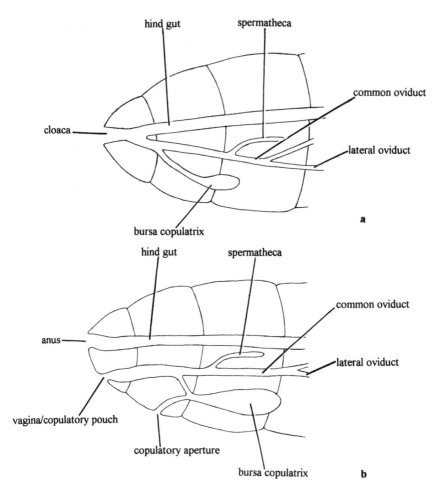

Fig. 10-3. Comparison of (*a*) the single reproductive aperture (in this case fused with the anus into a cloaca) of a monotrysian lepidopteran as opposed to (*b*) the dual reproductive apertures and anus of a ditrysian lepidopteran (Modified and redrawn from various sources).

Function

Oogenesis

Egg development begins in the germarium (*Fig. 10-2 a*) as the primordial germ cells give rise mitotically to oogonia and these, in turn, into oocytes and, in polytrophic and acrotrophic ovarioles, also to nutritive cells. As the oocytes mature and enter the vitellarium, they become arranged in linear fashion along the length of the ovariole, one behind the other, by the addition of newer oocytes issuing from the germarium. Each oocyte becomes enclosed within a one-cell thick *follicle* derived from the germarial mesoderm at the junction of the germarium and vitellarium. The entire ovariole soon

Function

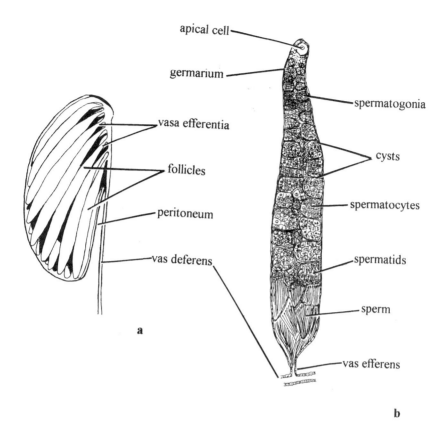

Fig. 10-4. (*a*) A testis of a generalized insect with many follicles and (*b*) an enlargement of one testicular follicle (*a* is modified and redrawn from Snodgrass, 1935; *b* is modified and redrawn from Wigglesworth, 1965).

becomes distended into a series of distinct follicles that, from front to back, are successively larger and older. The largest, most mature follicle then becomes encased within a *chorion*, or egg shell, secreted by the follicular cells. It is kept from entering the pedicel by a plug that eventually disintegrates, liberating it into the lateral oviduct.

Each oocyte and its cluster of nutritive cells within a polytrophic ovariole (*Figs. 10-2 c, 10-2 d*) consist of a clone of interconnected sister cells arising from successive, incomplete divisions of a single oogonium. The most caudal of the sister cells becomes the oocyte surrounded by follicular cells, and the remaining cells of the clone become interconnected nutritive cells that transfer nutrients to the oocyte which, because of its original placement, is always the most posterior cell. A major element of the oocyte consists of ribosomal messenger and transfer RNAs synthesized by the nurse cells and acrotrophic cells, except for panoistic ovarioles in which they are produced by the nucleus. The rest of the yolk elements either pass through or arise from the follicle cells. The nutritive cells disintegrate upon maturation of the oocyte.

Fully formed, chorion-enclosed oocytes are not yet mature eggs. They do not mature until *oviposition*, an act which usually awaits impregnation by sperm from the male. Following coitus, sperm

Chapter 10 / REPRODUCTIVE SYSTEM

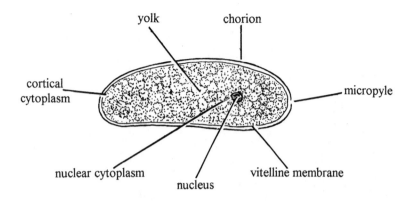

Fig. 10-5. Longisection of a common pterygote egg encased within its chorion, with yolk concentrated in the center around the nucleus and cortical cytoplasm at the periphery (Modified and redrawn from R. E. Snodgrass, 1935).

are introduced into the female tract via the copulatory pouch or bursa copulatrix, if any, and stored for a time within the spermatheca (*Fig. 10-1 a*). Then, as oocytes exit from the ovarioles into the lateral and common oviducts, they are propelled backward into the copulatory pouch by peristaltic contraction. They receive sperm at or near the *micropyle apertures* of the chorion (*Fig. 10-5*) as they pass the mouth of the spermathecal duct. This entry stimulates two oocyte maturation divisions culminating in fusion of the egg nucleus with a sperm nucleus, at which point they receive a coating of female accessory gland secretion that provides an embedding or protective covering.

Fertilized eggs are readied for oviposition in a manner depending on the species. As many eggs as there are ovarioles with fully formed oocytes may be deposited at one laying; eggs may be released singly, either in continuous or in discontinuous succession; oocytes soon to become eggs may be liberated first from one ovary and then from the other; or, in the event the calyx is dilated, oocytes may accumulate there, making possible egg masses.

Growth, Yolk Formation, and Related Processes

Oocyte growth and maturation depends on the uptake of nutrients from the bloodstream or fat body. This transfer is mediated by the ovarioles in a way that varies with structure. The oocytes of panoistic ovarioles (those without nutritive cells) (*Fig. 10-2 a*) rely in part on food that passes between or is elaborated by the surrounding follicular cells. An element of the oocyte is ribosomal m-RNA and t-RNA synthesized within the nucleus. The nutritive cells of polytrophic ovarioles (those with an alternating succession of nutritive cells and oocytes) (*Fig. 10-2 c, d*) synthesize and pass glycogen, lipids, and other substances to their oocyte until they disintegrate and are resorbed. The rest of the yolk elements pass through or arise within the follicular cells. The oocytes of acrotrophic ovarioles (those with nutritive cords extending between the oocytes and the apical nutritive cells) (*Fig. 10-2 b*) are nourished via the nutritive cords. The acrotrophic relationship lasts until the oocytes reach a determined size, whereupon the cords break, leaving the oocytes dependent on the surrounding follicular epithelium.

Function

Insect eggs are *centrolecithal*, their yolk being central and their cortical cytoplasm at the cell periphery. Yolk formation involves a massing of protein, lipid, and sometimes glycogen bodies that become prominent as the oocyte enlarges beyond its sister nutritive cells.

Chorion Formation

A fully formed oocyte (*Fig. 10-5*) consists of a nucleus surrounded by yolk composed largely of protein and fat globules, bounded by a delicate homogeneous vitelline membrane apparently synthesized by the follicular epithelium, and covered at the periphery by a chorion. The oocyte has its total yolk complement at this stage, soon to be used by the embryo to complete development up to hatching. But more is required. It must be protected from the desiccating terrestrial environment, yet it cannot be so shielded as to preclude sperm entry and gas exchange. These conflicting responsibilities are handled by the chorion and associated structures. The chorion is structurally similar to the cuticle covering the external body wall except for its non-chitinous composition. It retains impressions of the follicular cells that produced it everywhere on its surface, seen usually as hexagonal ridging. This so-called *chorionic sculpture* is often diagnostic of the species and may be used taxonomically. Sometimes it is provided with a detachable cap. The chorion forms a complete covering over the oocyte except at the micropyle/s and the gas apertures. The micropyles are one or more points where the chorion is closely adherent to, or absent from, the vitelline membrane, allowing entry of fertilizing sperm. The gas apertures open into an extensive network of spaces close to the embryo's vitelline membrane through which oxygen diffuses inward and carbon dioxide outward.

Hormonal Control

Growth and reproduction are mediated by the endocrine system, as discussed in Chapter 6.

Spermatogenesis

A testis (*Figs. 10-1 b, 10-4 a*) is composed of one or more *follicles*, each an elongate tube accommodating within its cavity the entire process of spermatogenesis. Hence, along the length of each testicular follicle (*Fig. 10-4 b*) occurs a succession of male germ cells in different stages of development. A zone of spermatogonia called the *germarium* is located at the follicular apex. Next is a *spermatocyte zone* where individual spermatogonia set free from the germarium divide into cysts, each a bundle of cells surrounded by mesodermal cells. Next is a *maturation and reduction zone* in which the newly formed primary spermatocytes divide twice meiotically to form haploid spermatids. Finally, there is a *transformation zone* in which spermatids elongate into flagellated spermatozoa.

The cells of a given cyst, each derived from a single spermatogonium, are at the same stage of development and, like those of single oogonia, represent an interconnected clone. They assume a radial position within their packet and for awhile remain centrally attached. The cysts are, at first, rounded in outline, but mutual pressure forces them into a polyhedral form as they grow and are displaced backward by new cysts and by the force of elongation of the developing germ cells until eventually mature spermatozoa rupture the cyst wall and are released.

Chapter 10 / REPRODUCTIVE SYSTEM

The transformation of spermatids into sperm involves several steps as a result of which each germ cell elongates and develops a head and a tail that are little different in diameter. There is no middle piece like that characteristic of vertebrate sperm.

Sperm Passage

Thousands of spermatozoa per cubic mm appear in the basal part of the follicle grouped within bundles coinciding with the cysts from which they arose. As they accumulate, others begin forming in the germarium at the opposite end, causing distension and an increase in length of the follicle. The mature cells remain close to their vas efferens for a time until they are expelled, either by elongation pressure or by the periodic lashing of their flagella. They pass first into the short vas efferens and then into the vas deferens on their respective side. Once within one of the vasa deferentia, they may be stored for a time within the dilated seminal vesicles where they occur in masses, sometimes with their heads still embedded in the epithelial wall and their tails projecting into the lumen or merely lodged there, unattached.

The male accessory glands (*Fig. 10-1 b*) often secrete a cream-colored *seminal fluid* or a gelatinous *spermatophore*. These are vehicles by which sperm may be passed to the female during coitus. Details of mating, spermatophore formation and exchange, and impregnation are dealt with in Chapter 16.

Trophic Function

All nutrition stems ultimately from the bloodstream to which the fully developed sperm lack direct access. Accordingly, they must obtain nutrients through intermediary structures, often a group of cells called the apical cell complex (*Fig. 10-4 b*). Developing spermatogonia cluster around and attach to this complex by protoplasmic extrusions that apparently serve a trophic function. The complex apparently breaks down and absorbs one set of spermatogonia to secrete absorbed nutrients to another set destined to become functional germ cells. This exchange takes place only during early stages of spermatogenesis. At later stages, the nutritive function is usually assumed by the cells of the follicle or cyst wall.

Suggested Additional Reading

Berry (1985); Chapman (1991, 1998); Davies (1988); Engelmann (1970); Gullan & Cranston (2000); Leather & Hardie (1995); Nation (2002); Patton (1963); Richards & Davies, vol. 1 (1977); Snodgrass (1935); Wigglesworth (1972, 1984).

Chapter 11

Metamorphosis

Today I saw the dragon-fly
Come from the wells where he did lie.
An inner impulse rent the veil of his old husk;
From head to tail came out clear plates of sapphire mail.
He dried his wings; like gauze they grew;
Through crofts and pastures wet with dew
A living shaft of light he flew.
— Tennyson

Except for the few that bear living young, most insects are "egg layers" that deposit their eggs in appropriate places where, after a given interval of time, the young hatch to take their place in nature as new individuals.

Hatching is egress from the egg. It is that major landmark in the insect life cycle indicating the end of embryonic development. *Reproduction* is a second major landmark, one normally restricted to adults. The two delineate a postembryonic segment of the life cycle during which individuals undergo a transformation spoken of as *metamorphosis*. This transformation consists of the collective, step-wise changes in form that the animal undergoes between hatching and adulthood. Table 11-1 puts into perspective the life processes and events both of insects with an incomplete metamorphosis (Hemimetabola) and those with a complete metamorphosis (Holometabola).

There are commonly five or six molts in insects. The number tends to be a species characteristic but is subject to other intrinsic variables such as sex and caste and extrinsic ones such as food and weather. The structure and condition of the insect between successive molts is termed the *instar*, and the time period between its molts is the *stadium*. Hence, an insect issuing from an egg is a first instar in its first stadium. The creature molts again after some days or weeks characteristic of the species and depending on temperature and available food. It is then a second instar that remains in its second stadium until its next molt and so on.

Types of Postembryonic Development

Some insects or hexapod relatives of insects have a *direct development* involving graded changes between young and adults. The young resemble adults in this instance except in size, lack of fertility,

Chapter 11 / **METAMORPHOSIS**

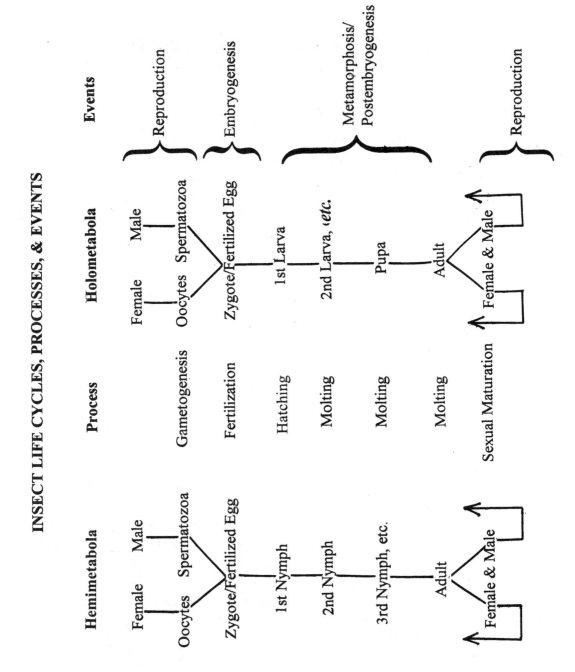

Table 11-1. Life cycles, processes, and events in incomplete (hemimetabolous) development as compared to those of complete (holometabolous) development.

Types of Postembryonic Development

and winglessness. Other insects have an *indirect development* involving pronounced changes in structure and function between young and adults. Still others have an indirect development that is intermediate between the two extremes. The following types may be distinguished from within this gamut.

Ametabolous Development including Anamorphosis

The primitive, wingless hexapod orders Collembola, Diplura, Microcoryphia, and Thysanura have a direct development. They are grouped as *Ametabola*, and their direct development is termed ametabolous. The word is a Greek derivative indicating *change in form without a change in form*. This contradiction expresses the fact that development is minimal and graded. They tend to undergo numerous molts, usually more than ten. The juveniles (*Fig. 11-1 a*) differ little among one another and from the adults of their species (*Fig. 11-1 b*) except in size, body proportions, and fertility. Another kind of direct development, *anamorphosis*, occurs in several arthropod classes including the telsontails of order Protura. In telsontails, the first instar juvenile has nine abdominal segments, to which an additional segment is added with each molt until, in the adult, there is a total of twelve abdominal segments counting the telson.

Hemimetabolous or Incomplete Development

Hemimetabola include those usually terrestrial insects, either winged or secondarily wingless, that lack the persistent molting of some adult Ametabola as well as the pupal stage of Holometabola (see below). The young of Hemimetabola, termed *nymphs*, or *larvae* (*Fig. 11-1 c*), resemble one another and also the adult (*Fig. 11-1 d*) in their general structure and mode of life but differ in their smaller size, different body proportions, and incompletely developed genitalia and wings. Grasshoppers and their allies, termites, and the other orthopteroid orders and bugs, hoppers, and their allies, and certain other hemipteroid orders belong to the Hemimetabola. Dragonflies and damselflies, mayflies, and stoneflies are also classified as Hemimetabola, but their aquatic nymphs, sometimes called *naiads* (*Fig. 11-1 e*), present a substantially different appearance from the terrestrial adults of their species (*Fig. 11-1 f*) and have tracheal gills and closed spiracles. Like other Hemimetabola, naiads develop *external wing pads* that, except in mayflies, do not become functional wings until emergence of the adult. Most Hemiptera: Homoptera are hemimetabolous, except for certain scale insects which have evolved a primitive holometabolous development. The secondarily wingless nymphs of biting lice and sucking lice lack the wing pads characteristic of the young of other Hemimetabola, which is doubtless an adaptive response to their ectoparasitic mode of life.

Holometabolous or Complete Development

The more advanced orders of insects, including neuropterans, beetles, butterflies and moths, flies, bees, wasps, and their relatives, comprise the *Holometabola*. They undergo an indirect development characterized by completely dissimilar larval, pupal, and adult stages.

The *larva* is the often soft-bodied, worm-like juvenile that intervenes between the egg and the pupa. Several types may be recognized, among them campodeiform larvae (*Fig. 11-2 a*) which are

Chapter 11 / METAMORPHOSIS

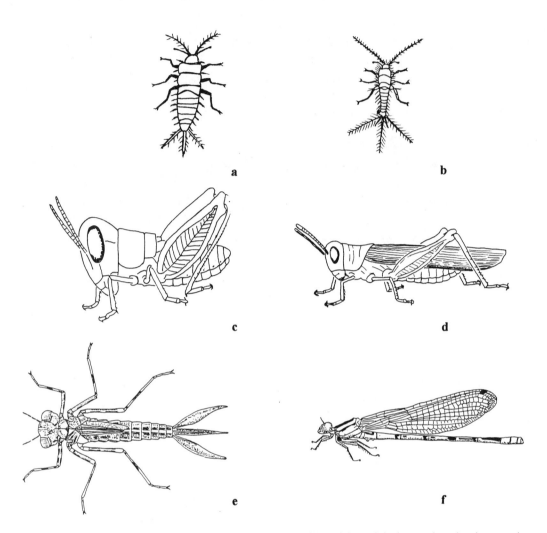

Fig. 11-1. Direct development in the firebrat, with *a* = juvenile and *b* = adult; incomplete development in a grasshopper, with *c* = nymph and *d* = adult; and incomplete development in a damselfly, with *e* = naiad and *f* = adult (Figures are not to scale, the early juveniles, *left side*, being a fraction of the size of adults, *right side* (*a* and *b* are modified and redrawn from C. L. Metcalf; *c* and *d* are modified and redrawn from V. B. Wigglesworth, 1972; and *e* and *f* are modified and redrawn from C. H. Kennedy).

elongate, often flattened, active juveniles with well-developed thoracic legs and antennae; scarabaeiform larvae (*Fig. 11-2 b*) which are thick-bodied, sluggish juveniles with a well-developed head and thoracic legs but without abdominal legs; apodous eucephalous larvae (*Fig. 11-2 c*) which are legless juveniles with a well-developed head; and eruciform larvae (*Figs. 3-4 a, 11-2 d, 19-19 d*) which are caterpillars with a cylindrical body, a well-developed head, thoracic legs, and abdominal legs.

Larvae increase in size with each molt but do not exhibit major structural changes. Whatever their form, they bear no resemblance to the adult of their species. They are, in general, actively moving juveniles specialized for feeding and growth; they often eat ravenously, using biting-type mouth-

Types of Postembryonic Development

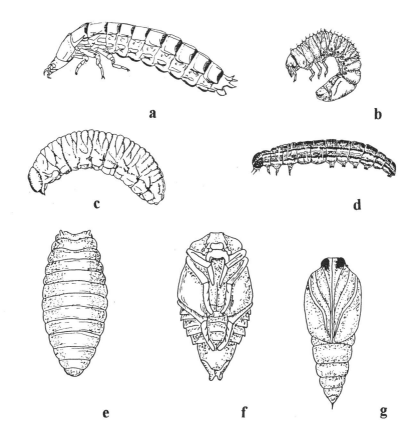

Fig. 11-2. Selected larval types including *a* = campodeiform, *b* = scarabaeiform, *c* = apodous eucephalous, and *d* = eruciform. Selected pupal types including *e* = adecticous obtect, f = decticous exarate, and *g* = adecticous exarate (*a* – *d* are modified and redrawn from H. Oldroyd, 1962; *e* – *g* are modified and redrawn from CSIRO, 1991).

parts and a digestive system completely different from that of the adult; they lack compound eyes and ocelli but may have stemmata; and they develop their wings internally from *wing discs*, as opposed to the external wing pads of Hemimetabola.

The *pupa* is the instar intervening between larva and adult in Holometabola, and the process of becoming one is known as *pupation*. The external features of adults are present in pupae, though not fully developed, so they appear substantially different from the larvae from which they arose. Several types may be recognized, among them adecticous obtect pupae (*Fig. 11-2 e*) with immovable, non-functional mandibles and appendages "glued" to the body; decticous exarate pupae (*Fig. 11-2 f*) with movable, functional mandibles and free legs; and adecticous exarate pupae (*Fig. 11-2 g*) with immovable, non-functional mandibles and free appendages.

The pupa is a so-called "resting stage," but it is hardly quiescent. It is really a period of intense structural, functional, and developmental activity as larval structures are being broken down, yielding materials for construction of new adult structures. The pupal stage lasts for intervals from a few days to many months. The latter is characteristic of species with a protracted *pupal diapause* (a period

Chapter 11 / METAMORPHOSIS

of arrested development and reduced metabolic rate correlated with the unfavorable season, either winter in temperate zones or the dry season in the tropics).

Pupae usually transform within some kind of protective covering. For example, the silkworm and many other Lepidoptera transform within an above-ground silken *cocoon* secreted by the larval salivary glands; certain beetles burrow into the ground where they construct earthen cells in which to pupate; and the larvae of cyclorrhaphan flies retain and harden the last larval skin called the *puparium* (*Fig. 19-18 h*) used in lieu of a cocoon. However, butterflies do not transform within protective cells or cocoons but merely suspend themselves from the abdomen by a hook fashioned from a remnant of the cocoon (*Fig. 3-4 c*).

Emergence of the new adult through the pupal skin occurs along a medial line of weakness, usually somewhere atop the head and/or thorax (*Fig. 3-4 c*). The insect escapes by spasmodically contracting its muscles and swallowing air to increase the body volume, sometimes facilitated by using the mouthparts or specialized spines, knobs, or other devices that weaken the cocoon locally. These forces cause the animal to burst through at the line of weakness, whereupon it pulls itself out of the *exuviae* and inflates its wings and body until the new exoskeleton hardens and darkens.

One might ask whether the pupa is merely a last larva or a new instar between the last larva and the adult. The former is probably the answer. It is essentially a last larval instar that bridges the profound structural and physiological gap between larva and adult. Why has it evolved? The answer could lie in the fact that holometabolous insects are *endopterygotes*, their wings developing as internal wing discs responsible for crowding within the limited confines of the body. Perhaps they solve this crowding by everting the wings. In this light, therefore, the pupa is the instar adapted both for wing eversion and for the profound larva-adult transformation. Understandably, this adaptation does not occur in juvenile Hemimetabola which are *exopterygotes*, having external wing pads and so not subject to the same degree of juvenile wing crowding as in Holometabola. It could also be that the pupal stage evolved to allow the insects to occupy a wider variety of environmental niches than would otherwise be possible.

Growth and Differentiation

Growth

Development involves two separate processes, growth and differentiation. *Growth* is an increase in the mass of living substance. *Differentiation* involves both physical and metabolic changes in cells from the embryonic to the adult state. The two processes go forward concurrently causing both an increase in size and weight and a change in body form.

The recognized types of insect development seem related to the degree of change between the juvenile instars and the adult. Most juvenile organs in non-holometabolous insects are carried over essentially unaltered into the adult, but the larval organs of Holometabola are generally destroyed and then replaced by adult structures.

Growth and Differentiation

Differentiation

A single zygote of an insect undergoes cleavage and forms a ball of more or less similar cells surrounding a central core of yolk. Certain cells lying along the venter differentiate into a plate destined to form the embryo, leaving the remaining cells to become extraembryonic tissue that is ultimately discarded and absorbed. A region of the embryo destined to become thorax then generates a wave of determination that spreads over the embryo and somehow establishes a larval role for certain cells. This is followed by a second wave of determination fixing a pupal and an adult role for other cells.

The first few cleavage nuclei are *totipotent*, each having the potential to become any part of the future insect, for each has a full double complement of all genes. The role of these early cells is determined, in part, by where they land in the developing blastoderm. Cells landing in one place are destined to become one part of the larva; cells in another place become another part. They differentiate rapidly, and the tiny embryo takes form. In the meantime, other cells determined to form parts of the pupa and the adult are set aside, often in the form of *imaginal discs*. These islands of inactive, undifferentiated tissue lie scattered throughout the larval body. Juvenile hormone temporarily ceases to be produced by the corpora allata during pupation or at least is produced in lesser amounts that no longer inhibit development of the imaginal discs. Now influenced by molting hormone from the prothoracic glands, the discs differentiate into pupal and adult organs using the broken-down tissues of the degenerating larva much as they would a culture medium.

Development in the Ametabola may be characterized as simple, progressive differentiation. These primitive hexapods increase in size and change slightly in body proportions with each molt, but their overall transition to adult form takes place without altering the metabolic activity of their cells. Epidermal cells continue to form the same kind of cuticle in each instar; muscle cells the same kind of muscle protein, *etc*. The only exception are the sex cells. Completely suppressed in juveniles, the sex cells of Ametabola do not complete their differentiation and become functional until adulthood.

The changes that occur when the last nymph of a representative of the Hemimetabola molts into an adult are more obvious than those that take place in Ametabola. The sex cells develop and become functional in much the same manner as in Ametabola, but the cuticle undergoes extensive alteration. The cuticle of nymphal assassin bugs, for example, is soft, thick, pigmented, and beset with a certain kind of sensory bristle, as opposed to the hard, thin adult cuticle provided with a completely different kind of bristle and a different pigmentation. The other nymphal organs of assassin bugs are either greatly or little affected during transformation. The muscles associated with the wings and external genitalia undergo extensive change at the same time as the tracheae, Malpighian tubules, and other organs remain little altered.

This trend culminates in the Holometabola, in which the organ systems undergo wholesale change during transformation. The larva is completely torn down and rebuilt into a new adult from cells stemming at least partly from imaginal discs held in reserve until pupation. A holometabolous embryo may be said, therefore, to possess two separate, latent forms. The first is the larval form which differentiates, changes, and grows, while the second, adult form remains as imaginal discs. Then the fully grown larva is more or less completely destroyed and replaced by the adult. This transformation is so profound that apparently it can only be accomplished within a transitional pupal stage.

Chapter 11 / **METAMORPHOSIS**

Suggested Additional Reading

Chapman (1998); Chu & Cutkomp (1992); Davies (1988): Gullan & Cranston (2000); Peterson (1948-1951); Richards & Davies, vol. 1 (1977); Snodgrass (1954); Wigglesworth (1954, 1959, 1972, 1984).

Chapter 12

Insect Natural History

*It is interesting to contemplate a tangled bank,
clothed with plants of many kinds,
with birds singing on the bushes,
with various insects flitting about,
and with worms crawling through the damp earth,
and to reflect that these elaborately constructed forms,
so different from each other,
and dependent upon each other in so complex a manner,
have all been produced by laws acting around us.*
— Charles Darwin

Environmental Factors

The environment is a composite of interwoven factors, both physical and biotic, that simultaneously impinge upon organisms. It is, in this sense, analogous to a rope composed of many fibers, yet more than the sum of its fibers. Confronted with this complexity, entomologists often resort to an analytical approach to study the environment, factor by factor, to deduce overall environmental relationships as they affect insects.

Environmental factors vary through a range of intensity usually greater than organisms can tolerate. An insect's optimal response to a particular factor tends to lie somewhere near the middle range of intensity of a bell-shaped curve (*Fig. 12-1 a*). A threshold is the minimal quantity of energy, such as heat, humidity, light, *etc.*, capable of producing a perceptible effect on behavior or development. Insects do not respond below threshold, but above it their response increases with intensity up to the point of optimal response. Above this level, they exhibit no further increase either because of fatigue or because the stimulus becomes deleterious. Insects exhibit a comparatively wide middle zone of optimal response with respect to most environmental factors, flanked above and below by physiologically stressful zones, each of the latter causing inactivation and eventual death. What are these environmental factors, and how do they influence insects?

Light

The sun, moon, stars, and bioluminescence are sources of light on earth, of which sunlight is the most important. Paramount among the sun's biological effects is its capture and transformation by green plants into the chemical energy of food. This process, *photosynthesis*, is beyond our present scope. However, there are other biological effects of light including vision, periodism, pigmentation, and protective coloration whose discussion is relevant.

Chapter 12 / **INSECT NATURAL HISTORY**

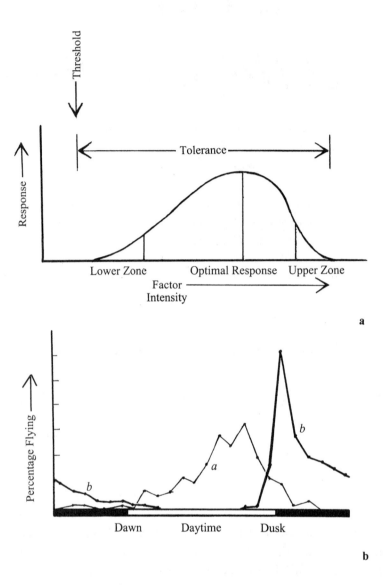

Fig. 12-1. (*a*) The idealized response of an insect to an environmental factor showing optimal response near the center of a bell-shaped curve and minimal response in the lower and upper zones of physiological stress; (*b*) the contrasting activity patterns of two different mosquitoes, *species a* (light line) being a day-active species and *species b* (heavy line) a night-active species (*Fig. 12-1 b* is modified and redrawn from Haddow).

The limits of visible light are fixed arbitrarily according to the human eye. However, insects' visual organs depart from the established human standard. Humans see light ranging from about 3,900 to 7,700 Å (angstrom units) which includes the colors of the rainbow from violet to red. Insects generally discriminate among greens, blues, and other shorter wave lengths well into the ultraviolet range though they are generally blind to reds and other longer wave lengths.

Vision is important in insects' orientation and movement. Their compound eyes are their most significant visual organs. The slightly blurred, coarse mosaic image of nearby objects that they produce is adequate for recognition of simple patterns and forms and is remarkably capable of perceiving movement. Even slight spatial displacements register immediately on one or more ommatidia. Insects may also have color perception.

Dawn ends the darkness of night and dusk the light of day. This familiar cycle imposes on insects an external rhythm by which to regulate their daily activities. The term *circadian* comes from the periodicity of such cycles which approximate a 24-hour period. Most insects are not randomly active throughout the entire 24 hours but exhibit a definite *periodicity*, defined as a temporal gradient of activity (*Fig. 12-1 b*). This periodicity is termed *diurnal* when they are active during daylight hours, as with dragonflies, grasshoppers, and butterflies, *nocturnal* when active at night, as with cockroaches, most katydids, and moths, and *crepuscular* when active at dusk or dawn, as with certain pomace flies. Each of these insects tends to forage, reproduce, and move about during its period of activity. It then becomes inactive and often hides during the alternative period during which it may be temporarily replaced in the community by species having the alternative periodicity.

Temperature, humidity, and many other environmental factors vary over a day. All may be involved in circadian behavior, but light is salient. The transition from light to dark or dark to light may act as a simple cue to switch insects from day to night activities and *vice versa*, or their periodic behavior may be more complex. A circadian rhythm may persist in some insects, at least for a time, when caged in total darkness or in continuous light. Such creatures may be said to have a *biological clock*, being influenced by two separate periodicities: an exogenous photoperiod of 24 hours and an endogenous circadian rhythm that only approximates 24 hours, the two of which are brought into daily synchrony by environmental clues such as dawn or dusk. Their internal biological clock does not merely start when, for example, the light cue is given and stop when it is withdrawn. It runs or, better, oscillates continuously but requires setting by environmental cues. Caged individuals of the nocturnally active American cockroach exposed to constant light or to constant darkness continue to be active at about the same hour, suggesting an endogenous circadian rhythm. Over time, however, they adjust this innate rhythm to a new photoperiod, resuming activity early when the onset of darkness is advanced or delaying it when it is retarded.

No less important than these daily rhythms of insects are their *seasonal rhythms*. A shift in hours of illumination relative to those of darkness follows the march of the seasons, and insects adjust their activities accordingly. For example, the shorter days of autumn trigger larval diapause in the pink bollworm which is not terminated until day length again increases in late winter.

Pigmentation, another light-related phenomenon, has evolved in insects to the point where their colors and color patterns rival those of birds and satisfy many biological needs. Colors may block harmful sunlight, help them gain or lose heat, help them find mates, or hide them from potential predators.

Ultraviolet rays do not penetrate the deep tissues of most insects, but excessive UV radiation inhibits enzyme action or RNA or DNA synthesis and may be deleterious to small, unprotected insects and their eggs. Infrared rays may be even more dangerous than UV rays because they carry heat, concentrated doses of which burn unprotected individuals.

Chapter 12 / INSECT NATURAL HISTORY

Protective pigmentation is a common adaptation shielding insects from excessive radiation, their pigment granules interrupting the harmful rays. Diurnal insects tend to be more heavily pigmented than insects that are customarily exposed to little or no light. Subterranean worker termites are pale, yet their sexual forms which mate in daylight are well-pigmented. Many cave crickets and other dwellers of deep caverns are colorless.

Dark colors absorb and light colors reflect heat, and insects may be pigmented accordingly. Snowfleas are warmed through their deeply pigmented cuticle, while the lightly pigmented desert and beach forms of certain insects are more likely to reflect heat than to absorb it.

Several types of protective and/or attractive coloration and color patterns occur in insects. Common among them are: (1) *cryptic coloration*, as with many plant-perching katydids whose green body color matches that of their homogeneously colored green environment, presumably making them inconspicuous to potential predators; (2) *disruptive coloration*, or camouflage (*Fig. 12-2 a*), as with certain grasshoppers whose blotched or mottled color pattern breaks up their outline when in their irregular, bare-ground or rocky habitat, and as with pine sawyers which are rendered inconspicuous when at rest on the bark of trees; (3) *object resemblance*, in which an insect takes on not only the color but also the form of some object within its environment, often combined with the ability to "freeze," motionless, for extended periods, as with various walkingsticks (*Fig 12-2 d*), leaf insects (*Fig. 19-5 c*), and the dead-leaf butterfly that mimic twigs or leaves, the exotic pink mantises that resemble flowers, and the slant-faced grasshoppers (*Fig. 12-2 c*) that resemble the grass on which they perch; (4) *aposematic*, or warning coloration, as with bumblebees whose bright colors and conspicuous color pattern associated with a sting, certain katydids that "rear up" in a threatening manner, exposing colorful mandibles and wings (*Fig. 12-2 b*), and the strikingly colored black and red milkweed beetles whose toxic substances ingested from the protected host plant they have eaten likewise protect them; (5) *mimicry*, in which an insect, or mimic, resembles in color and form an unpalatable or noxious species, or model, and is protected by this resemblance; (6) *industrial melanism*, as with the peppered moth of Britain whose urban populations are darker than are those of nearby unpolluted areas; (7) *color change*, as with certain walkingsticks whose body color is dark at night but light during the day; and (8) *sexual color dimorphism*, by which some insects recognize the opposite sex and aggregate for mating, as with many butterflies whose colorful, intricately patterned, sometimes lustrous males differ from the often drab females of their species.

Temperature

All ecology relates, directly or indirectly, to radiant energy from the sun. This energy translates into heat, which is our immediate concern, and light, which itself dissipates into heat.

Insects are *poikilothermic*, or "cold-blooded." They constantly gain heat from, or lose it to, the environment through convection, conduction, radiation, evaporation, and metabolism. Their body temperature fluctuates according to the environmental temperature as modified by the heat that they produce metabolically. Thus, their growth and metabolism fluctuate at a proportionate rate except at

Environmental Factors

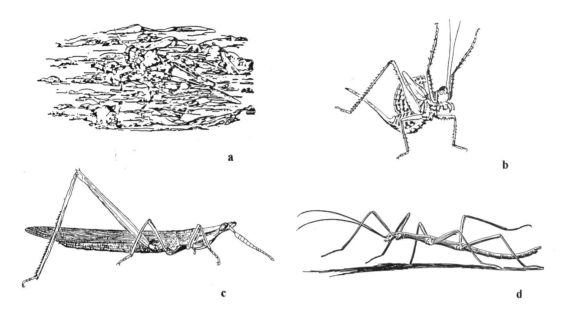

Fig. 12-2. (*a*) A disruptively mottled grasshopper well camouflaged in its bare-ground habitat; (*b*) an aposematic katydid in threatening defensive posture; (*c*) a slender slant-faced grasshopper that resembles in color and form the grass on which it perches; and (*d*) a twig-mimicking walkingstick (*a* is after L. Chopard, 1938; *b* is by courtesy of T. J. Cohn, 1965; *c* is modified and redrawn from Sharp, 1970; *d* is by courtesy of Illinois Natural History Survey).

extreme temperatures. This relationship can be demonstrated in the laboratory by plotting an insect's activity rate against increasing temperature. For example, the frequency of the snowy tree cricket's calling song varies so closely with external temperature that it can serve as a crude thermometer.

Some insects use their flight muscles to generate heat as well as to fly. Flight for most species is possible only at relatively high muscle temperatures, usually about 30 degrees C, so during cool weather they may precede flight by a period of pronounced shivering. This warm-up behavior is common among night fliers and in insects such as bumblebees during their early morning flights.

Insects tend to seek shelter and become immobile at low temperatures. Then they slip into cold coma as temperatures lower further though this state is usually not damaging. Eventually, however, at the point of low lethal temperature, their enzymes become inactivated, ice crystals form, and their cells burst, causing death. At high temperatures, insects become restless, hyperactive, and attempt to escape. Then they slip into heat coma and eventually, at the point of high lethal temperature, their body proteins coagulate and death ensues. Damage by lethal temperatures, either low or high, is progressive and depends on the intensity and duration of exposure. If milder conditions are restored before death, the rate of tissue and enzyme repair rises and that of destruction falls until they again become functional. However, if gradually exposed to extreme temperatures over a period of weeks, they may become acclimated and survive under the new temperature regime.

Most insects are active at temperatures between 12 to 35 degrees C though a few species are adapted for living under more extreme conditions of cold or warmth. Certain cavern-dwelling carrion

Chapter 12 / INSECT NATURAL HISTORY

beetles whose normal activities are in ice grottoes under near-freezing conditions, the alpine rock crawlers and the so-called snowfleas, both of which live on or near patches of ice and snow, and many temperate-zone stoneflies that mature, feed, and mate during the coldest months of the year are examples. In contrast, the firebrat lives in or near kilns, ovens, and other remarkably hot places.

Many insects adapted for life in cold regions survive the unfavorable season by undergoing *diapause* which is a programmed condition of arrested development before they resume normal development upon return of favorable conditions. *Univoltine species*, those having a single generation per year, experience an obligate diapause. In contrast, *multivoltine insects*, those having two or more generations per year, usually exhibit only a facultative diapause, their development proceeding according to local variations in environmental temperature.

Insects may be thought of as creatures of the *microhabitat*. They dwell within crannies beneath bark, under pebbles, between soil interstices, between leaves or flowers, *etc.*, where they are especially subject to *microclimate*. Air, which is the medium of terrestrial insects, is responsive to temperature and wind variations resulting from local differences in topography and cover. Summer air tends to lie in layers of decreasing temperature from ground level upward. This vertical heat gradient determines the perching situations sought by many insects and causes appropriate changes in their body physiology.

Noted earlier is the fact that light/dark is the critical environmental cue in most insects' circadian rhythms. However, temperature and humidity are involved also, though not so directly. The transition from day to night is accompanied by a drop in temperature and a rise in humidity. These microclimatic changes may be as important to individual insects as are light changes *per se*. A diurnal grasshopper, for example, usually spends the night, motionless, in a sheltered situation. Upon exposure to the warming rays of the sun the next morning, its metabolism becomes enhanced, allowing it to resume activity. It begins moving about and foraging as temperatures rise, and it continues to be active throughout the day until dusk. In contrast, a nocturnal katydid remains inactive during daylight hours but becomes active at night. It resumes activity at dusk, carries out most of its movement and feeding shortly afterward, and then becomes inactive toward dawn. The local thermal migrations of some insects are closely associated. For example, some grasshoppers move about on the ground until forced to ascend the vegetation during the heat of the day, as opposed to many desert katydids which spend the heat of the day under debris, deep at the base of plants, and migrate to the surface only at night.

Water

Water is critically important to insects because, being small, they have a high surface area/volume ratio. Water in the fluid state is the medium of aquatic insects; it offers terrestrial insects life-sustaining drink; and it supports growth of the vegetation upon which all insects depend, directly or indirectly, for food and shelter. Moreover, water in the form of atmospheric moisture determines the extent to which terrestrial insects may desiccate and thereby affects their geographical, ecological, and seasonal distribution.

Terrestrial insects' tolerance of moisture or dryness varies. Springtails, termites, cave and camel crickets, and certain other hygrophilous insects require a humidity close to saturation and desiccate under drier conditions. In contrast, mealworms, granary weevils, and certain other xerophilous insects are able to live under reduced levels of humidity, sometimes below 10% of saturation. However, most insects live under conditions intermediate between these extremes.

Insects supplement whatever natural tolerance they have for particular conditions of moisture or dryness by using specialized structures, functions, or behavior. Included are the Malpighian tubule adaptations that enable mealworms to produce dry, powdery excretions that occasion little water loss; the use of the so-called "water of metabolism" by powder-post beetles; adoption of nocturnal activity and diurnal hiding in sheltered places as in camel crickets and cockroaches; use of enclosed nests and runways as in termites; and drinking as in butterflies, honey bees, plague locusts, and many other thirsty insects.

Other Environmental Factors

Numerous other environmental factors influence insect distribution to a lesser extent, and many of them interact. For example, temperature modifies humidity levels and the rate of evaporation, and, in turn, evaporation and condensation modify the impact of temperature. Furthermore, high temperatures increase metabolism and growth which, in turn, may alter insects' need for food and water.

Populations

A *population* includes all individuals of a single given species living in a given area. It is an aggregation of potentially or actually interbreeding animals within that area. This definition fixes population relationships as intraspecific (within the species), as opposed to community relationships which are interspecific (involving interactions among different species).

Consequences of Aggregation

Over a century ago, Darwin noted that the struggle for existence is keener among individuals of the same species than among those of distantly related species. After all, the needs of conspecific individuals and the means by which they satisfy them tend to be most nearly similar. Animals need food, shelter, reproductive facilities, and other, often limited resources resulting in a competition that becomes increasingly severe upon population growth. The biological effects vary according to the point-of-view. Competition is harmful and even fatal to individuals, yet it may be either favorable or unfavorable to species because of populations' reliance on the selective pressure of competition in their evolutionary development.

The layman's conception of competition is of a dramatic confrontation between gladiators, whether insect or vertebrate. Such combat does occur among the pugnacious field crickets that fight over food and mates and certain large-bodied stag beetles that fight for mates. However, most competition among insects is more subtle. For example, populations of the confused flour beetle increase dramatically when food is abundant, but the beetles die off during food scarcity. Much the same happens in the common fruit fly, another favorite "test tube insect."

Chapter 12 / INSECT NATURAL HISTORY

Aggregation is an integral aspect of reproduction. Most insects reproduce by bisexual dioecious means and so must find mates if they are to perpetuate themselves. Mate location is obviously more difficult under reduced than under enhanced population levels. However, individuals may concentrate themselves locally using odors, lures, calls, *etc.*, enhancing the probability of finding mates.

Population Development

Insects differ in the number of individuals that hatch and enter a particular area, leave it, and die in a given unit of time. Accordingly, their populations may increase up to a maximum, maintain themselves at the same stable level, or decrease, sometimes to the point of extinction.

Natality is the rate of production of new individuals in the population (*Fig. 12-3 a*). It refers to the number of newly hatched individuals added to the population per unit of time and not to population growth, which depends on additional factors. *Mortality*, in contrast, is the rate of death within the population owing to aging, starvation, predation, disease, competition, accidents, or other negative factors. Natality less mortality represents the population's theoretical potential for growth under optimal conditions. This ideal rate, termed *biotic potential*, does not occur in nature over long periods because optimal conditions are never encountered. As population density increases, shortages of food or space occur as well as increased interference by other animals, and other negative environmental factors become limiting, causing a progressive increase in mortality, a decrease in natality, or both.

A population growth curve may be constructed by plotting the number of individuals within the population against time (*Fig. 12-3 b*). Starting with a small breeding stock unimpeded by regulation, the resulting growth curve is *logarithmic*. Such exponential growth is purely theoretical. Regulation invariably comes into play, the population quickly leveling off and reaching a self-limiting equilibrium. The resulting second kind of curve is *logistic*. Some forces underlying population growth can be appreciated by comparison of the logistic curve with a curve of population increment (numbers added to the population per unit of time) (*Fig. 12-3 c*). An *incremental* curve begins essentially the same way as a logistic one but quickly reaches a point of maximal increment called the *inflection point*. Thereafter, the increment declines progressively though the population as a whole may continue to grow, and the increment disappears altogether as the population reaches equilibrium. Mortality comes to equal natality during equilibrium, and the population ceases growing though it may remain large.

Mortality stems from the impact of two types of regulatory factors: (1) *density-dependent factors* that destroy an increasing proportion of individuals with increased population size, and (2) *density-independent factors* that are independent of fluctuations in population size. Among the density-dependent variables are supplies of food, oxygen, water, breeding sites, and other resources that may become increasingly scarce upon population growth. Among the density-independent ones are temperature, moisture, other weather-related variables, and pesticides which continue regulating regardless of population size.

Aphids, thrips, and certain other short-lived insects dependent on a single food source often follow a *J-shaped population growth curve* rather than a logistic one. Theirs involves almost exponential growth followed by an abrupt decline in numbers. The sharp initial increase in population size

Populations

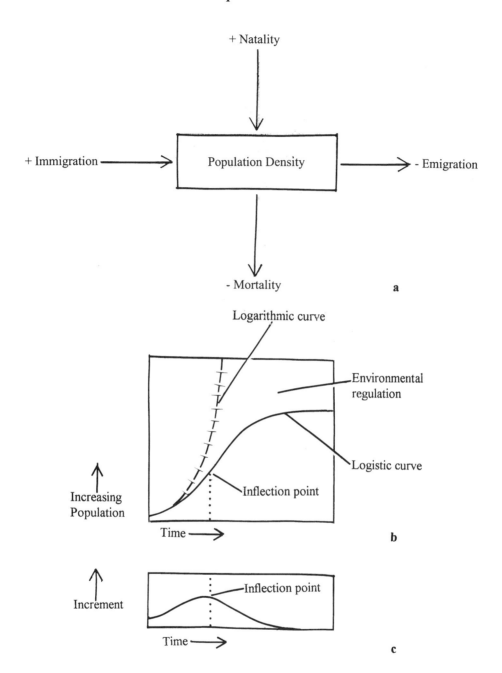

Fig. 12-3. (*a*) Positive (natality and immigration) and negative (emigration and mortality) determinants of insect population density; (*b*) logarithmic and logistic population growth curves relative to (*c*) a curve of population increment.

occurs during the brief favorable season and stems from the hatching of new individuals that, themselves, quickly become breeders. The decline that follows often stems from environmental regulation attributable to the seasonal dwindling of the food supply, accelerated attack by predators, *etc*.

169

Chapter 12 / INSECT NATURAL HISTORY

Spatial Relationships

Immigration, or dispersal of individuals into an area, and *emigration*, or dispersal of them out of it, also modify population development (*Fig. 12-3 a*). Natality and immigration combine to increase local population density, whereas mortality and emigration reduce it.

Vagility is the ability of organisms to disseminate, as with individuals of an old field dispersing across a marsh, woodland, or river to gain access to another field. *Adaptability* is their ability to adjust to changed conditions such as may be encountered upon arrival in the new habitat. Some adaptability is requisite to any immigrant because no two communities, even closely adjacent ones, are identical.

Insects' spatial needs vary with the species and are a function of available food, shelter, and breeding facilities. Several broad patterns of spacing or distribution may be recognized including *random* (*Fig. 12-4 a*) in which the statistical probability of an individual insect occurring in a particular place is as great as occurring in any other place within the species' range, *uniform* (*Fig. 12-4 b*) in which individuals are distributed according to some regularly repeating pattern, and *clumped* (*Fig. 12-4 c*) in which they cluster. Clumped distributions are common among insects, arising from their tendency both to form pairs, swarms, or other aggregations and to distribute themselves in accordance with the clumped vegetation they need for food, shelter, or perching. Granary beetles that live in a uniform flour environment are examples of insects that tend toward randomness. A uniform distribution is rare in nature, even among insects that exhibit territoriality in a relatively constant environment. There are, however, some pest insects that take on a semblance of distributional uniformity in response to mankind's horticultural or agricultural practices, as with the wheat stem sawfly responding to strip farming.

Population Age Structure

Pre-reproductive, reproductive, and post-reproductive age groups may be recognized within a population, and their relative proportions are useful in determining the population's status at a given time. An increasing, young population is characterized by more pre-reproductive individuals than reproductive ones, more reproductive individuals than post-reproductive ones, a high natality, and an exponential population growth. The age data of such a population, a young one (*Fig. 12-4 d, left*), plotted on successive layers of a bar graph take on a triangular appearance. A stable-age, mature population (*Fig. 12-4 d, middle*) has comparatively equal numbers of pre-reproductive, reproductive, and post-reproductive individuals and a bell-shaped graphic representation. A declining, old population (*Fig. 12-4 d, right*) has more reproductive and post-reproductive individuals than pre-reproductive ones, a reduced natality, and an inverted urn-shaped graphic representation.

These stages usually occur in seasonal sequence among insects. Insect populations tend to be young at the onset of the growing season, old at the end, and mature in between. However, different species vary widely with respect to their periodism and time of maturity. Many stoneflies of the northern hemisphere emerge, feed, and mate during winter, their old age being shunted toward the early

Populations

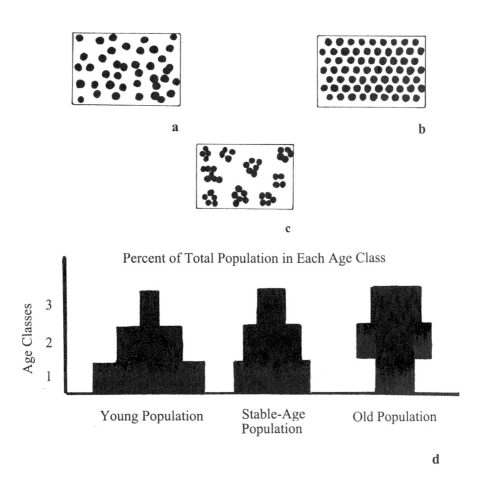

Fig. 12-4. (*Top figures, a-c*) Contrasting patterns of dispersion of individuals of a population (each dot representing an individual insect), with *a* showing random, *b* regular, and *c* clumped distributions. (*Bottom figures, d*) An idealized comparison of the population age structure of a young, increasing population (*left*), a mature, stable-aged population (*center*), and an old, declining population (*right*).

months of the calendar year. In contrast, most grasshoppers, even from the same region, emerge, mature, and mate in summer, their old age being shunted toward the late months of the year.

The duration as well as the chronology of insect life stages varies. Most species spend over half their life as pre-reproductive individuals and consequently have shortened reproductive and post-reproductive periods. Many temperate-zone insects spend eight or nine months in the egg stage and live only a few weeks after reaching maturity. Mayfly naiads require a year or more to go through postembryonic development, yet the adults live a few days at best and cannot even eat. Periodical cicadas have an even more disproportionately long juvenile existence. Theirs may last up to 17 yr, depending on the species, and is followed by an adult stage that lasts only about a month.

Chapter 12 / INSECT NATURAL HISTORY

Communities and Ecosystems

Communities consist of all species living together in a particular place, interacting with one another, and exhibiting at least a degree of mutual adjustment. Inasmuch as communities involve individuals of many different species (animals, vascular plants, fungi, bacteria, *etc.*), their relationships are interspecific, not intraspecific (unlike population ecology). Interspecific relationships sometimes may be beneficial, sometimes neutral, and sometimes deleterious; they may be permanent, lasting during the entire life cycle, or temporary, lasting only during certain life stages; and they may be *obligatory* and essential to survival, or *facultative*, being availed upon on occasion but not necessary. A convenient way to address these community relationships and their gradations is to divide them into arbitrary categories, as follows: (1) *mutualism*, a mutually beneficial relationship among individuals called symbionts belonging to two or more species such as termites and their intestinal flagellates and aphids, or so-called "honeycows," attended by ants (*Fig. 12-5 a*); (2) *commensalism*, a relationship beneficial to the commensal but neutral to the host, as with myrmecophiline crickets that live as "house guests" within ant nests, yet apparently provide nothing in return; (3) *competition*, a mutually harmful demand on the same limited environmental resources, as with different flour beetle species attempting to live within the same laboratory culture, resulting eventually in competitive exclusion of one or more of them; (4) *parasitism*, a relationship between species in which one, the parasite, lives on, in, or near the other, the host, deriving nourishment from it, usually without killing it, as with bed bugs and their human host; and (5) *predation*, a relationship between species in which one, the predator, attacks or traps, kills, and eats the other, the prey, as with praying mantises devouring other insects (*Fig. 14-2 c*).

The insects, other animals, and plants that live in a particular place are, therefore, more than just a casual aggregation of organisms. They constitute a coexistent, mutually adjusted assemblage of different species populations dependent on one another in complex ways. The term applied to this overall entity is *community*. It is an assemblage in which (1) each population finds at least its minimal life necessities provided by a tolerable local environment. It is an assemblage in which (2) such deleterious interactions as take place result in the death of many individuals though the populations as a whole survive them. For example, photosynthetic plants provide food, shelter, and protection to animals which eat the plants and plant products and avail themselves of the shelter afforded by plants, while decomposers break down the remains of the dead plants, animals, and their products, thereby supplying substances that the plants need for continuing photosynthesis. It is an assemblage in which (3) there is at least a degree of *integration*, or mutual adjustment, achieved in a context of tolerance and beneficial interaction. It is an assemblage in which (4) large numbers of producers, consumers, and decomposers are represented, their respective functions being essential to community self-sufficiency. Finally it is an assemblage in which (5) there are sometimes astoundingly large numbers of species populations, both plants and animals, and correspondingly complicated relationships.

A community consists, therefore, of the biotic environment of a given place, but this living component (fauna and flora) cannot be divorced from the physical, non-living environment around it, all tending to remain in a state of equilibrium. Therefore, the two are properly an *ecosystem*.

Communities and Ecosystems

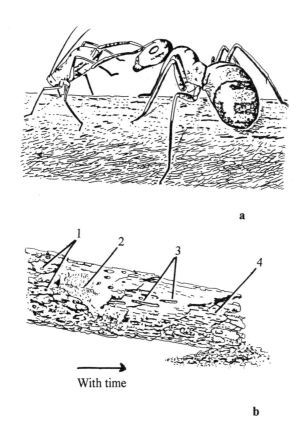

Fig. 12-5. (*a*) An ant obtaining honeydew from a "honeycow" that it protects and tends assuring a continuing supply of the aphids' sticky, sweet exudate; (*b*) a rotting pine log microsuccession, with *1* = holes from long-horned beetles, *2* = engraving by bark beetles, *3* = galleries by carpenter bees, *4* = damage by termites and carpenter ants (*a* is modified and redrawn from Wickler, 1968; *b* is modified and redrawn from T. C. Emmel, 1973).

Desert, alpine, salt lake, hot spring, tundra, and other severe ecosystems consist, in general, of few species each, as opposed to less austere communities which tend to have more numerous species. It is hard to generalize about them other than to comment on their species numbers. The best that one can say is that insects predominate in most communities, and the insect species represented tend to belong to different genera.

I have stressed communities' uniformity, but this is an oversimplification. Communities exhibit marked horizontal, vertical, and temporal gradients of condition creating additional places and time slots for occupancy by insects, other animals, and plants.

Zonation

The term *zonation* embraces the horizontal community gradients that occur in nature. It results from place-to-place changes in the physical or climatic conditions causing corresponding changes in the community. In southeastern Michigan, for example, an old-field community is not uniform, continuous grassland but may include grass-herbaceous upland, grass-lichen-moss upland, and other upland habitats as well as lush swales in local depressions where the water table approaches the surface. Zonation also takes place on a lesser scale, as with particular rock outcrops, rotting stumps, dung piles, ant hills, and other nests and burrows that interrupt the overall community, each constituting a *microhabitat* of its own and each occupied by representative lesser communities. Microhabitats are generally more important to insects than to vertebrates which explains why insects, with their small size and secretive habits, are often said to be "creatures of the microhabitat."

Stratification

Stratification is the vertical equivalent of zonation. It involves development of vertical gradients of condition, or *strata*. Those in a terrestrial community range in number from the single, ground stratum of rock ledges, to the three (subterranean, ground, and herb) strata of grassland, to the five (subterranean, ground-leaf litter, herb-shrub, low tree, and high tree) strata of deciduous woodland, and to the multiple strata of tropical rain forest. Stratification, like zonation, presents organisms with an ecological barrier that is often effective in separating them. For example, a recent study of oak-hickory forest showed 78 per cent of the insect, spider, and myriapod species to be restricted to a single stratum, with only a few of these taxa extending into more than two strata.

Periodism

Communities present temporal, in addition to spatial, gradients of condition that enable their inhabitants to restrict activity to one time period or another. This *periodism*, as it is called, occurs both on a seasonal (spring, summer, fall, winter) and on a daily (day, dusk/dawn, night) basis, allowing many more populations to avail themselves of the same limited community resources than if their activities coincided.

Succession

Communities appear static to one's eye but are subject to short-term progressive ecological change. This *succession*, as it is called, may be defined as the fixed, predictable sequence of communities that occurs in a particular area as a result of the impact of the local conditions of climate and soil. Succession involves these steps: (1) pioneering species, plants first and then animals, enter and dominate a newly cleared area, forming a closed community for awhile until eventually the local environment has been altered, favoring other species; (2) then these other species, both plants and animals, enter the area, out compete, and cause extinction of the pioneers, establishing themselves as a closed community for awhile until they too eventually alter the environment and make it suitable for still other communities; (3) one transitory community follows another until eventually succession leads to establishment of a permanently self-perpetuating *climax community* that cannot be displaced under the prevailing conditions of climate and soil.

Thus, abandoned farmland may be taken over by annual weeds, then by perennials and grasses, by shrubs, and finally by climax trees; a surface exposed by landslide is covered first by lichens and moss, then by grass, by shrubs and trees, and finally by climax woodland; a lake fills in and becomes a shallow lake, then a marsh, and eventually dry land that converts to climax woodland. Insects neither play a dominant role in these *macrosuccessions*, as they are called, nor are they characteristic of them though certain insects are common inhabitants of many successional stages. Some insects are actively involved in the dynamics of succession, but others are merely passive players though diagnostic of given stages. For example, sand dune succession in the Lake Michigan region starts with a community of cottonwood, beach grass, tiger beetles, grasshoppers, and wasps. This pioneering community is succeeded by several different forest communities, each with changing insect populations, and leads eventually to the region's climax community.

Insects play a more dominant role in development of the lesser successions that also occur in nature. These *microsuccessions* differ from the preceding in their smaller scale and lack of a climax community. Their ultimate stage is always incorporated into the microenvironment around them and disappears as an entity. This is illustrated by rotting log microsuccessions which vary according to the host trees' specific identity and the community, climate, and locality in which they occur. The tree dies, falls onto the forest floor, is invaded, and then gradually destroyed by successive communities of organisms and microorganisms. The example provided (*Fig. 12-5 b*) is initiated by long-horned beetles (#1) that attack the cambium and wood. Then bark beetles (#2), carpenter bees (#3), and termites (#4) burrow into the wood, carrying fungi and wood-rotting microorganisms with them, causing further disintegration. These wood-destroying insects attract numerous insect predators and scavengers that seek shelter under the peeling bark and course through the decaying wood. Then insect numbers decrease as bacteria increase during late stages of decomposition. Finally, the log becomes a decaying mound of humus on the forest floor in which mosses, ferns, and higher plant seedlings take root, now allowing insects of general occurrence to use it as home. Similar microsuccessions take place within carrion, dung, and some other substances.

Community Metabolism

A community harnesses solar energy and makes it available to insects and other organisms which use some of it for their metabolic processes and return some of it to the overall community until all of the energy is converted into heat and lost from the ecosystem. This energy flow is termed *community metabolism*.

As mentioned, the sun is the ultimate source of energy for life on earth. It releases radiant energy that green plants harness through *photosynthesis*, providing the community with the chemical energy of food. Herbivores, including most insects, ingest these plant tissues and, via molecular rearrangement, convert the plant chemical energy into potential mechanical energy. They either store the energy in this condition or subject it to the molecular rearrangements of *respiration*, resulting in its release as the kinetic energy of *metabolism*. Carnivores, including many insects, then eat the herbivores and obtain and metabolize the plant food second hand. Some heat energy is dissipated to the physical environment with each conversion.

Chapter 12 / INSECT NATURAL HISTORY

These transformations proceed according to the *Laws of Thermodynamics* which, in effect, hold that the potential energy incorporated into food is reduced, step-by-step, by successive levels of feeders until all has been dissipated as heat. It follows that the energy within communities undergoes a one-way flow, being lost continuously from the ecosystem and requiring regular replacement by new solar energy.

Scientists classify organisms according to how they receive energy and pass it on to others, as follows: *Producers* are the photosynthetic green plants of the first trophic (feeding) level. They use water, carbon dioxide, and minerals to make sugar and protein and thereby provide food for the entire biotic community (*Fig. 12-6*). *Herbivores* are second trophic-level vegetarians that eat producers. They include among their numbers over half of the insect species of the world. *Primary carnivores* are meat-eaters of the third trophic level consisting of many insects as well as vertebrates that consume small herbivores. *Secondary carnivores* are meat-eaters of the fourth trophic level. They consist of parasites and of large-bodied animals, especially vertebrates, and only a few insects. Finally *decomposers* of the fifth trophic level, chiefly bacteria and fungi assisted by a few insects, attack and break down the tissues of dead animals and plants, absorb some of the products for their own metabolic needs, and eventually release the other products to producers, allowing the cycle to begin anew.

Therefore, the ecosystem may be seen as a sequence of feeders, each operating within its own trophic level. The feeding sequences may be simple food chains involving only a few species but are usually complicated food webs. A *food chain* is a simple, linear series of foods and feeders involving at least a plant, an herbivore, and a carnivore. However, the total of foods and feeders is usually more extensive and involves many interrelated species, especially insects. Such complex, interlocking patterns are termed *food webs*. Species such as scavengers may consume food from several different trophic levels.

The total individuals of a community may be plotted according to arbitrary size groups (*Fig. 12-6*). For example, the numbers of individuals of small, medium-sized, and large-sized taxa may be determined, the data assembled into bar graphs, and the bars arranged on top of one another in general order of body size. The resulting *pyramid of numbers* usually takes on a step-wise, pyramidal form because of communities' tendency to have large numbers of small-sized individuals, moderate numbers of middle-sized individuals, and small numbers of large-sized individuals. The same result may be obtained by grouping the organisms of a community according to trophic level, resulting in a *pyramid of trophic level*. A recent study of the elm slime flux that arises where tree branches are torn free illustrates. One insect secondary carnivore was found supported by three insect primary carnivores that, in turn, were supported by one herbivorous mite and three herbivorous insects.

The principle behind such pyramids is that the rate of production must exceed that of primary consumption, the rate of primary consumption that of secondary consumption, *etc.*, in accordance with the *Second Law of Thermodynamics*. In short, a proper proportion of producers, herbivores, carnivores, and scavengers must be maintained within the community to assure the continuous existence of the food web. This equilibrium, the *Balance of Nature*, is most readily appreciated when disturbed by outside change. For example, grasshoppers exposed to unseasonably cool, wet weather may be

Community Metabolism

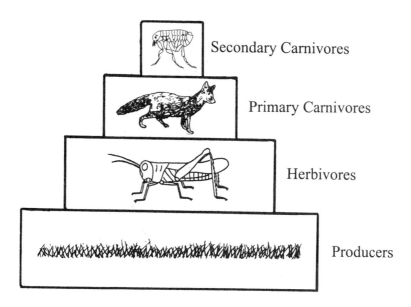

Fig. 12-6. An idealized pyramid of trophic level showing insects' typical role, in this case grasshoppers eating grass but being eaten by foxes which, in turn, are parasitized by fleas. Vertebrate secondary carnivores tend to be large-bodied and more powerful than the primary carnivores on which they feed. This relationship does not hold for insects. Insect secondary carnivores are either parasitic or specialized in some other way for their role in the food chain and, hence, dwarfed by the host on which they feed.

decimated by fungal disease which, in turn, drastically reduces the number of birds that can depend on grasshoppers for food and, as a consequence, starve. Conversely, favorable weather conditions may cause grasshopper irruptions to the point where their numbers temporarily exceed the environment's *carrying capacity*, or resources capable of supporting life. In this event, large numbers of grasshoppers will die by starvation and increased predation pressure, leading, in turn, to the eventual death of the increased numbers of birds and other carnivores that depend on grasshoppers for food. In short, animals, including humans, cannot survive an abundance that outstrips their food supply.

Suggested Additional Reading

Bursell (1974, 1974a); Davies (1988); Gullan & Cranston (2000); Huffaker & Rabb (1984); Norris (1991); Odum (1971); Price (1997).

Chapter 13

Insect Habitats

Well may we affirm that every part of the world is habitable!
Whether lakes or brine or . . . warm mineral springs . . . even
the surface of perpetual snow — all support organic beings.
— Darwin

The *environment* is the sum total of conditions affecting organisms in the general place (land, fresh water, or seas) in which they live. It provides clues to their life history and adaptations, but more is needed to understand insect occurrence. Entomologists also must turn to the *habitat*, or particular place where, and environmental conditions under which, organisms live. It includes an area of determinate biotic environment and physiognomy more or less distinguishable from surrounding areas.

About 71 per cent of the earth's surface is covered by ocean and only about 28 per cent by land. The oceans are also more continuous, more barrier free, and more environmentally stable than land, and they have been occupied by living things far longer than has land. On these bases alone, the oceans and seas of the world might be expected to be more prolific of insects than is land, but they are *not*. Insects are fundamentally terrestrial creatures whose adaptiveness has enabled them to take advantage of land's rich diversity of habitats to produce a staggering number of species. This is not to say there are no aquatic insects—far from it. Many insects have secondarily invaded fresh water, and a few have returned to the ancestral oceans and seas. The latter will be discussed first.

Marine Environment

Oceanic Zone

Certain water striders (*Fig. 13-1 a*) inhabit the calm, open waters of the *oceanic zone*, sometimes hundreds of km off shore, especially in equatorial areas with seaweed. These water-surface dwellers move over the water much as do their gerrid relatives on fresh-water lakes and ponds. They either prey or scavenge on small floating animals and lay their eggs on floating debris and vegetation.

Intertidal and Supratidal Zones

Other so-called "marine insects" are found not in open ocean but either in the *intertidal zone* between high and low tide or immediately above in the narrow *supratidal zone* wetted by the ocean spray.

Chapter 13 / **INSECT HABITATS**

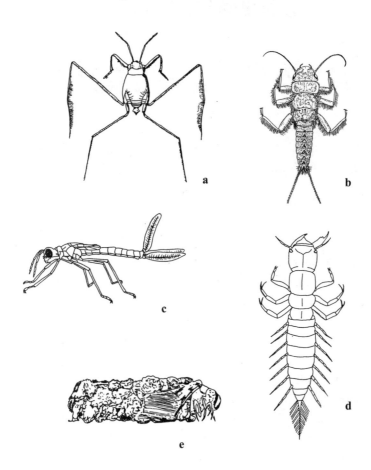

Fig. 13-1. (*a*) An oceanic water strider and selected fresh-water insects including (*b*) a stonefly naiad, (*c*) a damselfly naiad, (*d*) an alderfly larva, and (*e*) a caddisfly larva within its distinctive case (*a* is modified and redrawn from L. Cheng, 1976; *b – e* are modified and redrawn from various sources).

Among them are the insect dwellers of *rocky shores* and *cliffs*. They occupy a coastal environment of strong tidal action, powerful wind, fluctuating temperatures, and pools more saline than the very sea water that fills them. Included are the salt- and heat-tolerant larvae of shore flies, seaweed flies, and other Diptera, earwigs that hide within rock crevices or beneath debris during the day but emerge at night to scavenge, and springtails that crawl over the shore at low tide and cluster in submerged air pockets at high tide. Moreover, ants, ground beetles, and other terrestrial predators and scavengers frequent bare rock in this region, and dragonflies, butterflies, and other winged insects patrol the air above.

Other intertidal insects occupy *sandy beaches* that afford neither fixed attachment for seaweed growth nor crevices for hiding but have a gentler surf and fewer pools than rocky shores. Most sandy-beach dwellers are scavengers that live beneath seaweed or debris and converge on the drift line to feed as the tide ebbs, among them ants, ground beetles, earwigs, and many flies. Tiger beetles, robber flies, and other predators walk the sandy surface or fly over it in search of prey.

Still other intertidal insects frequent *muddy shores*, another habitat characterized by an unstable substrate. Muddy shores are typical of estuaries and bays where rivers deposit silt in an alluvial fan mixing fresh-water currents with oceanic tides, producing brackish conditions. Salt marshes and mangrove swamps, for example, have an emergent vegetation supporting marine, fresh-water, and terrestrial insects. Included are mosquito and midge larvae and dragonfly naiads found within the brackish waters and adults of these insects and of various beetles, ants, and bugs that either rest on or fly about over the vegetation.

Fresh-Water Environment

The ancestral arthropods apparently spread from the seas to land where insects evolved, flourished, and then secondarily invaded fresh water. Approximately 1 per cent of the earth's surface is covered by fresh water whose habitats are more varied in temperature, salinity, chemical composition, turbidity, current, and biota than are oceanic habitats though far less varied than terrestrial habitats. Fresh water is also geographically discontinuous. These factors have had an important bearing on the resident insect fauna.

Fresh water includes two basic environments, *standing-water* and *running-water*, between which there are no absolute differences except for the obvious disparity in current which is largely responsible for shaping the inhabiting insect fauna (*Fig. 13-1 b – d*) and flora.

Standing-Water Environment

This environment includes lakes, ponds, swamps, marshes, bogs, salt lakes, and certain subterranean bodies. *Lakes* are large standing-water bodies having well-defined shore, open water, and deep zones and, in temperate regions, thermal stratification during certain seasons of the year. *Ponds* are smaller, shallower standing-water bodies than lakes. The shore zone of ponds supports rooted vegetation over much of the bottom; the open-water and deep-water zones of lakes are lacking; and thermal stratification is negligible. Many ponds are temporary, being filled or empty according to season, but others are permanent, being water filled at all times. The fauna varies accordingly. The insects of temporary ponds belong to species able to mature quickly and reproduce during the ephemeral wet season and survive the dry season in dormancy. The standing-water biota includes plankton, benthos, periphyton, nekton, and neuston.

Plankton consist of the assemblage of small, submicroscopic or microscopic floating plants and animals largely incapable of moving against the current. Algae, bacteria, Protozoa, rotifers, Crustacea, and larval insects including mosquitoes (*Fig. 9-5 d*) and phantom midges are common plankters. Some occur in tremendous number, sometimes on the order of millions per liter of water. For this reason and because they include many photosynthetic species, they form the basis of the aquatic food chain. The plants among them are distributed near the surface where photosynthesis is effective, but the animals, being less dependent on light, often reach into the depths.

Chapter 13 / INSECT HABITATS

The *benthos* includes the assemblage of organisms attached to or associated with the bottom. They either crawl over the surface, burrow into it, or attach to it in some manner, so their occurrence is largely determined by depth, nature of bottom, and the presence or absence of rooted plants. The benthos of fresh-water shores consists of numerous submicroscopic animals as well as flatworms, annelids, rotifers, Crustacea, water mites, midge and caddisfly larvae (*Fig. 13-1 e*) and mayfly (*Fig. 19-4 b*) and flat-bodied dragonfly (*Fig. 19-4 d*) naiads. This great diversity within the benthos of the shore zone contrasts with the more limited benthos of the deep zone. The latter includes only certain clams and annelids, bloodworm larvae, and a few other animals able to withstand prolonged periods of reduced oxygen tension.

The *periphyton* is the assemblage of organisms that cling to rooted plants or brush projecting from the bottom. Among insects of this type are midge larvae and mayfly, damselfly (*Fig. 13-1 c*), and slender-bodied dragonfly naiads adapted for clinging and climbing.

The *nekton* is the assemblage of organisms able to swim effectively against the current. The nekton of the open-water and deep-water zones is poorly developed, but that of the shore zone includes fishes and insects that inhabit the zone permanently and amphibians, turtles, and snakes that occur there "part-time," being terrestrial the rest of the time. Many insects with a streamlined body form and oar-like legs including backswimmers (*Fig. 19-12 b*), waterboatmen (*Fig. 19-12 a*), predacious diving beetles (*Fig. 9-8 a*), and water scavenger beetles (*Fig. 9-8 b*) are representative.

The *neuston* is the assemblage of organisms that rest upon or swim in association with the water surface. Being dependent for support on the surface film, these animals are more common in the quiet waters of ponds than in lakes. They include water striders and water measurers that walk over the upper surface, pushing against it with their water-repellent legs as if it were land, and whirligig beetles that swim in association with the under surface, immersed except for their water-repellent dorsum.

Running-Water

Rivers, creeks, springs, and subterranean water courses are the main running-water environments. The first two are arbitrarily separable according to stream width. A stream of more than a few m across may be termed a *river* and one of less than that a *creek* or *brook*. A *spring* is a point where water flows from the ground. It is often a stream source rather than a separate running-water environment, but occasionally it is a seepage area without apparent run-off.

Stream habitats are determined by the nature of the bottom and the rate of water flow. The stream gradient is steep in places, the bottom composed of boulders, trapped rocks, and debris, and the partly obstructed water flow is swift and capable of carrying gravel, sand, and silt. This zone is the called the *rapids* or, on a lesser scale, the *riffles*. The stream gradient is lesser and the current correspondingly slower at other points downstream. Here in the *sand-bottom pool* occurs deposition of heavy water-carried gravel and sand. Current is slower farther downstream in the *mud-bottom pool*, allowing for deposition of silt. These habitats, from riffles to pools, occur repeatedly along the length of a stream, together with endless microhabitats from stone to stone, *etc.*, and they determine the occurrence of the insect inhabitants.

The current is too swift for plankton and most nekton and neuston in riffles, but the habitat supports a characteristic assemblage of plant-clinging and bottom-dwelling organisms. The insects among them generally have a means of attachment and often a streamlined or flattened form enabling them to withstand the onrushing water. Included are net-spinning caddisworms, black fly larvae that sometimes hold fast at the very brink of falls, flattened, ovoid water-penny beetle larvae found appressed to the rocks of swift-flowing streams, and riffle beetle larvae and adults (*Fig. 9-8 c*), the latter having a velvety pile that traps air and functions as a mechanical gill.

The sand and mud substrate of the running-water pool zone is too soft for establishment of a surface fauna, and the current rules out the presence of plankton except in backwaters. However, pools have a burrowing bottom fauna, a nekton, and a neuston. Among them are case-making caddisworms (*Fig. 13-1 e*) found in ponds and slow-moving streams, certain mayfly naiads found in mud burrows except at night when they emerge to search for food, burrowing-type dragonfly naiads that are more cylindrical-bodied and shorter-limbed than the flattened, sprawling-type dragonflies that move about on the bottom of standing water, streamlined backswimmers, water boatmen, predacious diving beetles, and water scavenger beetles that use oar-like legs to move with facility through the water, and the whirligig beetles mentioned previously with respect to the standing-water neuston.

Subterranean Environment

The subterranean environment is a specialized part of the terrestrial environment, yet it has more in common with water than with land. Subterranean life offers its inhabitants relative protection from rapid changes in temperature, humidity, and light, leaving them affected only by substrate, aeration, chemical constituents, pH, *etc*. The subterranean environment, like the aquatic environment, is water logged at the water table and near the stream edge. It includes two radically different habitats, *soil* and *caves*.

Soil

Soil is a mixture of sand, silt, and clay, together with organic matter. Soil particles range from coarse gravel, through sand, to silt, and to fine clay, even the finest particles of which are separated by *soil cavities* comprising perhaps half the volume of average soil. In general, the coarser the soil, the larger the soil cavities, the more space there is for soil organisms, the more readily they can dig into it, the greater its oxygen content, and the more subject the soil is to percolation, preventing water logging. Thus, increased porosity favors development of a soil fauna.

Interstitial organisms are weakly moving, small or minute creatures that occupy soil cavities. They include, among others, telsontails (*Fig. 19-1 a*), springtails (*Fig. 13-2 a*), japygids (*Fig. 13-2 b*), thrips, and the juveniles of various pterygote insects associated with the bottoms of streams or pools or with land. They sometimes number in the millions of individuals per acre of soil, being most abundant in forest soil and least in desert sands. They are most common in the upper layers and only occasionally are found below 1 m in depth. They differ strongly from the fossorial forms that share the soil habitat with them.

Chapter 13 / INSECT HABITATS

Fossorial insects tend to be colorless or pale creatures with a streamlined, smooth body and eyes and wings that are often reduced or lacking altogether. Examples include the burrowing cockroaches and crickets that use their legs to push dirt behind them, termites and ants that remove stones, gravel, and debris with their mandibles and thereby construct extensive, mound-like nests, digger wasps that excavate with their legs and mandibles, nymphal cicadas that use enlarged fore legs to dig to the roots of trees on which they feed, and mole crickets. Mole crickets (*Fig. 13-2 c*) are the fossorial insects *par excellence*, being elongate, smoothly pubescent, mole-like creatures with broad, powerful, spade-like fore legs that they use to burrow almost effortlessly in soft, friable, moist soil.

Caves and Burrows

Cavernicoly pertains to the occupancy of caves, caverns, rock spaces, mines, cellars, and other subterranean places. These habitats are often strongly geographically isolated from one another and are sometimes remarkably constant in their prevailing physical and climatic conditions. There is no light and little or no temperature variation in deep caves, humidity is uniformly high, and air currents are negligible. The insects of deep caves range from facultative inhabitants usually found underground but capable of moving to the surface to obligatory inhabitants restricted to life deep within the caves. Deep caves have distinct dark, twilight, and parietal zones. Among the inhabitants of the *dark zone*, deep within caves, are certain springtails, Diplura, and camel and cave crickets (*Fig. 19-6 d*). In absence of green plants on which to base a food chain, they eat food that stems from the surface, directly or indirectly, being scavengers, predators, or parasites that eat in-washed wood, leaves, and seeds, fungi, small living or dead invertebrates, and/or the bodies or excreta of vertebrates. The *twilight zone*, near the cave mouth, and the *parietal zone*, immediately outside the cave, are also frequented by cave-dwelling insects, especially at night and during damp weather. These two habitats are similar, and at times insects that frequent them may leave the cave and come into contact with an expanded assortment of food. Included are crickets, camel crickets, cockroaches, ground beetles, darkling beetles, and many flies. Some of them are true cave inhabitants; some merely use caves as a retreat; some hibernate within them; and some are only occasional inhabitants.

A few insects take advantage of the relatively constant temperature and humidity of prairie dog, mole, and other mammal burrows to live as scavengers on accumulated dead materials and waste. Various flies and camel crickets are representative of these commensals. There are also various *inquilines*, or true guests, including rove beetles and myrmecophiline crickets, that live in the nests of social termites and ants and provide nothing in return.

Parasitic Environment

Numerous insects live in parasitic association with other animals, among them the ectoparasitic head and body lice of humans (*Fig. 19-10 a*), the human crab louse (*Fig. 19-10 b*), and the human cat (*Fig. 19-19 a*) and dog fleas. They live on the body of their particular host which offers them an environment in its own right though they remain affected by the external environment. Others, such as ox warble fly

Terrestrial Environment

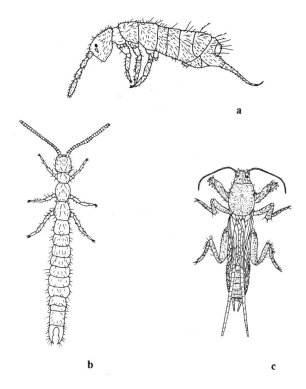

Fig. 13-2. Selected insect representatives of the subterranean fauna including (*a*) an interstitial springtail, (*b*) an interstitial japygid, and (*c*) the fossorial mole cricket (*a* and *b* are modified and redrawn from various sources and *c* from L. Chopard, 1951).

larvae, live endoparasitically within the host whose body provides the environment of immediate concern to them.

Terrestrial Environment

The terrestrial environment is the harshest of all. Land inhabitants are exposed to wide fluctuations in temperature, light, humidity, and wind, both daily and seasonally. They are also subject to minor differences in local topography and cover that translate into *microclimate*. Inasmuch as insects are creatures of the *microhabitat*, the terrestrial forms among them are especially subject to, and must be precisely adapted for, withstanding these pronounced fluctuations.

The following terrestrial *biomes* are among those inhabited by insects.

Tundra is the Arctic zone north of the tree line in the northern hemisphere up to the region of perpetual ice and snow, but it also occurs along the margins of Antarctica. It is a harsh, treeless zone of reduced precipitation, cold temperatures, short growing season, and pronounced seasonal changes

Chapter 13 / INSECT HABITATS

in day length. It consists mostly of flat or rolling, poorly drained land interrupted by innumerable lakes, ponds, marshes, and bogs. It is frozen over much of the year but in summer thaws to a depth of a few cm. It supports an open type of cold-tolerant vegetation consisting of mosses, lichens, sedges, rushes, grasses, a few herbs, dwarf willows, and shrubs.

The resident insect fauna of tundra consists of certain true Arctic forms along with some wide-ranging immigrants from more temperate climes. Predacious diving beetles, water scavenger beetles, and midges are common aquatic inhabitants, and among the terrestrial ones are snowfleas and other springtails, ichneumon wasps (*Fig.19-20 i*), and bumble bees (*Fig. 19-20 l*). Among the many flies are hordes of biting species including mosquitoes, black flies, and deer flies. There are comparatively few Lepidoptera and practically none of the other orders characteristic of more temperate climes. The insect fauna of Antarctica is even more impoverished than is that of the Arctic, consisting of only a few midges, springtails, and certain bird and seal ectoparasites.

Taiga is a vast circumpolar belt of boreal forest extending southward from the Arctic tundra to the temperate region as well as along certain mountain ranges. Winters are harsh, the growing season short and cool, and rainfall moderate. The dominant trees include spruce, pine, fir, and other conifers that grow in stands as far as the eye can see, interrupted here and there by lesser communities of bogs, lakes, or grassland. The fallen conifer needles decay slowly, resulting in a soil that inhibits development of an herb-shrub stratum and limits the insect fauna mostly to forms associated with conifers.

Certain bark beetles, sawyers, borers, weevils, sawflies, horntails (*Fig. 13-3 a*), ants (*Fig. 13-3 e*), termites (*Fig. 13-3 d*), the spruce budworm, and other moths, all associated directly or indirectly with conifers, are common North American representatives.

Temperate deciduous forest is well developed in North America east of the Mississippi, through much of western Europe, and in parts of the Far East. This zone has cool winters and a warm, humid, extended summer growing season. The North American forest consists of hickory, maple, beech, walnut, basswood, and other trees that are leafless in winter over much of the range. There is usually a thick leaf litter on the ground offering insects additional abode. The herb-shrub stratum is well developed. The overall biome is a rich, diverse mosaic of deciduous woodland and successional communities including ponds, pastures, and other open fields as well as conifer stands.

The insects of the North American deciduous forest are so numerous that one can only list a few. For example, on the forest floor are springtails, rove beetles (*Fig. 19-14 e*), ants (*Fig. 13-3 e*), and larvae of various orders; on tree trunks may be found metallic wood borers and powderpost beetles, and in decaying wood live wireworms and a few termites (*Fig. 13-3 d*); on foliage in the shrub-tree strata may be seen leafhoppers (*Fig. 19-11 c*), scale insects, aphids (*Fig. 13-3 c*), katydids, crickets, walkingsticks (*Fig. 13-3 b*), fireflies (*Fig. 19-15 i*), leaf beetles, scarabs, the gypsy moth (*Fig. 16-3 b, c*), tent caterpillars, various leaf miners, noctuids, cankerworms, swallowtails, fritillaries, ichneumon wasps (*Fig. 19-20 i*), gall wasps, midges (*Fig.19-18 e*), gall gnats, and mosquitoes (*Fig. 19-18 c*).

Temperate grassland occurs in North America from the Mississippi Valley westward to the Rocky Mountains, in central Europe through the Russian steppes into the Far East, and in parts of Africa and South America. This dry grassland is to be distinguished from *savanna*, a tropical grassland interrupted by scattered trees and shrubs, as in East Africa. Temperate grassland is common in the interiors of con-

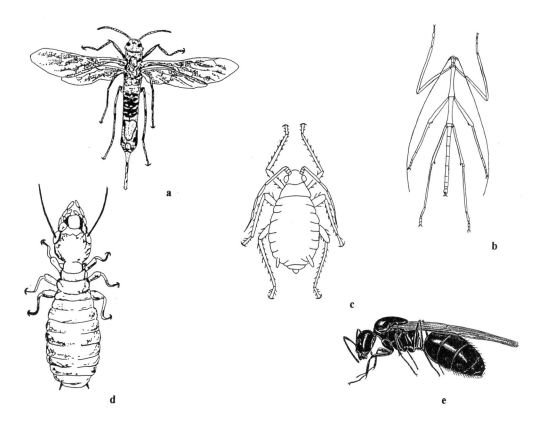

Fig. 13-3. Selected insect representatives of the deciduous forest fauna including (*a*) a horntail, (*b*) a walkingstick, (*c*) an aphid, (*d*) a termite worker, and (*e*) a carpenter ant (*a – d* are modified and redrawn from various sources; *e* is after Back, U. S. Bureau of Entomology, 1937).

tinents where winters are severe, summers hot, and rainfall seasonal, being too light to support forest, yet sufficient for a growth of grass that out-competes desert or scrub vegetation. The evaporation rate is high. The terrain is flat or rolling, and the soils high in humus content. The dominant plants are grasses, ranging from tall to short and growing either as clumped "bunch grass" or more uniformly spaced. Interspersed among them are numerous broad-leaved herbs, especially composites.

Insects occur in grassland in such number and variety that the best that can be done is to list a few representative kinds, as follows: grasshoppers (*Fig. 13-4 a*), crickets, leafhoppers (*Fig. 13-4 c*), spittlebugs, aphids, leaf bugs (*Fig. 13-4 b*), stink bugs, ground beetles (*Fig. 19-14 c*), leaf beetles, ladybirds, snout beetles (*Fig. 19-15 n*), grass moths, noctuids, and other moths, fritillaries, swallowtails (*Fig. 13-4 e*), and other butterflies, bees, wasps, ants, and parasitic Hymenoptera such as ichneumon wasps and braconids, deer flies and horse flies (*Fig. 13-4 d*), robber flies, flower flies (*Fig. 19-18 g*), and flesh flies as well as many other Diptera.

Desert is arid land with sparse, highly modified, xerophytic vegetation. Hot desert occurs at low elevations in certain widely separated locations, notably in northern Mexico through southwestern United States, North Africa through Arabia into central Asia, and South America, southwestern

Chapter 13 / **INSECT HABITATS**

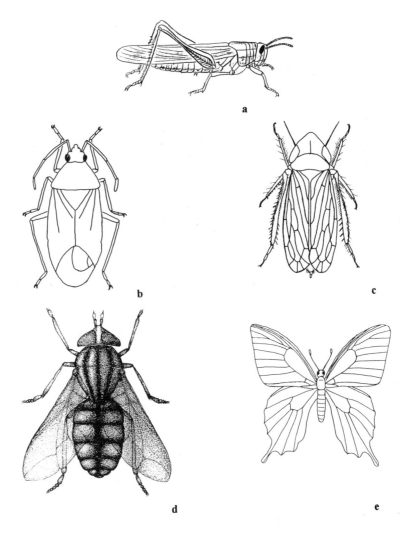

Fig. 13-4. Selected insect representatives of the grassland fauna including (*a*) a grasshopper, (*b*) a leaf bug, (*c*) a leafhopper, (*d*) a horse fly, and (*e*) a swallowtail butterfly (*d* is by courtesy of M. Debabneh; the other figures are modified and redrawn from various sources).

Africa, and central Australia. The soil is rocky or sandy and highly subject to erosion by wind and, during the infrequent flash floods, by water. Solar radiation is strong and continuous throughout the year. Summer and daytime temperatures are comparatively high and winter and nighttime ones low, causing pronounced daily and seasonal heat fluctuations. Precipitation is sparse, seasonal, and erratic, generally less than 25 cm per year, water retention is reduced, and the rate of evaporation high. The vegetation that grows under these harsh conditions includes scattered clumps of perennial cacti, yuccas, euphorbias, and other succulents that store water, have a tough cuticle, and are often thick-leaved, small-leaved, or leafless, and often thorny. The few desert annuals are adapted for rapid growth and flowering during the ephemeral rainy season.

Terrestrial Environment

The insect inhabitants of desert tend to be nocturnally active creatures that burrow or hide during the day in recesses in the ground or within vegetation. They include, among others, bristletails, grasshoppers, crickets, katydids, termites, leaf bugs, leafhoppers, scale insects, tiger beetles, darkling beetles, leaf beetles, dung beetles, various moths and butterflies, and ants and wasps, as well as flower flies, bee flies, house flies, and other Diptera.

Rain forest occurs in certain low equatorial areas, notably in northern South America through Central America, in western Africa, New Guinea, parts of Australia, and the Orient. Rainfall is heavy and either distributed throughout the year or restricted to distinct, alternating rainy and dry seasons. Evaporation is minimal. Proximity to the equator assures an elevated mean temperature that tends to be comparatively uniform day and night and through the seasons and a day length that varies little seasonally. The soils are heavily leached by rainfall and largely unsuited for agriculture. The typically lush, dense vegetation is either evergreen or deciduous. The numbers and diversity of plant species is extraordinarily great. The flora includes broad-leaved trees, twisted woody vines, bromeliads, orchids, and other plant-perching epiphytes as well as ferns, mosses, numerous herbs, and a wealth of other plants. The tall trees are often buttressed in compensation for their shallow root system and stratified into several distinct canopies sometimes producing a dense shade beneath them. Much of the biota is concentrated in the canopies, so the forest floor herbage may be reduced except at the edge of clearings and adjacent to streams. Leaf litter is negligible because of rapid decomposition and the irregular, rather than seasonal, leaf drop.

The insect fauna of rain forest is great in numbers and diversity because of the richness of the food resources and the relative climatic constancy of the forest. The best that one can do to characterize it is to list some representative kinds, as follows: springtails, walkingsticks, leaf insects (*Fig. 19-5 c*), mantises, cockroaches, katydids, termites, cicadas, planthoppers, many bugs, butterflies and moths, beetles, ants and other Hymenoptera, and flies.

Mountains are characterized by marked variations in temperature, pressure, oxygen, humidity, solar radiation, and wind according to elevation and exposure. Their difficult terrain, steep slopes, and discontinuous distribution present strong physiographical barriers that most lowland insects cannot surpass, and they isolate on their peaks the relatively few alpine insects able to establish themselves. Mountains provide a series of *vertical life zones*. The lower levels may be forested by deciduous woodland similar to that of the surrounding lowlands; coniferous forest replaces the deciduous woodland at higher levels up to the tree line; and an alpine zone begins beyond the tree line in mountains that are sufficiently high.

Corresponding changes take place in the montane insect fauna. For example, arboreal and wood-boring insects are necessarily restricted to the lower elevations and excluded above the tree line, meadow insects occur in the expanses of grass and broad-leaved herbs beyond the tree line but below the alpine barrens, and insects are generally absent from the barrens except for certain wind-borne stragglers and a few specialized alpine occupants such as springtails. Except for the few true alpine katydids, grasshoppers, and one primitive genus of rock crawler (*Fig. 19-7 d*), Orthoptera are best developed in the lower zones. Lepidoptera occupy both lower elevations and the high alpine zone, occurring in the latter as ground-dwelling caterpillars beneath stones and as adult moths, including

Chapter 13 / INSECT HABITATS

noctuids, geometers, various microlepidoptera, and even swallowtails, cabbage butterflies, and fritillaries. Among the alpine Coleoptera are ground beetles, leaf beetles, snout beetles, and darkling beetles, as well as others, many of them wingless scavengers or predators found beneath stones, on flowers, or within dung. The alpine Diptera range in habit from scavengers, to predators, to parasitoids, and to flower-feeders. They include crane flies, mosquitoes, midges, mountain midges, black flies, robber flies, flower flies, and blow flies. Finally, the common alpine Hymenoptera include ants, sawflies, ichneumon wasps, and bumble bees.

Suggested Additional Reading

Cheng (1976); Gressitt (1967); Huffaker & Rabb (1984); Lehmkuhl (1979); Mani (1962, 1968); Marshall (1981); Merritt & Cummins (1996); Wallwork (1976); Ward (1992).

Chapter 14

Insect Feeding Behavior

"And what does it live on?" "Weak tea with cream in it."
"Supposing it couldn't find any?" she suggested. "Then it
would die, of course." "But that must happen very often,"
Alice remarked thoughtfully. "It always happens," said the gnat."
— Lewis Carroll

A million or so insects occur in nature. They come in all sizes, shapes, and colors and exhibit a wealth of adaptations. They live virtually everywhere and sometimes reach staggering numbers of individuals. Their exceptional variability, ubiquity, and tremendous numbers in combination with the wide assortment of foods that nature makes available to them contribute to their almost infinite variety in feeding. They consume virtually anything organic including tissues of animals, higher plants, fungi, bacteria, wood, feathers, hair, feces, carrion, wax, *etc*. Despite this variety, careful analysis discloses the existence within insects of generalized trends, or *food-habits*, that lend themselves to description. Selected examples are noted below though there remains a residuum that overlaps categories and defies ready characterization.

Carnivory

Carnivores consume animals. Some of them live in close association with, and feed upon, the host parasitically, without killing. Others kill in the process of preying on the host. Still others scavenge upon it when it is dead or merely partake of its products or secretions.

Parasitism

Parasitism is the carnivorous-type relationship between organisms in which one, the *parasite*, usually the smaller-bodied, lives on, in, or near the other, the *host*, from which it derives nourishment, usually without killing. This definition may be qualified by noting that the parasite may be host-specific; its relationship with the host may or may not be obligatory; and the physical association between the two may last for an extended period of time.

Parasites have evolved numerous behavioral and structural adaptations to aid them in their life style. They may undergo profound, often degenerative changes in body structure and function. These changes may involve loss or reduction in the powers of locomotion, development of clinging or other attachment devices, and modified mouthparts. Metamorphosis may be reduced or lost, in which case both young and adults have a similar parasitic relationship. In other cases, the parasite retains its

Chapter 14 / INSECT FEEDING BEHAVIOR

metamorphosis, resulting in juveniles exquisitely adapted for parasitism and adults little modified from their free-living relatives. A convenient, albeit artificial classification recognizes *ectoparasites* that live on or near the host's external surface and *endoparasites* that live within the host's body cavity, gut, or tissues. The following examples illustrate.

Sucking lice (*Fig. 14-1 a*) are highly specialized, wingless, flattened mammalian parasites with a poorly developed metamorphosis, eyes reduced or wanting, legs consisting of a single tarsal segment and a claw for clutching hair, and highly modified piercing-sucking mouthparts. Their ectoparasitism involves a *continuous association* with the host, to the fur of which they cling tenaciously by their legs. They cannot long survive without the host's body warmth. They breed continuously, anchoring their eggs directly onto the host's fur. They drink blood sucked through the beak and are so host-specific that, for example, a monkey-feeding louse cannot take blood from a human.

Mosquitoes, in contrast, are parasites with a *discontinuous association* with the host and a complete metamorphosis. They oviposit in water. The hatched larvae (*Fig. 9-5 d*) feed on microorganisms and organic debris in the water; the pupae (*Fig. 9-5 c*) are likewise aquatic; the terrestrial adults (*Fig. 14-1 b*) are not flattened, and they have well-developed eyes, legs, and wings. Adult males eat nectar and other plant substances, but those females associated with mankind enter human dwellings or encounter people in nature where they inflict irritating wounds through their "bites." Some species are so completely dependent on blood-feeding that they cannot ripen their eggs in absence of a blood meal. Others suck blood but do not need it to produce eggs. Some mosquitoes are host-specific, but others attack mammals and birds somewhat indiscriminately.

Stylopids are among the most aberrant endoparasites. Stylopid females (*Fig. 14-1 e*) are legless, wingless, eyeless larviform creatures without developed mouthparts. They parasitize bees and certain other insects within the body of which they remain, their head and thorax exposed between the host's abdominal sclerites. The winged, free-living males (*Fig. 14-1 f*) mate with them in this partly exposed position. The eggs develop and hatch within the female body which literally becomes a sac of young. The agile first-instar larvae, or triungulins (*Fig. 14-1 d*), escape through a slit in the exposed maternal cephalothorax, attach to and enter a new host individual, usually a larva, before molting into additional legless, quiescent, blood-feeding larvae until maturity. They bulge their head and thorax outward through a tear between the host's abdominal sclerites when fully grown and transform into the adult while still within the larval skin. If males, they escape to the outside; if females, they remain within the host and await fertilization by an incoming male.

Parasitoidism

Parasitism grades into predation. The area of overlap is recognized as a pattern of its own, *parasitoidism*, characteristic of larval stages only, not of adults. Parasitoidism in insects is a relationship in which, unlike in parasitism, the larva eventually kills the host, and it involves a relatively large larva, sometimes almost as large as the host on which it feeds. (It cannot be larger, of course, because it must be able to complete development on the tissues of that single individual.) Adults of parasitoid species are typically free-living individuals whose females (*Fig. 14-1 c*) oviposit in, on, or near the host. The eggs hatch and the larval parasitoids slowly consume the host's blood, fat body, or other tissues, often

Carnivory

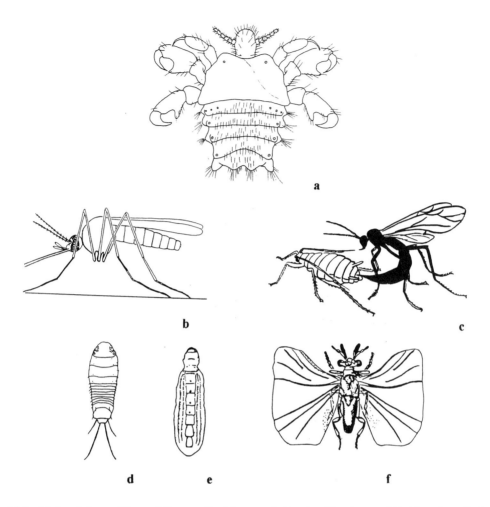

Fig. 14-1. (*a*) A crab louse, (*b*) an adult mosquito, (*c*) a parasitoid ovipositing in an aphid, (*d*) a triungulin larva, (*e*) an adult female stylopid, and (*f*) an adult male stylopid (*a* is modified and redrawn from *Turtox Key Cards*; *b* is modified and redrawn from Jones, 1978; *c* is modified and redrawn from Crossman; and *d – f* are modified and redrawn from Essig, 1942).

eating non-vital organs at first but eventually turning to vital organs, causing the host's death as the larva reaches maturity. Parasitoidism is common in ichneumonoid and chalcidoid Hymenoptera, in tachinid Diptera, and in Strepsiptera, but it occurs in some other groups too. Most parasitoids attack Coleoptera, Lepidoptera, Orthoptera, Hemiptera, or individuals from within their own order.

Predation and Cannibalism

Predators constitute the other major group of carnivores. Except for trappers, they differ from parasites in that typically they are larger and more powerful than their host animals, or *prey*, and they attack a series of them. They are generally not host-specific. They eat the prey, usually consuming vital tissues

Chapter 14 / INSECT FEEDING BEHAVIOR

and all, causing death. They sometimes attack and eat other individuals of their own species, the act of *cannibalism*. Predation and cannibalism are closely associated behavioral patterns, and many predators turn to cannibalism when the opportunity permits.

Insects' predatory behavior takes three forms: hunting, ambushing, and trapping. *Hunters* either search for and capture prey on the wing or overtake it and capture it on foot. Dragonflies (*Fig. 19-3 c*) and robber flies are examples of the former and certain predacious katydids (*Fig. 14-2 d*) an example of the latter. *Ambushers* including mantises (*Fig. 14-2 c*), mantispids (*Fig. 19-13 f*), and ambush bugs await the approach of prey and then, when it moves close enough for capture, pounce on it using specialized legs. *Trappers* build traps or other devices to capture prey. Antlions (*Fig. 14-3 a*), for example, dig pitfalls in the ground (*Fig. 14-3 b*), and certain larval caddisflies construct silken aquatic nets reminiscent of spider webs by which to ensnare prey that floats by.

Predators, like parasites, are usually highly adapted for their mode of life. The fore legs of hunters and ambushers are often powerful grasping organs armed with sharp spines that impale and hold struggling prey until it can be eaten. Giant water bugs (*Fig. 14-2 a*), waterscorpions, ambush bugs, mantises, and mantispids have raptorial fore legs of this type. Hanging scorpionflies (*Fig. 14-2 b*) reverse the pattern. They have raptorial hind legs, freeing the fore legs for support. Other strong predators including robber flies, dragonflies, and predacious katydids (*Fig. 14-2 d*) lack raptorial fore legs but use all three pairs of their powerful, spiny legs to hold the prey until it can be dispatched by the mouthparts.

The mouthparts of predators are as well-developed and distinctive as are their leg adaptations. Mantises, predacious katydids (*Graph 14-1, # 2*), and many predacious beetles have elongate mandibles with sharp cusps used to cut and tear flesh. This pattern is analogous to the "fang pattern" in the mammalian order Carnivora. Other predators have evolved specialized sucking mouthparts. Included are larvae of the common lacewings and antlions (*Fig. 14-3 a*) and those of certain predacious diving beetles whose paired, elongated, sickle-shaped mandibles pierce and suck the body juices of prey.

Most other predators with sucking mouthparts use a single powerful beak rather than paired piercing mandibles. Robber flies, giant water bugs (*Fig. 14-2 a*), and assassin bugs (*Fig. 19-12 e*) are examples. They simply thrust their beak into the prey and suck. This activity may be facilitated by injection of powerful digestive enzymes into the wound, helping disintegrate the prey's viscera and musculature.

Animal Scavenging

Scavenging involves the eating of dead tissues, plant or animal. Animal scavengers eat dead animals or their parts ranging from fresh, newly dead carcasses to decomposed, long-dead tissues. Fresh carrion is hardly different from living tissues except the food does not fight back, and it is much the same as eating physiologically weakened, newly molted, or moribund animals which are likewise unable to defend themselves. Mantises and certain other rapacious predators generally refuse any prey animal that does not move, but other predators willingly accept carrion whenever it becomes available. There are entire guilds of carrion feeders, many of which specialize on a particular stage of decay. For example, blow flies, skin beetles, and some other carrion feeding insects select host tissues

Carnivory

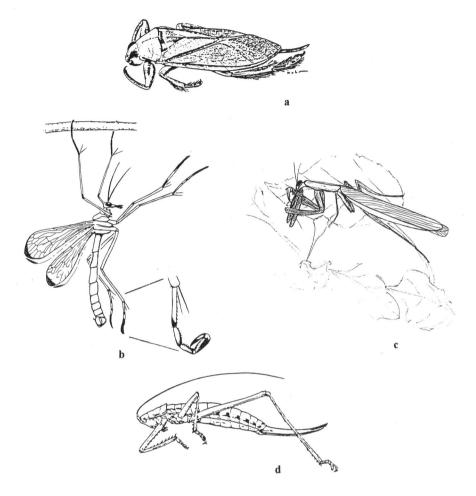

Fig. 14-2. (*a*) A giant water bug showing its sharp suctorial beak and powerful raptorial fore legs; (*b*) a hangingfly with its raptorial hind legs extended at left and enlarged and flexed at right; (*c*) a mantis consuming insect prey; and (*d*) a predacious katydid (*a* is by courtesy of Illinois Natural History Survey; *b* is modified and redrawn from Thornhill, 1980; *c* is by courtesy of Visual-Auditory Services, Wayne State University; and *d* is modified and redrawn from Chopard, 1951).

according to a definite, fixed *microsuccession* determined by comparative decomposition from fresh, to partly decomposed, to decomposed to the point of liquefaction, and to dried skin and bone. Only a few insects, among them carrion beetles (*Fig. 14-4 b*), are true carrion-feeders. They are so restricted in their diet that entomologists collect them by looking, not for them, but for the carcasses upon which they feed.

Dung constitutes a rich food source in its own right and also harbors microorganisms that are the real food of many dung feeders. Most dung feeders are general scavengers that take advantage of any feces they come upon, but a few are dung specialists. Among them are tumblebugs (*Fig. 14-4 c*), a type of dung beetle which molds, rolls, and then consumes vertebrate stools, and larval fleas which

Chapter 14 / INSECT FEEDING BEHAVIOR

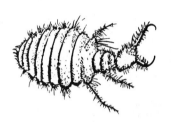
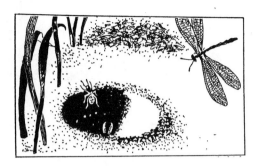

a   b

Fig. 14-3. (*a*) A larval antlion, (*b*) an antlion pitfall, with the gaping jaws of the larva shown at the bottom, an ant (*upper left of pitfall*) scrambling to escape the trap, and a flying adult (*upper right*) (*a* and *b* are modified and redrawn from Buchsbaum, 1938).

eat insect feculae and other debris. Tumblebugs are so dependent on feces that entomologists who wish to collect them do so by searching, not for them, but for dung. Dung, like carrion, changes progressively with age and decomposition. This change is responsible for an interesting microsuccession of dung-dwelling and dung-feeding arthopods.

Omnivory

Omnivory, or eating both animal and plant substances, is a habit essentially intermediate between carnivory and phytophagy, and the distinction between it and scavenging is even more tenuous. Many omnivores scavenge, and many so-called scavengers are, in fact, omnivores. The difference lies in the fact that an omnivore chooses with comparative indifference between the living and the dead, but scavengers prefer the dead. Insects exhibit few distinct adaptations for omnivory other than a full complement of digestive enzymes. This lack of specialization is perhaps explained by their nondescript mode of existence.

The Mormon cricket is an omnivore that partakes of most food substances, plant or animal, that it encounters in nature. It has been recorded eating hundreds of plant species, preying or scavenging on numerous animals, eating dung, and it is also highly cannibalistic. Cockroaches (*Fig. 14-4 a*) are, however, the omnivores *par excellence*. A few notorious genera include domestic, cosmopolitan species whose individuals frequent kitchens, restaurants, bakeries, grocery stores, hospitals, and other places with warmth, humidity, and organic debris to eat. They prefer soft, starchy foods but consume almost anything edible, either fresh or decayed or animal or plant in origin. They even gnaw cardboard, glue, paste, clothing, book bindings, hair, leather, garbage, and the secretions and wastes of vertebrates.

Phytophagy

ADAPTIVE RADIATION AS SHOWN IN MANDIBLES OF SELECTED ORTHOPTEROIDS

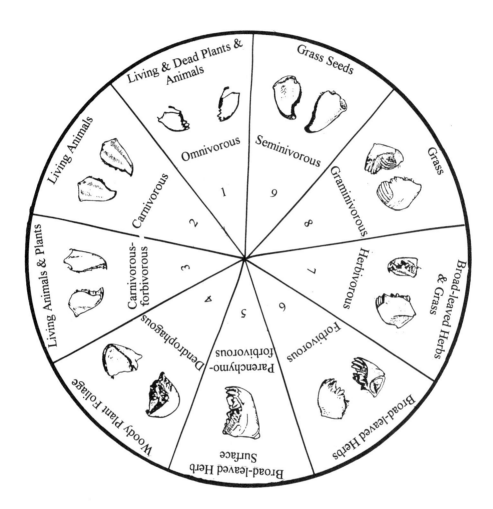

Graph 14-1. Adaptive radiation as shown by the mandibles of selected orthopteroid insects. The name of each food habit appears in the middle of the pie slice, and the foods for which they are adaptive are at the periphery; the numbers refer to particular species of feeder; the illustrations are of left and right mandibles of each of the species except for # 5 (*right mandible only*) and # 6 and # 8 (*left mandible only from external and internal surfaces, respectively*).

Phytophagy

Phytophagy, or plant-feeding, is insects' most common food-habit. It embraces behavior ranging from eating herbs, grasses, shrubs, and trees and often involves specialization on specific plant parts such as leaves, flowers, fruits, stems, trunks, roots, buds, *etc*. Selected examples from among this continuum are provided below.

Chapter 14 / INSECT FEEDING BEHAVIOR

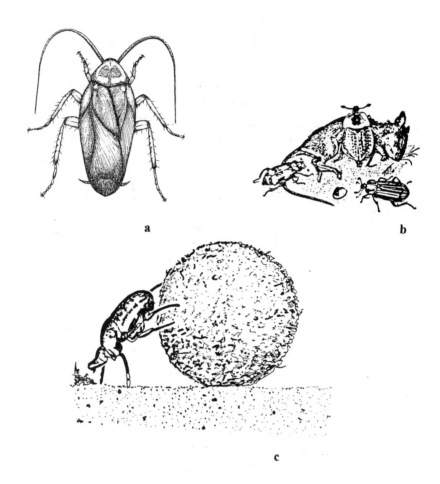

Fig. 14-4. (*a*) The American cockroach, (*b*) individuals of three different species of carrion beetle eating a dead mouse, and (*c*) a tumblebug, having shaping a ball out of dung, now rolling it to a burial place for oviposition and feeding (*a* is by courtesy of M. Dababneh; *b* is from S. W. Frost, 1959, after Walton; and (*c*) is modified and redrawn from Heinrich & Bartholomew, 1979).

Plant-feeders, like carnivores, may be highly adapted to their mode of life. Their purchase on plant tissues is often enhanced by claws and other attachment devices; their locomotion is often less well-developed than that in other insects; and their behavior is often more sluggish, particularly in juveniles.

Plant-Herbivore Co-Evolution

Insects have fed on plants for millions of years, attacking palatable ones and rejecting others that have defensive chemicals or tough or spiny tissues presenting mechanical barriers to feeding. As plants evolved potent chemical and physical defense mechanisms against prospective feeders, this put pressure on herbivorous insects to respond. The end result has been a plant-herbivore co-evolution in which the plants are chemically and/or mechanically defended from insect feeders except for those insects that evolved with them and whose adaptations circumvent the plant defenses. For example, few insects can eat milkweed with its milky sap and poisonous cardiac glycosides except for a small number of narrow specialists such as the monarch butterfly, the milkweed bug, the milkweed beetle, and the milkweed aphid. Each plant genus has been said to be like an island, carrying with it a small set of co-evolved, adapted herbivores restricted to, and "trapped"on, this particular host.

Phytoparasitism

As noted, animal parasites tend to be host-specific and smaller-bodied than their host and do not necessarily cause its death. Some insects have evolved a similar relationship, not with animals, but with plants. Many bugs, aphids, and thrips as well as some springtails and flies fulfill all criteria demanded of ectoparasites except their chosen host is a plant, not an animal. They are properly *phytoparasites*. Suctorial mouthparts are a common adaptation. Aphids (*Fig. 14-5 a, b*), for example, force their mouthpart stylets into the host plant's phloem, and sap rises through the beak by a combination of sap pressure and feeder suction. The often-sluggish phytoparasites need only to be near the appropriate host plant or to have powers of locomotion sufficient to seek it out. Once in contact, they need some kind of anchoring device, often nothing more than well-developed tarsal claws and the beak itself. The insect remains practically motionless as it feeds, its mouthparts buried deep within the host tissues.

Dendrophagy and Forbivory

Dendrophagy, or eating woody plant foliage (*Fig. 14-5 c, d*), overlaps *forbivory*, or eating broad-leaved herbs. The distinction between the two rests on the hosts' comparative content of lignin, or wood. Woody plants have a comparatively high and forbs a low lignin content. Both dendrophagy and forbivory are important, but, for the sake of brevity, I limit the following discussion to forbivory.

Forbivory is so widespread that it is almost easier to list orders in which it does *not* occur than to list those in which it does. In general, it is well developed in grasshoppers and their allies (*Fig. 14-6 a*), larval moths and butterflies (*Fig. 14-6 b*), beetles, and ants, all with biting mouthparts. The sucking bugs, cicadas, and their relatives could also be included, except they are herein classsified as phytoparasites. The degree of damage that forbivores inflict on the host varies. The plant is often lightly eaten and damage restricted to a particular part such as leaves, flowers, stem, roots, *etc*. Under more concentrated attack, however, the host may exhibit temporary wilting, and occasionally it is so severely defoliated that it dies. This happens in the case of certain flea beetles eating barberry. (This intense kind of forbivory differs from predation only in that the host is a plant rather than an animal). The tooth-like mandibular cusps of forbivorous grasshoppers are sharp and well-defined (*Graph 14-1, # 6*), never ridged grinding organs as in grass-feeders or fang-like as in carnivores.

Graminivory

Grasses, sedges, and their monocot allies are among the most ubiquitous, abundant of plants. They share linear leaves, a parallel venation, and often a high silica content, making them tough to eat. Successful *graminivory*, or grass-feeding, requires adaptations largely determined by these attributes. Among the primary feeding adaptations of grass-feeding grasshoppers, for example, are mandibles

Chapter 14 / **INSECT FEEDING BEHAVIOR**

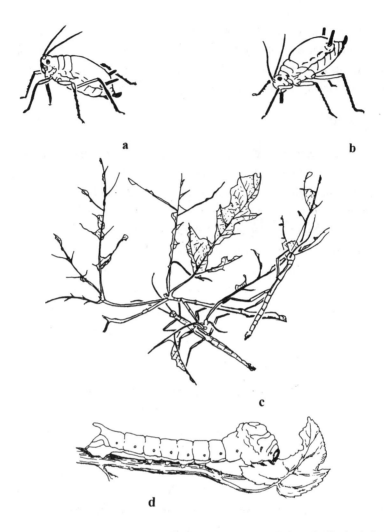

Fig. 14-5. (*a*) An aphid engaged in initial contact with the host plant and (*b*) inserting its beak deeply into the host tissues and sucking; (*c*) two walkingsticks eating oak foliage; and (*d*) a silkworm larva eating mulberry (*a* and *b* are modified and redrawn from Snodgrass, 1935; *c* is modified and redrawn from Graham, 1937; *d* is modified and redrawn from Verson).

armed with parallel grinding ridges (*Graph 14-1, # 8*) rather than the separate, distinct cusps of other plant feeders. They also have a straight, poorly valved digestive tract that is relatively uniform in diameter. Only a straight, unobstructed gut can pass successfully the elongate morsels into which grass splinters when eaten.

Leaf Mining

The circumstances of *leaf mining* are so special that the behavior is best treated as a separate food-habit. Mining is carried out by certain larval insects that burrow between the upper and lower sur-

Phytophagy

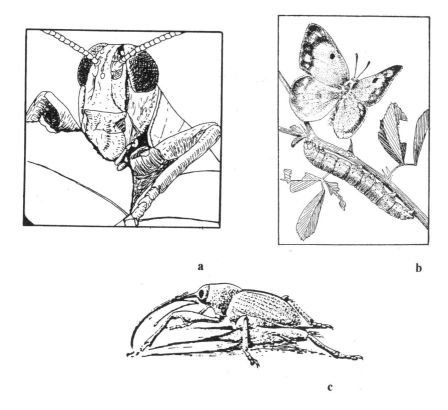

Fig. 14-6. (*a*) A grasshopper consuming clover, (*b*) an alfalfa caterpillar, together with the adult butterfly, on the host plant, and (*c*) a weevil showing its long beak used both for feeding and for oviposition (*a* is modified and redrawn from Illinois Natural History Survey; *b* is after USDA, 1952; *c* is by courtesy of Illinois Natural History Survey).

faces of leaves, consuming the internal tissues and leaving the surfaces unbroken (*Fig. 14-7 e, g, h*). They gain both sustenance and shelter ensconced, as they are, between the two leaf surfaces. They eat palisade tissue immediately below the leaf's upper epidermis, parenchyma cells above the lower epidermis, both tissues, or sap and, in doing so, excavate tunnels often identifiable to species. These tunnels vary from linear, to ovoid, to serpentine, and to irregular and may be further characterized by cast larval skins or deposits of feculae that stand out darkly against the translucent excavation.

Leaf miners include, among others, a few beetles, some sawflies, Agromyzidae and certain other true flies, and certain caterpillars including Gracillariidae and Nepticulidae, all adapted for occupancy of a specialized, rigorous environment. They share some features with fossorial insects, those which burrow within soil. Leaf-mining larvae are typically flattened insects with reduced legs, antennae, and eyes. Their sharp mandibles are borne on a wedge-shaped head rotated into horizontal position. Adults of these insects, as opposed to their larvae, are little different from other insects except for their tendency to select particular host plants for oviposition. Sometimes they use the ovipositor for direct insertion of eggs, or they gnaw deposition holes using the mouthparts, or they merely place the eggs on the leaf surface, leaving the hatched larvae to chew their own way into the leaf tissues.

Chapter 14 / **INSECT FEEDING BEHAVIOR**

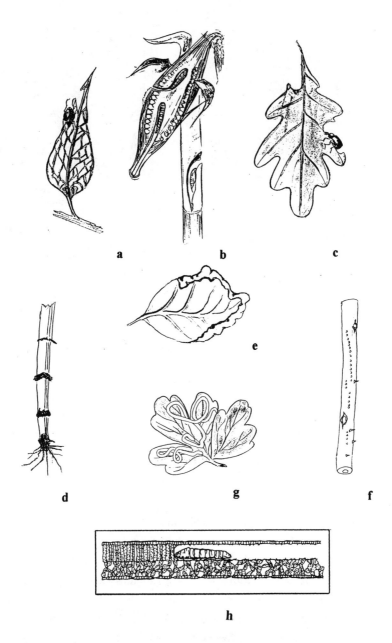

Fig. 14-7. Insect damage to selected plants, with *a* and *c* showing foliage feeding, *b*, *d*, and *f* showing boring, and *e*, *g*, and *h* showing mining (*a* – *g* are modified and redrawn from USDA Yearbook, 1952; *h* is modified and redrawn from Atkins, 1978).

Boring

Mining converges with *boring*, but a miner eats immediately beneath the leaf surface, and a borer works deep within other plant parts. Representatives of many different insect taxa derive food and shelter from boring (*Fig. 14-7 b, d, f*), for which their larvae are highly modified. They are generally cylindrical in cross-section, the exoskeleton often roughened, the head well-sclerotized, telescoped into the thorax, and provided with strong jaws, and the antennae and legs are reduced or lacking. These attributes are characteristic of the borers that excavate sound wood and are more pronounced than those of borers of soft, decaying wood. Most adult boring insects are specialized for selection of particular host plants for oviposition. Horntails (*Fig. 19-19 g*), parasitoid wasps (*Fig. 14-1 c*), and other boring Hymenoptera use a long, often needle-like ovipositor to insert eggs deep into host tissues; weevils (*Fig. 14-6 c*) lack an ovipositor but use the elongate proboscis for this purpose as well as for eating; and most boring flies and moths merely drop their eggs on or near the host, leaving the hatched young to work their own way into the plant tissues.

Gallivory

Galls are abnormal masses of hypertrophied tissue apparently produced through the interaction of the host meristematic tissues with animal, fungal, or other secretions, chiefly irritants of insects. They are purely plant contributions, but they depend for their development on the gall maker's continued stimulation. They provide the gall maker with food and shelter. Those gall makers with biting mouthparts eat the host tissues outright, but those with sucking mouthparts imbibe fluids from them. Adult gall-making insects are usually host specific in their feeding, almost always on a flowering plant, on or within the tissues of which they oviposit. Leaves are a common target and, to a lesser extent, petioles, buds, roots, stems, flowers, and fruit. Distinctive galls may be produced by different insects on the same host species (*Fig. 14-8 d, e*).

There are two types of galls: open and closed. *Open galls* (*Fig. 14-8 a, c*) are the numerous, small, often conical galls of certain Homoptera. These gall makers feed externally at first, stimulating leaf tissue to grow around them and eventually to form a chamber that envelops them. They mate while still ensconced within the gall, and the young escape to the outside through the gall aperture. *Closed galls* (*Fig. 14-8 b*) consist of a shallow external epidermis, an extensive central parenchyma, and an internal nutritive layer that surrounds and is progressively destroyed by the feeder ensconced within. Closed galls are usually the product of Cecidomyiidae, or gall midges, or of Cynipidae, or gall wasps.

Feeding Latitude

One system of classifying insects involves their relative specificity of food choice. This system recognizes three overlapping categories: *monophagy*, eating only a single host species, *oligophagy*, eating a few closely related foods, usually of the same genus or family, and *polyphagy*, eating many unrelated foods. True monophagous and oligophagous insects are so closely adapted to their particular food that they may starve in its absence though surrounded by large supplies of other foods. Some, however, turn to alternative foods when under starvation pressure. Polyphagous species, in contrast, feed so broadly that disappearance of a particular host poses no problem. They quickly turn to others, often those belonging to completely different phylogenetic groups.

Chapter 14 / **INSECT FEEDING BEHAVIOR**

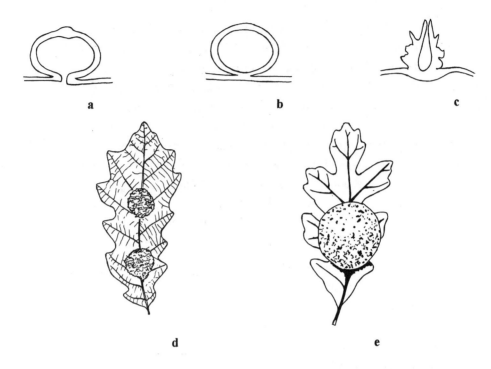

Fig. 14-8. Diagrammatic illustrations of open (*a* and *c*) and closed (*b*) galls; *d* and *e* are detailed views of selected galls (*a – c* are modified and redrawn from various sources; *d* and *e* are modified and redrawn from Atkins, 1978).

Most predators probably fall into the polyphagous category; gall makers, miners, borers, and most parasites into the restricted category; and most other plant feeders into a position intermediate between the two extremes. Carnivorous katydids (*Fig. 14-2 d*) and other large, powerful predators are usually polyphagous to the point of eating almost any prey they can catch and overpower, whereas smaller predators such as tree crickets (*Fig. 19-7 b*) necessarily confine their prey to small, weak, blundering insects. Most plant feeders eat a more specialized diet than carnivores. This generalization holds with respect to the herbivorous representatives of orders Coleoptera, Hymenoptera, and Lepidoptera (in which monophagy and oligophagy prevail), but it does not hold with respect to most Orthoptera (which are polyphagous).

Feeding Regulation

Climatic Factors

Air- and ground-level temperature, humidity, light, and wind act separately and together on insects and on the plants or animals that they eat. These climatic agents are subject, of course, to modification by topographical features and by the density of plant cover determining the degree of shelter afforded, and they act in control of behavioral rhythms including insects' feeding periodicity.

Feeding Periodicity

Insects vary widely in the timing of their meals. Mayflies and a few other insects do not eat as adults. Mosquitoes are among those that feed only once as adults. Still others, including certain grasshoppers, feed during several intervals each day, whereas some caterpillars feed nearly 24 hrs per day. The time of day when this periodic feeding takes place may be either diurnal, nocturnal, or some combination of the two. In contrast are many insect inhabitants of the dark zone of caves. They are *aperiodic* in their feeding in apparent response to the climatic constancy of their environment.

Food Palatability

Feeding in herbivorous insects is properly studied within the mosaic of plant-herbivore co-evolution noted earlier. Plants have been engaged through the ages in what has been called an "evolutionary arms race" with the herbivores that eat them. Part of plants' metabolic budget is spent in the form of physical and chemical defense mechanisms. They manipulate their organic environment through evolved chemicals, so-called *secondary substances*, which affect their relative palatability, or the lack thereof, with respect to other organisms. Herbivores that would eat these protected plants are obliged, therefore, to devote part of their energy to host-location and host-attack mechanisms to counteract the defenses raised against them.

Palatability, whether in a host plant or a host animal, is determined by the prospective feeder's response to the host's physical and chemical characteristics. Among the physical attributes are shape, visual patterns, movement, relative possession of spines or pubescence, hard, thick leaves or body parts, and tough, fibrous tissues which make the host difficult or easy to locate, approach, bite into, and eat, especially by early-stage juveniles, with their small, weak mouthparts (*Fig. 14-9 a, b*). Among the chemical attributes are the presence or absence of *succulence* (water content) and of secondary substances including alkaloids, essential oils, resins, and milky or colored saps which contribute as much or more as do physical attributes to the host's relative attractiveness or repulsiveness. Secondary substances are often highly effective in light of insects' acute sense of taste (which is on the order of 200 times that of humans) and their even more acute sense of smell. Insects may be strongly repelled by certain chemical substances called *deterrents*, or phagorepellents, but they are attracted by other substances, so-called *attractants*, or phagostimulants.

Food Availability

An acceptable host plant or animal may be present or absent, abundant or scarce, and generally distributed or localized within a given community. An abundant host is often more likely eaten than a scarce one and a generally distributed host more likely eaten than a localized one. This is explained by the fact that generalized feeders tend to eat the first acceptable food they encounter rather than seeking more preferred foods. Thus, insect food selection may be based as much on host availability and chance

Chapter 14 / INSECT FEEDING BEHAVIOR

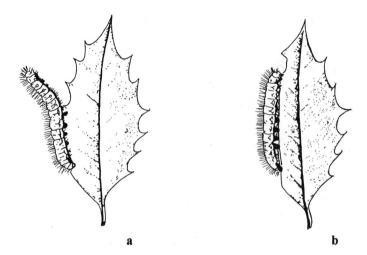

Fig. 14-9. (*a*) Holly is physically resistant to feeding by tent caterpillars unless (*b*) its serrated leaf edge is trimmed with scissors allowing the caterpillars to eat (Modified and redrawn from Ehrlich & Raven, 1977).

encounter as on palatability. However, food availability is not static. It changes with season as host plants and prey grow, mature, and die, as others take their place, and as particular parts such as flowers and fruits appear and then disappear.

Feeder Behavior

A prospective feeder encountering a plant touches it with the palpi, apprizing itself both of the plant's physical characteristics and of chemicals on its surface. This behavior may lead the insect either to make one or more exploratory nibbles or to reject the food outright. If the food provides appropriate attractant chemicals and lacks deterrent ones such as might inhibit feeding, the insect begins eating the acceptable food and may continue doing so until its crop is full. However, it begins searching for another food if it loses contact with the food or if the plant's attractant/repellent balance is inappropriate.

Feeder Hunger, Thirst, and Crowding

Hunger and thirst combine to weaken insects' physiological state. *Hunger* is dependent on the length of time since, and size of, the last meal, the rate of digestion, and the cumulative effect of past meals. Insects that are typically fastidious in their choice of food often select more widely when hungry. Some grass-feeding grasshoppers, for example, turn to forbs when starved. *Thirst* is important in that death by starvation may be delayed under conditions of high humidity or access to drinking water and, conversely, accelerated under conditions of dryness or lack of water. *Crowding* occurs naturally in insects that increase to outbreak populations. They often engage in cannibalism and other types of altered behavior stemming from food shortage and/or close proximity to one another. Individuals of the Mormon cricket, for example, become cannibalistic and also eat plants that they normally reject, and the desert locust and other plague locusts may turn from their normal host plants, sometimes eating virtually anything green, as their populations reach plague dimensions.

Feeder Injuries and Disease

Numerous afflictions beset insects and affect their feeding. Unfortunately, the mode of action of these agents is difficult to assess except when the injury is severe or the disease advanced. Loss of legs is common but has little apparent effect on insects' feeding. However, loss of the antennae or palpi that house most of the chemoreceptors involved in food selection may be more significant. Innumerable bacteria, protozoans, and fungi infect insects and may influence their feeding. They may display a lessened appetite and other symptoms as the condition becomes debilitating, and death may result as the body functions become deranged.

Feeder Development, Maturation, and Senescence

Insects undergo profound physiological change as they grow, molt, mature, mate, and become senescent, and these changes alter their feeding. Newly hatched and newly molted insects are unable to feed at first, owing to the initial need to harden their sclerites to resist the pull of muscles and the weakness of their mouthparts; male and female insects may turn to reproductive activities and spend less time eating upon attainment of sexual maturity; and old age is often characterized by decreased attention to food until feeding ceases, generally a few days before death. Insect larvae often have different diets, different mouthparts, and different habits from adults of their species. Not surprisingly, the adults of mayflies, some bot flies, and a few other insects that lack developed mouthparts are unable to feed though they carry sufficient energy over from their immature stages to fuel their adult sexual activities.

Suggested Additional Reading

Balduf (1939); Bernays & Chapman (1994); Brues (1946, 1952); Chapman (1974, 1990); Chapman & de Boer (1995); Clausen (1940); Davies (1988); Dethier (1963); Felt (1965); Gangwere (1972, 1991); Gullan & Cranston (2000); Norris (1991); Rodriguez (1972); Southwood (1972).

Chapter 15

Insect Locomotion

Fly, white butterflies, out to sea,
Frail pale wings for the winds to try;
Small white wings that we scarce can see,
Fly.
— Swinburne

Insects use almost every locomotor mechanism known from the Animal Kingdom. They hop or walk over the earth's surface, dig into it, swim within its waters, and even invade the atmosphere above it. Sclerotized insects use their skeletointegumentary system as a framework for internal muscle attachment that moves the legs and wings that are ultimately responsible for locomotion. In contrast, soft-bodied insects rely upon telescopic, syringe movements of the body as a whole under somatic muscle control independently of any involvement by appendages. These movements obey the laws of physics. A motionless insect has *inertia*, a tendency to stay at rest. It moves its legs, wings, or overall body to overcome this inertia, causing a change of position relative to, and exerting a backward force against, the substrate or medium. The substrate or medium then exerts an equal and opposite force against the organism, setting it into motion. Once in motion, the animal has another kind of inertia. It now tends to stay in motion until its forward force is arrested by *friction* which is maximal in burrowers, minimal in fliers, and intermediate in swimmers and walkers.

A terrestrial animal needs a body covering to hold its organs and body fluids in place, a skeleton to provide leverage, and muscles to lift and move the skeleton. It may move by dragging the body over the substrate, resulting in great friction, so few insects take this approach. It may also use legs to lift the body off the ground, minimizing friction. Most terrestrial insects take this second approach. Their legs have evolved in accordance with the physical principle that weight increases by the cube of body dimensions and strength of support by the square. Consequently, insects find delicate limbs sufficient for their purposes, as opposed to elephants and other large vertebrates with their disproportionately heavy limbs.

Turning to the aquatic environment, we find that the *sinking rate* of a body heavier than water is directly correlated with the ratio between friction and the difference between the body's specific gravity and that of the surrounding water. Insects being small, they have comparatively great surface with respect to volume, so their friction (one of the forces resident in surface) is enhanced, minimizing the tendency to sink. They also have a natural *buoyancy*. An object's buoyancy is equal to its weight in relation to the weight of displaced water. Accordingly, aquatic insects have adaptations of size and shape, strength of support, and flotation which, together with the supportive power of water, offset whatever tendency they have to sink. Water's *coefficient of viscosity* is 60 times greater than that of air

Chapter 15 / INSECT LOCOMOTION

at the same temperature, so resistance to movement in the aquatic environment is correspondingly greater than in air. Consequently, insects that move rapidly through water have evolved a streamlined form that minimizes the resistance of water flowing past them, and they take advantage of water's viscosity to use broad, paddle-like legs for propulsion.

The aerial environment, in contrast, affords minimal density, viscosity, and supportive power necessitating adaptations on the part of flying insects to keep them in the air against the pull of *gravity*. Enlarged wings that function as an airfoil, powerful or high-frequency wing vibrations, and a small body size providing great surface with respect to volume are among insects' flight adaptations.

Terrestrial Locomotion

Walking

Insects' walking, like that of vertebrates, is *out-of-phase* meaning they use the legs of a pair alternately. There is an obvious difference in the number of legs used, being six in insects and either four or two, respectively, in quadrupedal and bipedal vertebrates. The use of six limbs forces alteration in leg-use pattern but contributes to stability. There is, however, a more fundamental difference between movement in insects and that in vertebrates. Bipedal vertebrates' *fall-and-recovery* out-of-phase locomotion is not structurally feasible in animals with six legs. Insects substitute instead a *push-pull method* involving *traction*, *support*, and *propulsion*. All six legs perform each of these functions, but the legs of different segments vary in the degree to which they perform them. Fore legs are usually the most tractive, mid legs the most supportive, and hind legs the most propulsive.

Insects' coordination of leg movements during walking results from interaction between a basic pattern of movement generated by the central nervous division and performance feedback from sense organs on the legs.

An insect's pattern of leg use at most speeds may be termed *out-of-phase tripoding*. It moves forward on a tripod (*Fig. 15-1 a*) consisting of the fore and hind legs of one side and the mid leg of the other, as the opposites of each pair are lifted off the ground, extended, and placed down in a more forward position (*Fig. 15-1 b*). It then uses the opposite legs as a second tripod (*Fig. 15-1 c*) on which to move forward as the first-used legs are lifted. In doing so, the creature's *center of gravity* falls within an area of support provided by the legs in contact with the ground. This suggests that the legs of the tripod are moved simultaneously. Though the fore legs have a slightly higher probability of being lifted before the other two of the tripod, there is variability in which leg is lifted first, especially during fast walking. However, the two legs of a pair always move alternately with respect to one another.

Walking may depart from the above in different species and even in the same individual animal moving at different speeds. It is also different in insects that walk with less than six legs. For example, mantises usually lift the raptorial fore legs and walk with the mid and hind legs; grasshoppers, katydids, and other leaping Orthoptera may lift the leaping hind legs and walk with the fore and mid pairs; and

Terrestrial Locomotion

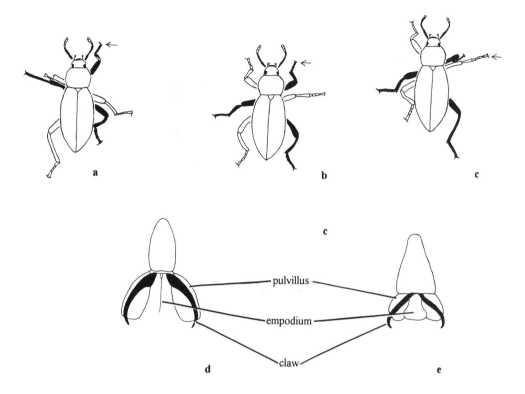

Fig. 15-1. Tripoding in a beetle is shown above, the insect supporting itself by the three darkened legs as it (*a*) lifts, (*b*) moves forward, and (*c*) places down the three unshaded legs in a more forward position from which they, in turn, form a supportive tripod allowing for forward movement of the darkened legs. Shown below are the pretarsal structures of (*d*) a robber fly and of (*e*) a house fly (*Figs. a–c* are modified and redrawn from Daly *et al.*, 1978; *d* and *e* are modified and redrawn from Borror *et al.*, 1989).

insect amputees lacking a leg on one side necessarily have an altered gait from expected. Walking in all such cases generally follows the out-of-phase, *diagonal pattern* of quadrupedal vertebrates. The walker moves ahead on the forward-most usable leg of one side and the hind-most leg of the other side as the opposite limbs are lifted. Then it moves forward on the opposite legs as the first-used legs are lifted.

House flies and certain other insects scale precipitous inclines and even walk, inverted, on the underside of smooth ceilings. The organs involved in this behavior include arolia, pulvilli, empodia, plantulae, and other *adhesive organs* and tarsal claws, tibial spines, and spurs (*Figs. 2-15 b, 15-1 d, e*). The mechanism behind the use of adhesive organs is controversial, but they seem instrumental in climbing smooth surfaces. Tarsal claws apparently substitute for them on rough surfaces.

Running

Most insects do *not* have a distinct running gait. Running cockroaches, for example, use a rhythm of leg movement that is the same as in basic tripodal walking except at extremely slow speeds. This kind of running is merely fast walking, being accomplished by changes in stride length. Technical difficulties

Chapter 15 / INSECT LOCOMOTION

owing to insects' small size and special habits have made difficult measurement of velocity attained. Nevertheless, reliable data on the walking speed of more than a dozen species are now available. Certain cockroaches are known to reach speeds of almost 5 km per hr at temperatures above 30 degrees C. This velocity seems slow, but a cockroach running at a speed of 5 km per hr covers a distance of about 40 times its body length per sec. This performance far exceeds that of horses, for example, which cover distances of only about six body lengths per sec.

Leaping

Most leaping in insects involves a simultaneous, violent *in-phase extension* of the hind legs against the ground hurtling the leaper into the air. Selected examples of the many insects adapted for leaping include flea beetles, certain planthoppers, fleahoppers, toad bugs, booklice (*Fig. 19-9 c*), fleas (*Fig. 19-19 a*), grasshoppers (*Fig. 15-2 a*), katydids, and crickets. In all, the hind legs are longer than the fore and mid legs, the hind femora are enlarged to accommodate increased leaping musculature, and the femoral extensor muscles of the hind legs are more highly developed than the flexors.

A leaping grasshopper rises on its fore legs and squats on its flexed hind legs (Fig. 15-2 a); it co-contracts its hind leg extensor and flexor muscles which allows the extensors to develop maximal force before the insect launches itself and assures a longer jump; the jump is initiated not by an excitatory signal to the extensors but by an inhibitory signal to the motor neurons activating the flexors; it extends the hind tibiae forcefully using its powerful femoral muscles to thrust itself forward and upward into the air (Fig. 15-2 b) and then either catches itself with the fore legs and lands upright at the end of its trajectory or undergoes a headlong spin and lands upside-down on its back. However it lands, its light weight assures that it is not hurt. The muscles involved are as powerful as any in the animate world. A locust weighing 2 gm can lift a 20-gm weight fixed to the hind leg. Several adaptations are involved (Fig. 5-7). The individual muscle fibers are short and packed in herringbone-like fashion along the entire inner femoral surface evenly distributing the load; they insert on a powerful tendon extending to the tibia; and they are subject to graded control by fast and slow nerve fibers.

Springtails leap by means of two medial organs derived by fusion of paired segmental appendages: a clasp-like catch, or *tenaculum*, arising from the third segment and a spring, or *furcula*, from the fourth (*Fig. 2-19 b*). The furcula, normally flexed beneath the abdomen, is held in place by the tenaculum. As the tenaculum releases, the pent-up tension of the furcula extends the appendage forcibly against the substrate (*Fig. 15-2 c, d*) thrusting the animal forward and upward, sometimes more than 50 times its body length.

Creeping

Soft-bodied insect larvae use *telescopic, syringe movement*s that vary according to the presence or absence of thoracic and abdominal legs and of backwardly directed spines. Legless larvae use the peristaltic movement of somatic muscles within the body wall. The waves of muscular contraction usually move along the body in the same direction as progression. However, crane fly and other larvae that force their way through soil do so by narrowing and elongating the anterior part

Terrestrial Locomotion

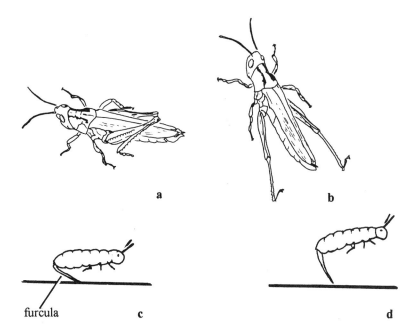

Fig. 15-2. Above is a leaping grasshopper, with (*a*) showing the insect rising on its fore legs and squatting on its hind legs preparatory to a leap and (*b*) showing it with the jump underway. Below is a catapulting springtail, with (*c*) showing the insect in an early and (*d*) in a later stage of leaping by means of the extending furcula (*Figs. a* and *b* are modified and redrawn from Matagne; *c* and *d* are modified and redrawn from L. J. & M. Milne, 1978).

of the body (*Fig. 15-3 a*) followed by progressive contractions that work backward (*Fig. 15-3 b*). As the wave reaches the posterior end (*Fig. 15-3 c*), now drawn into a new forward position, that end acts as a temporary support for the anterior segments which relax, push forward (*Fig. 15-3 d*), and expand laterally to enlarge the burrow. The cycle is then repeated.

Larval Lepidoptera have well-developed thoracic legs like those of other insects as well as fleshy abdominal legs called *larvapods*, or prolegs, equipped with hook-like crochets. The larvapods operate in conjunction with turgor and locomotor muscles (*Fig. 15-3 e*). The turgor muscles which extend across the body folds maintain body pressure. Their relaxation at a given point allows that surface to balloon outward. The locomotor muscles include longitudinal, dorsoventral, and transverse bands. The dorsal longitudinal muscles of a given abdominal segment contract simultaneously with the dorsoventral larvapod retractor muscles of the next posterior segment and with the ventral longitudinal muscles of the following segment. The body of the caterpillar arches at this point as the middle of the three pairs of larvapods is lifted, moved forward, and placed down again. Each new foothold involves relaxation of the dorsoventral retracting muscles accompanied by the ballooning downward of the relaxed larvapods. These events occur in a wave of movement that sweeps forward from the last pair of larvapods, each carried a step forward in its turn.

Chapter 15 / **INSECT LOCOMOTION**

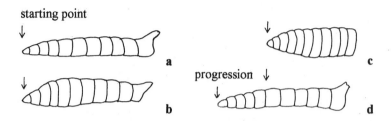

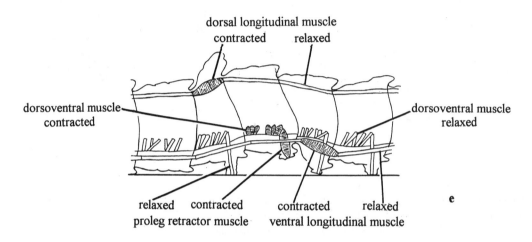

Fig. 15-3. Above are successive stages (*a, b, c*, and *d*) of creeping in a crane fly larva. (*e*) Below is a stage of creeping in several abdominal segments of a caterpillar, the wave of contraction beginning at a selected point, sweeping forward from there into the next segment, then into the next, *etc.*, producing forward movement as a whole (*Fig. a – d* is modified and redrawn from Kevan; *e* is modified and redrawn from Hughes).

Aquatic Locomotion

Water places its indelible stamp on the organisms that dwell within it, all of which have special adaptations to live there. The aquatic medium is fluid rather than gaseous providing great overall support, pronounced current effects, and reduced heat fluctuations compared to conditions encountered in air, but oxygen concentration within the medium is a fraction of that in the terrestrial environment.

Walking

There are two types of aquatic walking, each involving push-pull use of the legs in contact with the substrate. The first is characteristic of dobsonfly larvae (*Fig. 19-13 b*), creeping water bugs, and many other insects that walk over the aquatic vegetation or along the floor of their fresh-water habitation much as they would if they were terrestrial animals exposed to air. Nothing more need be said about them. The second kind of aquatic walking, *surface striding*, deserves more attention.

Aquatic Locomotion

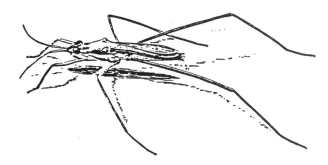

Fig. 15-4. A water strider atop water (Courtesy of Illinois Natural History Survey).

Water striders (*Fig. 15-4*) and some other insects spend their life in association with the surface film of water. Their body surface is water repellent, and they are sufficiently light for their size that they can glide about on the water surface. Though their weight tends to drag them down through the water, the contact angle they have established with the surface film is such that they are supported by the water's surface tension. Hence, they walk upon it as if supported by a thin, rubbery membrane against which their tarsi make impressions but do not break through. They use these depressions as footholds to push forward.

Creeping

Some legless larvae move across the surface of aquatic plants and over the bottom of streams and ponds by *creeping* carried out in much the same manner as do their terrestrial counterparts on land. Nothing additional need be said about them.

Swimming

Most swimming insects are secondarily aquatic air-breathing, streamlined forms with a specific gravity less than that of water, so they float to the surface and breathe when not swimming beneath water. They have oar-like legs often augmented by fringes of articulated setae that further expand the leg surface. The legs are extended fully, turned widthwise, and the setae spread to maximize resistance on the backward power stroke (*Fig. 15-5 a*). Each power stroke consists of backward and sideward components. The backward components push against the water whose inertia acts with an equal and opposite force to drive the swimmer forward. The sideward components, being delivered in opposite directions, cancel out one another. The legs trail behind and are turned edgewise, or "feathered," to reduce water resistance on the recovery stroke (*Fig. 15-5 b*).

Water boatmen (*Fig. 19-12 a*) and backswimmers (*Fig. 19-12 b*) use the hind legs for swimming. Giant water bugs (*Fig. 19-12 c*) and, to a degree, predacious diving beetles (*Fig. 9-8 a*) use both the mid and the hind legs for this purpose, and each uses the two legs of a pair simultaneously, or in-phase, so their pattern of movement consists of distinct forward bursts. If both the mid and hind pair are used in swimming, they alternate strokes. The fore legs are not used in swimming except by water scavenger

Chapter 15 / INSECT LOCOMOTION

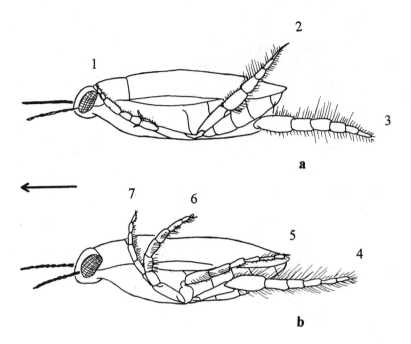

Fig. 15-5. The sequence of swimming movements of the mid legs of a predacious diving beetle, (*a*) showing stages of the backward power stroke using fully spread setae and (*b*) showing the forward recovery stroke using feathered setae, with numbers indicating successive positions. The fore and hind legs have been eliminated from the figure for the sake of simplicity (Modified and redrawn from Nachtigall).

beetles (*Fig. 9-8 b*) which row using all three pairs. In doing so, they use the legs of a pair alternately, or out-of-phase, to achieve a relatively smooth, continuous forward locomotion.

Many terrestrial insects including grasshoppers and cockroaches can swim as well. The interesting thing about swimming in the American cockroach is that it uses the alternating tripod coordination scheme even in water, whereas other terrestrial insects tend to swim by in-phase stroking with the two legs of each pair in synchrony.

Other Means of Aquatic Locomotion

Other methods of aquatic locomotion include: (1) *sculling* which is analogous to the tail of a fish swung alternately from side to side, for example, damselfly naiads' (*Fig. 13-1 c*) awkward abdominal movements enhanced by leaf-like caudal tracheal gills; (2) *undulation*, involving eel-like movements of an elongate body whose curves push against the water, as with biting midge larvae; (3) *wriggling*, a whip-like motion of the body first one way, then the other, causing a slight jerking movement each time, as with mosquito (*Fig. 9-5 d*) and midge larvae; and (4) *jet propulsion*, as with dragonfly naiads (*Fig. 19-4 d*) which, upon disturbance, bring the legs close to the sides of the body, dilate the rectal chamber to accommodate an increased volume of water, and forcibly eject it through the anus propelling them a distance of some inches.

Aerial Locomotion

Birds, bats, and insects are the only true fliers in the Animal Kingdom. They have conquered the air independently of one another. Insects' evolution has favored large numbers of species of comparatively small body size, and they have tended toward occupancy of limited habitat niches within which to gain shelter from enemies and avoid adverse climatic effects. Their existence in these restricted places depends on effective dispersal for food, mates, oviposition sites, and the like assured especially through flight. The flight mechanism that flies, bees, beetles, and other higher insects have evolved is unique. Their wings are new structures superimposed on the ancestral motor equipment, and they are powered by *indirect muscles* (*Fig. 15-6 a*). However, dragonflies, cockroaches, locusts, and certain other insects use *direct flight muscles* (*Fig. 15-6 b*) either exclusively or to supplement the indirect wing musculature.

Gliding

Wings may be likened to a flattened plane exposed to an air current. If oriented parallel to the air stream, the airfoil tends to be carried downstream by *drag*, or wind resistance. However, if the airfoil is gently curved and its leading edge is raised slightly, producing an angle with respect to the air stream, there occurs a differential pressure, and air then flows faster over the upper surface than over the lower. The effect is to *lift* the airfoil and drag it backward. Finally, if the lift and drag forces are adjusted to equal the weight of the animal, or *gravity*, it flies. However, the kind of flight that is manifested depends on the wings. If the wings are fixed, the animal can only coast from a higher to a lower place. Such limited flight is termed *gliding*. In this case, the air flow resulting from the insect's forward motion produces both lift and *thrust*, and forces instrumental in the thrust are completed before gliding begins except as supplemented by updrafts. Only a few insects glide effectively, among them certain wide-winged swallowtail butterflies and monarchs, dragonflies, and plague locusts which can soar great distances with the wings held stationary. Most insects are too small and too light to maintain the velocity needed for directed gliding.

Flying

Flying insects have evolved means of moving the wings, so they do not depend on gravity for lift. They flap their wings rapidly, creating a region of lowered pressure in front of and above them into which they are "sucked" in a continuous, self-propelled flight that can be maintained for periods of time. The flapping wings of these insects furnish both lifting and propulsive forces, whereas the stationary wings of airplanes provide only lift, leaving thrust to be provided by propellers or jets. Insects' upstroke/downstroke, except in dragonflies, certain Orthoptera, *etc.*, is caused by contractions of indirect muscles (*Fig. 15-6 a*) that do not insert on the wings themselves but warp the resilient thorax. Each wing pivots against a dorsolateral thoracic wing process (*Fig. 2-14 c*). Contraction of the tergosternal muscles depresses the notum (*Fig. 15-7 a*) and carries each wing base downward against the wing process forcing the wings to pivot upward. Then the dorsal longitudinal muscles contract (*Fig. 15-7 b*) upon relaxation of the ter-

Chapter 15 / INSECT LOCOMOTION

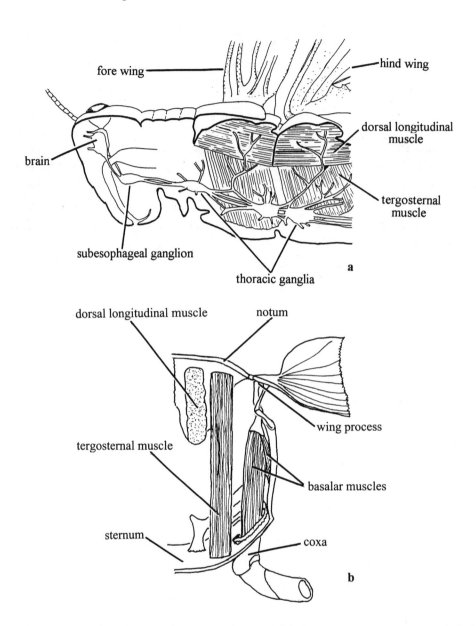

Fig. 15-6. (*a*) Above is a sagittal section of the fore part of a locust body showing relationships among the indirect (tergosternal and dorsal longitudinal) flight muscles, wings, and nervous system. (*b*) Below is a diagrammatic cross-section of the left side of a pterygote winged segment showing both indirect (tergosternal and dorsal longitudinal) and direct (basalar) wing muscles. The tergosternals depress the notum causing wing upstroke; the dorsal longitudinals warp the notum upward causing wing downstroke; the basalar muscles arising from the episternum depress the costal margin and extend the flexed wing. (The subalar muscles arising from the epimeron are hidden from view behind their antagonists, the basalars; they depress and extend the wing) (*a* is modified and redrawn from Wilson, 1968; *b* is modified and redrawn from Snodgrass.)

Aerial Locomotion

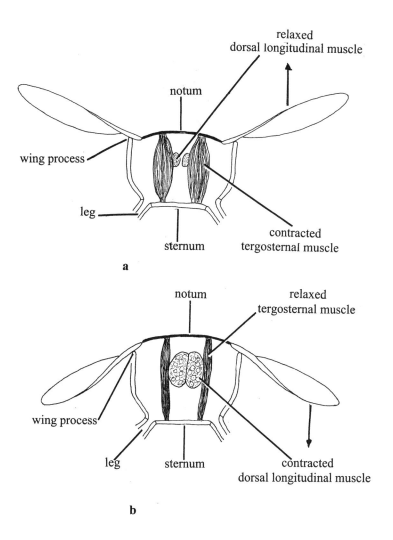

Fig. 15-7. The flight mechanism in an insect using indirect muscles to produce wing flapping is shown. Cross-section *a* shows the tergosternal muscles contracted, the dorsal longitudinal muscles relaxed, the notum depressed, and the wings forced upward against the wing process; *b* shows the tergosternals relaxed, the dorsal longitudinals contracted, the notum arched upward, and the wings pivoted downward (Modified and redrawn from Snodgrass).

gosternals. The dorsal longitudinal muscles insert on the phragmata at the fore and hind margins of the thoracic segment (*Fig. 5-6 a*), so they restore curvature to the notum and, in doing so, carry each wing base upward causing wing downstroke.

Upstroke/downstroke is more complicated than indicated as demonstrated by an anesthetized house fly. If one depresses the fly's mesonotum with a blunt probe, simulating tergosternal muscle contraction, nothing happens until the pressure reaches a certain level. Then suddenly the wings flip upward. Similarly, if one forces simultaneously against the fore and hind margins of the mesonotum, arching it

219

Chapter 15 / INSECT LOCOMOTION

to simulate the action of the dorsal longitudinal muscles, nothing happens until the pressure reaches a given point. Then suddenly the wings flip downward. Apparently, the resilient thorax is stable in only two positions, wings up and wings down. In short, insect wings operate through a click mechanism analogous to the action of an ordinary electrical switch.

Rotation, or twisting of the wings on their long axis, and their forward-rearward movements are joint products of the so-called direct wing muscles (*Fig. 15-6 b*) located on either side of the thorax. These muscles insert on the alar sclerites in front of and behind the wing process (*Fig. 2-14 c*) and also pull on the axillary sclerites (*Fig. 2-16 b*) at the wing base. Their torsion is enhanced in flight by the wings' differential flexibility. The heavier, stronger wing veins are concentrated near the costal margin leaving the posterior surface more flexible. Consequently, during simple upstroke/downstroke, there is torsion owing to changing air pressure on the flexible posterior surface of the wings. This torsion, together with differential contraction of the direct wing muscles, causes the wing's leading edge to be deflected downward during downstroke and upward during upstroke (*Fig. 15-8 a*).

Each wing behaves like a vibratory propellor drawing air in from above and in front, creating a partial vacuum, and forcing it backward in a high pressure airstream that constitutes thrust. Insect flight is, therefore, the resultant of the wings' lift force directed upward and in front, their drag force directed backward, and the force of gravity acting downward.

The flapping wing tips of a living insect fixed to a pin describe an inclined figure eight (*Fig. 15-8 a*), as seen from the side. The wing path is downward and forward, with downward deflection of the leading edge in forward flight. Then, it becomes upward and backward, with upward deflection of the leading edge. However, the inclination of the figure eight varies according to flight pattern. The inclination approaches vertical (*Fig. 15-8 b*) in the rapid forward flight of certain wasps and flies, nearly horizontal (*Fig. 15-8 d*) in backward flight, and intermediate in position (*Fig. 15-8 c*) during hovering.

Sideward steering and roll of the insect about its longitudinal axis involve different mechanisms. Insects cannot steer by altering wing-beat frequency on one side or the other because of the resonance of their wing-thorax system, and, unlike birds, they cannot do so by altering the wing area to change air speed on one side or on the other. Instead, they use differential rotation of the wings about the long axis to vary the lift provided and thereby steer. They simply flatten the angle of rotation on a side that is advancing too fast. Bees and certain other insects steer by altering the range of wing vibration from 100 or fewer degrees of arc to over 180 degrees. They maintain the same wing frequency of vibration on both sides but alter the amplitude on one side or the other producing a sideward movement away from the side of greater amplitude. They control roll by increasing, on downstroke, the angle of attack of the wing on the side toward which the insect is rolling.

Insects' wing-beat frequency varies from the slow beat of butterflies and moths, often less than five cycles per sec, to the moderate beat of most insects (usually 20-100 cycles per sec), to the more rapid beat of certain flies and wasps. Flies of one genus of biting midge have the fastest recorded beat, one exceeding 1,000 cycles per sec. Such a high rate poses problems to physiologists who would explain it. One might suppose that, in insects, as in other flying animals, the wings beat in direct response to individual contractions of the muscles that drive them which take place

Aerial Locomotion

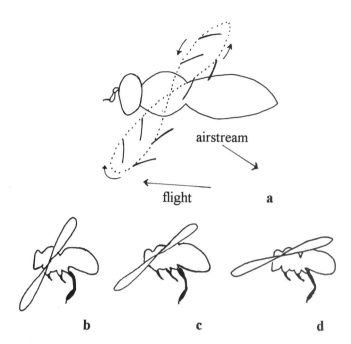

Fig. 15-8. (*a*) Above shows the figure-eight downstroke-upstroke of a wing viewed from the side, with airfoil orientation and directions at different points of the cycle indicated by arrows. Below compares the inclination of a wing figure eight during (*b*) forward flight, (*c*) hovering, and (*d*) backward flight (*a* is modified and redrawn from various sources; *b* – *d* are modified and redrawn from Wigglesworth).

only as fast as single nerve impulses can reach them. This hypothesis is applicable to butterflies, grasshoppers, and other insects with a comparatively slow wing beat. Their *synchronous wing muscles* contract in direct response to individual nerve impulses (*Fig. 15-9 a*). However, flies, wasps, and other insects with a more rapid wing beat cannot have muscles of this type. No muscle ever described can undergo nerve-stimulated contraction followed by relaxation at frequencies exceeding 100 beats per sec. Muscles such as these, their rate of contraction exceeding the rate of nerve stimulation, are termed *asynchronous* (*Fig. 15-9 b*).

It seems that the kinetic energy of wing motion in most insects, particularly small ones, is provided only partly by muscle contraction. The remaining energy stems from the inherent elasticity of the cuticle, muscles, wings, and resilin pads at the wing base. These elastic structures operate as a mechanically resonant unit that vibrates most readily at a particular frequency dependent on size. In general, the smaller the insect, the higher its resonant frequency and, hence, the more rapid its wing beat. Flight in these small insects proceeds by nerve impulses delivered to muscle in an active state causing contraction of the dorsal longitudinal muscles that warp the notum upward, forcing the wings downward until they "click" into the stable down position, and stretching the tergosternal muscles. This stretching of the tergosternals induces their spontaneous contraction, flattening the notum and forcing the wings upward until they "click" into the stable up position, stretching the dorsal longitu-

Chapter 15 / INSECT LOCOMOTION

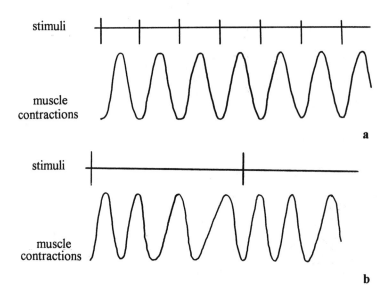

Fig. 15-9. Comparison of the relationship between nerve stimulation and muscle contraction in (*a*) synchronous as opposed to (*b*) asynchronous flight muscle (Modified and redrawn from Chapman).

dinals. Stretched this way, the dorsal longitudinals contract, with or without being stimulated by a nerve impulse. The cycle is repeated again and again, and the wings flap many times in succession for each arriving nerve impulse. However, a steady stream of nerve impulses is required to keep the muscles in an active state.

Insect wings vary in shape, deflection, and frequency of beat, so it comes as no surprise that insects' flight velocity differs. Butterflies and other broad-winged insects that fly with comparatively few strokes per sec are obliged to keep the airfoil almost horizontal during their fluttering, uneven flight, whereas dragonflies, horse flies, certain hawk moths, and other swift-flying insects have narrow, powerful wings that are deflected nearly vertically during each stroke. Some insects reach flight velocities in excess of 50 km per hr, but most fly at speeds of only a few km per hr which, at face value, is not impressive. However, the little insect that travels only 11 km per hr covers a distance of perhaps 300 times its body length per sec, a performance far exceeding that of any bird or mammal.

Some flying beetles, thrips, and wasps that are less than 0.1 mm long share a remarkable adaptive convergence in wing design despite their diverse origin. Their wings each consist of a club-like stem with a fringe of articulated "hairs" (*Fig. 15-10 a*) Inasmuch as the drag:lift ratio increases disproportionately at wingspans of such reduced magnitude, these insects cannot fly using the standard high lift:low drag method of other insects. Instead, they use a low lift:high drag system. Being so small and so light, they find air to be highly viscous relative to their weight, so, like a dust speck, they fall only slowly, and their mass may be offset altogether by the seemingly negligible lift provided by "rowing" with the wings, much as aquatic beetles swim in water. Their articulated "hairs" spread during each power stroke, providing greater drag, and "feather" during each return.

Aerial Locomotion

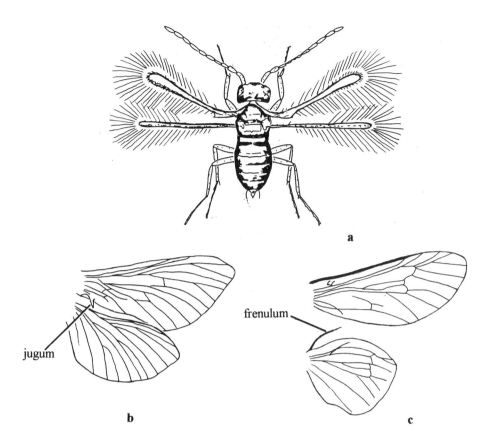

Fig. 15-10. Above shows the club-like, fringed wings of a diminutive fairfly wasp. Below illustrates two wing-coupling devices in moths: (*b*) a jugum and (*c*) a frenulum (*a* is modified and redrawn from Melanotti; *b* and *c* are modified and redrawn from Comstock).

Flight Adaptations

Insects' dorsal longitudinal and tergosternal indirect muscles provide thrust during both upstroke and downstroke. This constant propulsion is lacking in birds and bats whose flight mechanism is based almost entirely on downstroke.

Flies, bees, and wasps fly by means of asynchronous flight muscles whose use requires the click mechanism and an ability to apply a quick stretch to the flight muscles during their contraction cycle.

Insects begin to fly when properly stimulated, initiating wing flapping followed by leg retraction. This fact may be demonstrated by a tethered insect that begins flight movements immediately upon withdrawal of tarsal support. Conversely, reestablishment of substrate contact promptly stops wing vibration.

Insects have evolved specialized structures to monitor their flight adjustment and provide them with the "feel" of the air. Among these structures are tactile setae and bristles that respond to the rush of wind and mechanoreceptive campaniform sensilla and chordotonal organs that respond to wing torsion. These devices are widespread among insects. Other more specialized structures include

Chapter 15 / INSECT LOCOMOTION

the aristae and halteres of certain Diptera that function as air-speed indicators and gyroscopes, respectively, when provided a flow of air over their surfaces enabling them to maintain wing vibration without deviating from the flight path.

The earliest hexapods presumably belonged to subclass Apterygota. These primarily wingless relatives of insects have relatively distinct thoracic segments movable with respect to one another, but all true insects, past and present, are referable to the winged subclass Pterygota. Pterygotes' thoracic segments are specialized as an attachment for the legs and wings and as a housing for the muscles that power them. Not all pterygotes are winged, however. Lice (*Fig. 19-10 a – c*) and fleas (*Fig. 19-19 a*), for example, have lost their wings through adaptation to parasitism. Some other pterygotes including most walkingsticks (*Fig. 19-5 d*) and the worker and soldier castes of termites (*Fig. 17-4 c, b*) and of ants have lost their wings through adaptation to specialized habits. Winglessness in still other pterygotes including scale insects (*Fig. 19-11 f*), louse flies (*Fig. 19-18 a*), and stylopids (*Fig. 19-16 b*) is associated with sex, females being wingless and males fully winged. And, of course, all pterygote juveniles lack functional wings except for mayflies' subimaginal instar.

Insect evolution has proceeded in the general direction of wing simplification. The primitive condition presumably consisted of two pairs of similar, uncoupled wings with a net venation, the two pairs of which are moved independently during flight. In today's more advanced insects, however, one pair of wings is usually emphasized to the partial or complete exclusion of the other, and there is an accompanying decrease in the thoracic segment associated with the de-emphasized wings. Such wing reduction is noticeable in some Hemiptera: Homoptera, in which the fore wings are more leathery than the hind wings, and in Hemiptera: Heteroptera (*Fig. 19-12 g*) in which the fore wings are modified wing covers, or *hemelytra*, whose base is leathery and apex membranous. In both of them, as well as in Lepidoptera and Hymenoptera, the fore wings are better developed and more effective in flight than the hind wings. Diptera have extended fore-wing dominance to the point of exclusive use, the hind wings having been modified into gyroscopic *halteres*.

The opposite tendency, hind-wing emphasis, occurs in Orthoptera (*Fig. 1-1 b*) in which the fore wings are narrow leathery *tegmina* and the hind wings are expansive, membranous, pleated organs of flight. Both pairs are used in flight, but the fore wings are less effective than the hind ones. Coleoptera (*Fig. 19-14 d*) continue hind-wing emphasis with their sclerotized wing covers, or *elytra*, that are held extended but not usually flapped during flight. Male stylopids (*Fig. 19-16 a*) carry hind-wing dominance to its logical conclusion, the fore wings being reduced to haltere-like organs.

There is one other recourse to reduce slip stream, and many insects have adopted it. They have evolved a wing-coupling device that locks the fore and hind wings of each side into a single functional unit. Included among these coupling devices are the digitiform jugum (*Fig. 15-10 b*) and the bristle-like frenulum (*Fig. 15-10 c*) of Lepidoptera, the numerous hook-like hamuli of bees, and similar devices in other groups.

Suggested Additional Reading

Chapman (1991, 1998); Delcomyn (1985); Gray (1953); Gullan & Cranston (2000); Hughes (1952, 1958); Hughes & Mill (1974); Nachtigall (1974); Pringle (1975); Wigglesworth (1972, 1984).

Chapter 16

Insect Mating and Reproduction

They often repeat the form of their progenitors.
— Lucretius

Dioecious Bisexual Reproduction

Insemination

Insemination, the transfer of sperm from male to female, is usually via a membranous, sperm-filled sac called the *spermatophore* secreted by the male accessory glands (*Fig. 10-1 b*). It is used in apterygotes and in most pterygotes except scorpionflies and their relatives, most flies, and bugs. Male insects that lack a spermatophore may compensate by development of an elongate intromittent *aedeagus* able to reach or nearly reach the female's spermatheca where the ejaculate is deposited. A copulating male does not actually introduce his spermatophore into the spermatheca, deep within the female's genital tract, but into her copulatory pouch or bursa copulatrix (*Figs. 10-1 a, 10-3 b*), the latter in cases in which a bursa is developed. *Sperm* may be released by pressure of the spermatophore case on the ejaculate, by the female gnawing on the spermatophore, by rupture of it by bursal spines, or by dissolution of it by means of proteolytic secretions. However released, the sperm migrate on their own into the female's spermatheca. Certain male crickets and katydids remain alongside the females after mating, guarding them by antennal contact and sometimes stridulation if they attempt to move or if rival males approach.

Fertilization

The spermatheca acts as a reservoir until the sperm can be used in fertilization to produce a *zygote*. Sperm may be kept alive in the spermatheca for long periods of time in the case of females that mate infrequently or only once. In honey bees, for example, they may be kept for some years. Whatever the duration of storage, sperm release is caused by some unknown mechanism triggered immediately before the egg is laid.

Chapter 16 / **INSECT MATING AND REPRODUCTION**

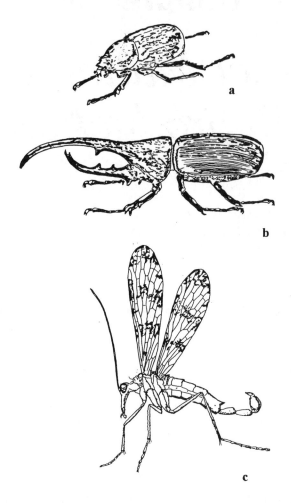

Fig. 16-1. (*a*) Female and (*b*) male of the hercules beetle, a species that violates the rule that female insects are the larger of the two sexes; (*c*) a male scorpionfly showing its scorpion-like terminalia (*a* and *b* are modified and redrawn from Essig, 1942; *c* is modified and redrawn from Borror *et al.*, 1989).

Sex Recognition and Attraction

Most insects are *solitary*, shunning others of their species as they go about their daily activities. They reverse this isolation during courtship using specific visual, auditory, vibratory, and chemical mating cues. Whatever combination of signals used, they make known to others of the same species their sex, location, and readiness to mate. Readiness is also dependent on certain internal physiological factors in absence of which a female receiving a male's invitation cannot respond positively. Her characteristic reluctance to mate must first be overcome. Insect males are typically competitive and females "choosey." The male must "woo" the often larger, stronger female before mating can occur. He makes her compliant through a courtship involving various mating signals that may be indicative of prospective mate quality.

Dioecious Bisexual Reproduction

Dimorphisms in body size, structure, and *color* are common mating signals. Male insects are usually smaller than females of their species (*Fig. 16-3 b, c*), sometimes by a factor of two-to-three times. Exceptions to this small-male : large-female relationship occur in certain beetles whose large aggressive males fight for their smaller female mates (*Fig. 16-1 a, b*). Not only is there a customary size disparity between the sexes, there is often a pronounced structural dimorphism aside from differences in the external genitalia. Male hercules beetles (*Fig. 16-1 b*), for example, are renowned for their powerful head and thoracic protuberances, male stag beetles (*Fig. 19-14 f*) for their powerful, antler-like mandibles, male dobsonflies (*Fig. 19-13 a*) for their elongate, sickle-shaped jaws, male scorpionflies (*Fig. 16-1 c*) for their scorpion-like, recurved terminalia, and female pelecinid wasps for their elongate, tubular abdomen. Wings may be sexually dimorphic. Male pterygotes tend to be long-winged and fully capable of flight, as are most females, except for the females of scale insects (*Fig. 19-11 f*), stylopids (*Fig. 19-16 b*), and some other insects that are wingless. In this event, the females' loss of flight is compensated for by life in secluded microhabitats, leaving the task of mate-seeking to the more wide-ranging, flying males of their species. There also may be pronounced color differences between the sexes. For example, the contrast between the sometimes colorful, intricately patterned, occasionally lustrous males of butterflies and moths (*Fig. 16-3 b*) and their often drab female counterparts (*Fig. 16-3 c*) may be extraordinary.

Sounds, odors, "dances," vibrations, and other enticements are other common mating signals used by insects.

Mythology is replete with tales of sirens wooing men with song. Much the same occurs among insects though on a more factual basis and with reversed roles. Here, males are usually the "seducers" and females the "seduced." Insect males are typically competitive and females "choosey." Most of insects' *organized calls*, or "songs," as they are often called, are directly or indirectly related to mating. Chorusing in cicadas, for example, may bring about dense heterosexual aggregations that facilitate mating. More commonly, as in crickets (*Fig. 16-2 a, b*) and katydids, calling brings together individual males and females sometimes from distances of many meters to the point of visual and physical contact.

Calling is one of a number of mate-acquisition tactics of which insects may avail themselves. Females are sometimes attracted to male songs as indicative of a possible source of nutrition including spermatophores, secretions, body parts, or other objects provided by males that the females eat during mating; song may also provide females access to male-held environmental resources including feeding sites or burrows usable for oviposition or for maternal care; and the songs themselves may indicate to the females something of the genetic quality or fitness of prospective mates. Males, in turn, may use mutual calling to establish reproductive dominance; to exploit the calls of rivals as a means of intercepting incoming females; and to attract proportionately greater numbers of females than could a single caller on his own.

Bioluminescence is neither phosphorescence, nor is it incandescence because phosphorus is not involved, and the reaction is independent of previous illumination. It is a cold light generated by oxidation of the substrate luciferin in the presence of the enzyme luciferase. The reaction is extraordinarily efficient, over 98% of the energy used being emitted as light. The light produced ranges over a

Chapter 16 / INSECT MATING AND REPRODUCTION

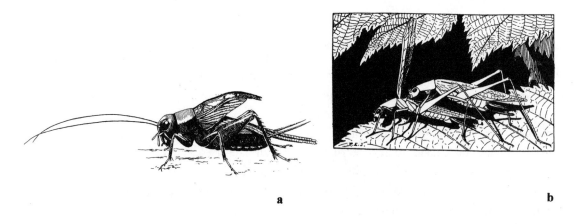

Fig. 16-2. (*a*) A male field cricket with its wing covers lifted in singing position; (*b*) a pair of mating tree crickets, the smaller male (*below*) elevating his wings into singing position, thereby exposing his metanotal gland to the female (*above*) which eats the exudate as they copulate (*a* is by courtesy of R. D. Alexander; *b* is modified and redrawn from Snodgrass, 1967).

continuous, narrow spectrum usually visible as green, blue, red, or orange according to the species. Certain insect larvae are luminescent on an occasional basis owing to the presence within them of luminescent bacteria or food. Truly bioluminescent insects, in contrast, generate light as a result of their own metabolism. They produce it through light organs consisting of a deep layer of reflective cells consisting of urate crystals, an intermediate layer of photogenic cells, and a superficial "window" of translucent cuticle. The reaction takes place within the photogenic cells, and the chalky white reflective layer reflects the light outward through the window. The chemical energy is supplied in the form of ATP generated by the photogenic cell mitochondria in the presence of oxygen, water, and salts, all fed into the luciferin-luciferase system controlled by the central nervous system.

The term *glow-worm* applies to certain wingless, worm-like beetle larvae and adult females that, from the ground, produce a soft, green light that attracts the winged, more beetle-like males. Equally familiar are fireflies (*Fig. 16-3 a*) whose males flash on the wing from within their characteristic habitat using the flash pattern and flight path typical of their species. The firefly females respond with a flash from their perch, usually atop low vegetation. If the females' answer is delivered according to the species' proper signal, the males approach, flash again, eventually land, and copulation ensues. However, the females of certain species have evolved the ability to respond to, and discriminate among, the flashes of different males of their species. The females of still other species have evolved the ability to respond to and attract the males of other species as well as their own, in which case the alien insects are liable to be eaten upon landing.

Insects may also use highly potent chemical substances with characteristic odors or tastes for excitation, inhibition, or control. These *pheromones*, as the substances are called, pass from one animal to another of the same species producing a change in the second's behavior, growth, or development. Included, among others, are the sex attractants bombykol and gyplure which attract the

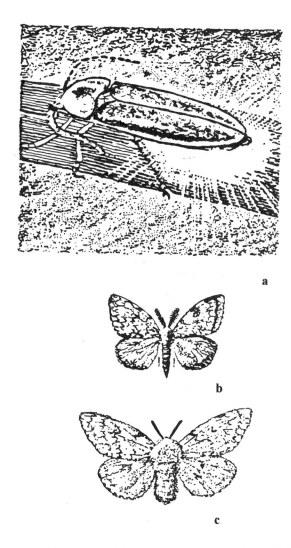

Fig. 16-3. (*a*) A flashing firefly; (*b*) a male and (*c*) a female gypsy moth (*a* is modified and redrawn from USDA, 1952; *b* and *c* are modified and redrawn from Wilson, 1963).

silkworm moth and the gypsy moth (*Fig. 16-3 b, c*), respectively. Their potency varies on the fantastic. Only 0.01 mg of gyplure can excite more than a billion male gypsy moths. The antennae are insects' primary organs of smell, so they are the major organs of pheromone detection. The antennae of certain male moths are bushy, bipectinate structures differing from the sparsely branched antennae of females of the species which correlates with habits. Females tend to be large-bodied, weak-flying, or flightless creatures that seldom move far from the point of pupal emergence. Their chances of mating are dependent on the ability of the strong-flying males to locate them by scent.

Insects share a remarkable habit, the proffering of a "*wedding gift*," with certain other arthropods, birds, and mankind. Males with this habit feed their potential mates prey, glandular secretions,

Chapter 16 / INSECT MATING AND REPRODUCTION

or even body parts. The gift-givers are sometimes predators that take advantage of almost any available prey of proper size, even their own species. Inasmuch as the courting males are usually smaller and more vulnerable than their potential mates, they could be eaten unless able to divert attention though it should be noted that, to date, avoidance of cannibalism garners relatively little overall support as an acceptable explanatory hypothesis. There is, however, empirical evidence on behalf of paternal investment and/or insemination guarantee hypotheses. Females of certain scorpionflies discriminate in favor of those males that provide them the largest wedding gift. Dance flies provide several interesting examples. Their males fly up and down in aerial swarms attractive to females. They may capture prey at this time but not devour it, sometimes wrapping it in a silken web or a frothy balloon before offering it to the females which eat the "gift" during coitus. Males of some other species of dance fly offer an empty balloon. Here, the gift exchange is an empty ritual that nonetheless serves as a necessary prelude to mating.

Males themselves occasionally serve as the ultimate gift. They may be sacrificed when females gain access to them during coitus. Such *husband cannibalism* occurs in antlion adults, praying mantises, and certain other highly predacious insects. Tree crickets (*Fig. 16-2 b*) obviate this difficulty through the males' habit of giving females access to a pheromonal gland at the wing base. The female cricket is first attracted to the male's call and then, on approaching him, to his pheromone, whereupon she mounts his back to eat the secretion as copulation ensues. Certain male tree crickets and katydids call from perches without obvious environmental resources with which to lure females, so many of them offer courtship gifts as an incentive for approaching and mating. Their calls vary from evenly spaced in some species to clustered in species that produce synchronized choruses. There are several possible explanations. Each male could have a proportionately higher mating success clustered than if he signaled alone; or the "superior" calls of certain males could provide opportunity for nearby "inferior" males to intercept attracted females; or perhaps it is nothing more than the mutual attraction of males to sites offering superior broadcasting opportunities. Other male katydids and crickets call from concentrations of food, oviposition sites, burrows, or other environmental resources around which attracted females cluster. These males may aggressively push, kick, and bite in an attempt to repel rivals and gain exclusive access to the lured females.

Male hangingflies (*Fig. 14-2 b*) catch a prey insect by the prehensile hind legs and then produce a volatile abdominal pheromone. They use this prey to attract a female which, if the gift is accepted, leads to copulation. She breaks off copulation, however, if the offered prey is too small or is otherwise undesirable. The amount of sperm transferred is proportionate to the time the female spends consuming the prey, so it is to her advantage to choose a mate with a large, desirable prey, and it is to his advantage to provide her with a prey that maximizes his gene contribution to the next generation.

Mayflies, midges, and some other insects are among those whose males "*dance*" in dense aerial swarms perceived visually by the females which then move individually toward them. These dances over lakes, ponds, and streams are occasionally in such number that their massed flight resembles a veritable insect snowstorm. The dancers prove to be males belonging almost entirely to a single species, the females of which are perched on nearby foliage. The dance that attracts the females consists of a gentle, upward mass flight alternating with a brief sinking. The females fly headlong into the swarm of dancers which cluster about them until clasping pairs are formed. Each pair falls to the ground and copulation ensues.

Courtship and Mate Choice

Pair formation in singing crickets and katydids provides the first opportunity for females to assess their prospective mates. Further opportunity is provided during the ensuing courtship as the males signal their availability by song, vibrations, odors, and other modes of communication. Females may continue evaluating prospective partners even after finding an available conspecific male. Female preference for a particular strength or quality of signal may give them access to the superior environmental resources that these males provide or defend; it may indicate a genetically superior male; or, through female preferences for nearby males, it may result in reduced mating effort. Whatever form the female preference takes, it imposes on males a strong sexual selection. Females relying on sound frequency as an indicator of male size and vigor often seem to prefer the low-frequency calls of large males because a large body size is positively correlated with nutritional and other resources or with male competitive superiority.

Pair formation without song takes place in some crickets. Perhaps the benefits that accrue to singers that broadcast their location are outweighed in this instance by avoidance of the parasitoids and predators that tend to attack singers. It could be that those silent males that merely search for females achieve a mating success comparable to that obtained by singers. Pair formation without song is facilitated in cases of high population density in which the chances of random encounter are enhanced.

Complex Patterns of Sex Recognition and Attraction

Sexual recognition in insects involves a number of subtle visual, auditory, chemical, and tactile stimuli delivered at a speed and frequency and in a sequence unique to the species. This does not mean that the male recognizes every detail of a prospective mate's structure, odor, color, and behavior. He often responds in an oversimplified way to one or more potent sensory impressions and ignores others. These relevant signals, so-called *sign stimuli*, are distinguishable from other stimuli that impinge on them in their environment. Females of the grayling, for example, reconnoiter aerially as males watch from the ground. Grayling males that perceive the presence of a possible mate fly to meet it, maneuver around it, and cause it to descend to the ground, if a female, or to withdraw, if another male. Female size, shape, and color are not distinguishing attributes at this early stage because the grayling males seem attentive only to flight pattern. Certain damselflies distinguish one another mostly on the basis of wing color and transparency. A male pomace fly that perceives a female scent approaches the "caller," taps her with the fore tarsi, circles her, and eventually faces her from one side or the other. He then "opens" and "closes" his wings horizontally, vibrates a wing causing air disturbance over her antennae, and licks her genitalia. Thus pacified and stimulated, she opens her genital plates and allows him to mount. Copulation ensues unless she is unreceptive or he delivers the signals incorrectly.

The complex preliminaries of courtship have been cleared away. Male and female are now together and mutually receptive. Their union is consummated through an elaborate mating behavior that is varied from group to group and often specific. Selected examples are provided below.

Chapter 16 / INSECT MATING AND REPRODUCTION

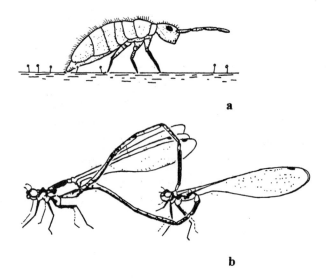

Fig. 16-4. (*a*) One method of fertilization practiced by springtails, showing a female squatting over a spermatophore left previously by an anonymous male; (*b*) a pair of damselflies mating in tandem, with the male in front and the female behind (*a* is modified and redrawn from Alcock, 1975; *b* is modified and redrawn from Walker, 1953).

Apterygote Mating

Springtails often mate with minimal effort. Their insemination may be purely impersonal, not involving copulation. In this instance, males merely deposit spermatophores on the ground, each on its own slender stalk. Females may encounter these stalks during their daily activities. When one does, she may squat over it and exude upon it a drop of vulval fluid (*Fig. 16-4 a*) causing it to rupture and release sperm into her genital chamber. In this instance, the two partners of the union do not meet, and paternity is anonymous. In other springtails, males either ride on the females and pull them over the spermatophore, or they use oral insemination, smearing sperm on their mouthparts prior to licking the females' genital aperture.

Male Thysanura usually deposit spermatophores on the ground much as do springtails except they deposit them near receptive females which are guided closer by silken threads. Courtship in the silverfish, for example, occurs in the buildings occupied by these domestic animals. A given male spins silken signal threads at an angle between floor and wall, deposits a spermatophore, and induces a female to move forward beneath the threads until her restricted movements bring her into contact with it whereupon she can make the "pick-up."

Pterygote Mating

Mating in pterygote insects differs from that in apterygotes in that it involves true *copulation*, a clasping of the sexes during which the male inseminates the female by means of an intromittent *aedeagus*, with or without use of a *spermatophore*. The copulators' position with respect to one another varies. They face forward with the male (*dark figure*) astride the female (*light figure*) (*Fig. 16-5 a*) in stoneflies, walkingsticks, and mantises; they face forward with the female astride the male (*Fig. 16-5 c*) in fleas, sucking lice, biting lice, and mayflies; they face forward with their undersides together (*Fig. 16-5 d*) in mosquitoes; they join end-to-end facing opposite directions (*Fig. 16-5 e*) in termites and earwigs; and in other insects they assume various combinations of the above with the male sometimes above (*Fig. 16-5 b*), the female sometimes above, or the two in side-by-side or end-to-end positions. Any question as to which position is primitive remains unresolved.

Dragonflies and damselflies exhibit an interesting variation. Their mating takes place aerially in tandems of coupled insects that circle back and forth over land and water. The odonate genital aperture, like that of other insects, is at the apex of the abdomen. Nothing is different here. However, the males' copulatory organs are developed, *not* at the abdominal apex, but on the underside of the second or third abdominal segments, well forward of the genital aperture. This lack of contiguity of the two genital complexes requires that the male bend the tip of his abdomen downward and forward to charge his copulatory organs with sperm from his caudal genital aperture. Once this pre-copulatory charging has been accomplished, he can mate. He does so in a copulatory flight with the female during which the two partners face the same direction and fly together as one (*Fig. 16-4 b*). The male uses his cerci to grasp the female by the neck, apparently pulling her along as they fly in tandem, and she bends her abdomen downward and forward until her genitalia come into contact with his on the second or third abdominal segment. Finally she lays her eggs in or near water, either after separating from him or occasionally while still in tandem with him.

Oviposition

Egg laying is the usual climax to the reproductive act. It is generally accomplished by means of an *ovipositor* (*Fig. 2-20 a, b*) originating from certain of the female's modified abdominal appendages. This organ varies more according to oviposition habits than to phylogeny. Sometimes the female lacks a developed ovipositor. In this event, she may use her specialized, elongate abdominal segments to lay eggs, or she may simply drop them unceremoniously without attempting placement.

Oviposition site selection is important in view of the limited means of locomotion characteristic of most larval insects. Oviposition tends to be on or near a supply of the right kind of food in a particular habitat niche. This commonality of habitat selection, food selection, and oviposition site selection is illustrated by the behavior of one species of swallowtail butterfly that eats only rutaceous and umbelliferous plants. Chemical analysis of the insect's host plants indicates that each plant has its own combination of essential oils always including methyl-nonyl-ketone, a substance eaten by the caterpillars and, therefore, attractive to the ovipositing females.

The mechanics of oviposition differ. Most walkingsticks merely drop their eggs indiscriminately on the forest floor; grouse locusts (*Fig. 16-6 a*) lay their eggs in soft soil; grasshoppers place their frothy egg pods in the soil; mantises and cockroaches produce purse-like egg cases within which the eggs are arranged in orderly rows; many katydids use their blade-like ovipositor to place eggs flat between the upper and lower surfaces of leaves; crickets use their needle-like ovipositor to penetrate

Chapter 16 / **INSECT MATING AND REPRODUCTION**

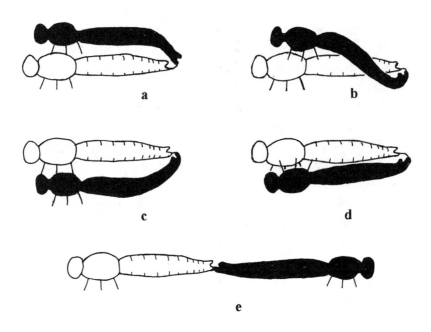

Fig. 16-5. Diagrammatic views of the copulatory position assumed by various insects comparing the female position (*light figures*) with the male position (*dark figures*) (Modified and redrawn from Chapman, 1982, after Richards).

host stems or soil; sawflies use their stout, serrated ovipositor to saw woody twigs or other plant tissues within which they place their eggs; lacewings (Fig. 16-6 b) attach their eggs to individual, slender stalks on leaves or other objects; warble flies attach theirs to mammalian hair with lightning-like speed, whereas sucking lice deposit their nits on hair in the more leisurely manner permitted by permanent residence on the host.

"Birth" vs. Hatching

The reproductive pattern discussed to this point is a general one in which an oocyte is invested with an egg shell, fertilized, and deposited somewhere to develop and hatch of its own accord. Sometimes it is placed in a special nest, and occasionally it is tended for a time by one or both parents. Whatever the case, the hatching insect issues directly into the environment through the ruptured egg shell. This overall procedure involving egg laying is termed *oviparity*. However, some insects do not seem to hatch from eggs but are apparently "born," or extruded alive, from within the mother's body. Such relationships include two fundamentally different patterns: *ovoviviparity* in which the eggs are retained, develop, and hatch within the maternal genital tract without benefit of nutrient exchange, and *viviparity* in which the embryonic eggs or hatched larvae are nourished for a time within the maternal genital tract before expulsion as living young. Ovoviviparity is merely a modification of oviparity except hatching occurs within the mother's genital tract rather than outside in nature. Where it occurs, as in tachina flies, it gives rise to living young, so it superficially resembles true birth but

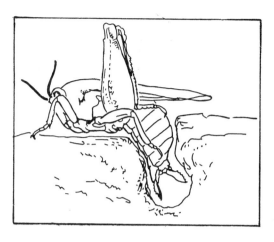
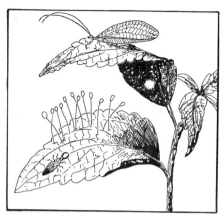

Fig. 16-6. (*a*) A grouse locust using its ovipositor to lay eggs in soft soil; (*b*) oviposition in lacewings, showing (*lower leaf*) a larva and many eggs attached to individual stalks and (*upper leaf*) a cocoon (*white object*) and an adult lacewing (*a* is modified and redrawn from Chopard, 1938, after Hancock; *b* is modified and redrawn from Comstock, 1948).

it is *not* because the young are not nourished before birth. They depend solely on the yolk allocated them during oogenesis. The relationship is different in true viviparity. In the viviparous Pupipara, for example, the fly larvae hatch within the uterus (which is a specialized, enlarged vagina) and are nourished there by special glands and breathe through the maternal genital aperture. They pupate immediately upon discharge to the exterior.

Other Methods of Reproduction

The insects that reproduce by means other than dioecious bisexual reproduction include a few hermaphrodites and many parthenogenetic forms. Both suffer from a reduced genetic variability though the genetic loss occasioned by hermaphroditism may be minimized by enforced cross-fertilization and that caused by parthenogenesis may be minimized by alternation of parthenogenetic with dioecious bisexual generations.

Monoecious Bisexual Reproduction

Those uncommon animals called *hermaphrodites* possess both testes and ovaries as well as the associated ducts, so are neither males nor females but both. Theoretically capable of self-fertilization, they usually avoid this most intimate kind of in-breeding. Instead, they resort to cross-fertilization. Only one insect, the California cottony cushion scale (*Fig. 16-7 a*), is known to be a functional self-fertilizing hermaphrodite. Its populations consist of a few males and numerous

Chapter 16 / INSECT MATING AND REPRODUCTION

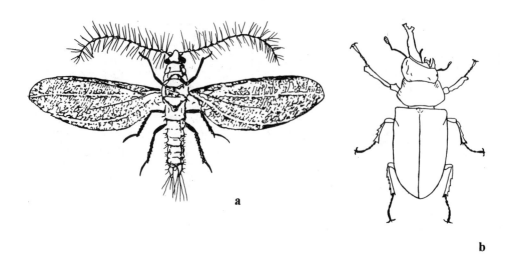

Fig. 16-7. (*a*) A haploid male of the hermaphroditic cottony cushion scale; (*b*) a gynandromorphic stag beetle showing male characteristics on the left side and female ones on the right (*a* is modified and redrawn from Richards & Davies, 1977, after Riley; *b* is modified and redrawn from Wigglesworth, 1972, after Lengerken).

hermaphrodites that correspond in appearance to the females of related scale insects. There are no true females in this cushion scale. The diploid hermaphrodites develop from fertilized eggs within the central cells of the first-stage nymphal gonad. These eggs undergo reduction to the haploid chromosome number and form a testis that, upon maturation, gives rise to sperm. The peripheral cells of the nymphal gonad become ovarioles that produce oocytes. Most oocytes are fertilized by sperm from the same animal, so the resulting diploid embryos become functional hermaphrodites. However, some eggs escape fertilization and develop parthenogenetically into haploid males. Upon maturity, these males may copulate with the diploid hermaphrodites.

Unisexual Reproduction

Parthenogenesis in insects exhibits a variety of form and number unrivaled in the Animal Kingdom. It ensures an increase in number that is as rapid as possible during the brief favorable season and obviates searching for mates. This is advantageous in small populations with feeble powers of dispersal in which mates may be in short supply. As mentioned, however, parthenogenesis occasions reduced genetic variability though the defect may be offset by alternation of parthenogenetic with dioecious bisexual generations.

Parthenogenesis occurs in nearly every insect order except Odonata. It also takes place sporadically in normally bisexually reproducing insects such as grasshoppers and moths. They produce eggs that, on occasion, remain unfertilized, yet somehow develop and hatch. Many other insects utilize parthenogenesis as a normal feature of the life cycle. Such *obligatory parthenogenesis* is found in certain mantises that lack males throughout their range and in those walkingsticks that, in some places, have males and females but, in other places, only females. Aphids (*Fig. 19-11 d*) and gall

Other Methods of Reproduction

wasps are phylogenetically dissimilar insects that share *cyclic parthenogenesis*, a regular alternation of parthenogenetic with dioecious bisexual generations. They avail themselves, therefore, of both the rapid increase in numbers afforded by parthenogenesis and the genetic variability inherent in dioecious behavior. The honey bee (*Fig. 17-7*) does much the same, its females being produced by bisexual dioecious means and its males by parthenogenetic means. This *haplodiploidy*, as the reproductive device is called, involves fertilized eggs that develop into diploid females and become either queens or workers depending on the larval diet and unfertilized eggs that become haploid drones whose sole purpose is to mate with the queens.

Special Modes of Reproduction

The preceding indicates something of the diversity of insect reproduction, but the coverage is hardly exhaustive. It ignores *paedogenesis*, or reproduction by immatures, as in certain gall midges and *polyembryony*, or division of a single egg into two-to-many embryos each, as in certain wasps. It also ignores *gynandromorphism* (*Fig. 16-7 b*), not a means of reproduction but the sporadic occurrence of sex mosaics. Treatment of these subjects is necessarily left to more comprehensive works than this one.

Suggested Additional Reading

Alexander (1964); Brown & Gwynne (1997); Chapman (1991, 1998); Gullan & Cranston (2000); Leather & Hardie (1995); Norris (1991); Snodgrass (1935); Thornhill & Alcock (1983); Wigglesworth (1972, 1984).

Chapter 17

Insect Aggregations and Societies

> *For so work the honey bees,*
> *Creatures that by a rule in nature teach*
> *The act of order to a peopled kingdom.*
> *They have a king and officers of sorts,*
> *Where some, like magistrates, correct at home,*
> *Others, like merchants, venture trade abroad,*
> *Others, like soldiers, armed in their stings,*
> *Make booty upon the summer's velvet buds*
> *Which pillage they with merry march bring home.*
>
> — Shakespeare

A few parthenogenetic insects lead isolated lives. Most other insects are solitary but at least come together to mate. Some of them mate in distinct aggregations and place their eggs or young on, in, or near the proper food. Others produce swarms during one or more periods of the life cycle. Still others form societies that may be composed of thousands of individuals whose relationships reach a high level of complexity. The task at hand is to characterize these diverse aggregations and discuss factors involved in their formation.

First, some definitions are in order. *Solitary* refers to individuals of a species that live alone. This condition is the opposite of *gregarious* which refers to individuals that live in groups, sometimes for extended periods, but do not cooperate in brood care. Some gregarious insects that share a nest are better termed *communal*. Sometimes one or both parents remain with and care for the young in which case the relationship is termed *subsocial*. In still other cases, that of the truly *social*, the individuals live together, are variably dependent on one another, and exhibit both brood care and division of labor. The social relationship is often confused with the *colonial*. Colonial forms such as corals are functionally independent animals that nonetheless are physically bound to, and inseparable from, other individuals of their species. Social insects are not physically attached to one another, so one cannot speak of a "colony" of them. However, such strict application of terms leaves missing a word to describe the individuals of a given nest, and the problem becomes acute in army ants in which there are great masses of individuals but no nest. Therefore, use of the terms *colony* and *nest* to mean the occupants and not the structure itself is incorrect though often practiced, as I shall do in the following pages.

Chapter 17 / **INSECT AGGREGATIONS AND SOCIETIES**

Insect Aggregations

Oriented Responses

The oriented responses of certain insects to specific environmental stimuli may lead them to form aggregations. For example, the box elder bug (*Fig. 17-1 c*) enters houses and other shelters in number during the fall season; domiciliary cockroaches (*Fig. 19-5 a*) sometimes inhabit kitchens in alarming number; and May beetles (*Fig. 17-1 a*) aggregate around artificial light. These insects gather together, not because of any "liking" for one another, but because of their mutual attraction to particular environmental stimuli, whether warmth, food, light, or shelter.

Mutual Attraction

The African subspecies of the migratory locust has a dual range. It is always found in its *residual breeding area* in northern and central Africa through Arabia into southwestern Asia, but it also extends into an invasion area far beyond. It is represented in the breeding area by distinct solitary (*Fig. 17-2 a*) and gregarious phases (*Fig. 17-2 b*) together with intermediate forms differing in behavior, color, shape of pronotum, wing length, *etc*. Individuals of the two phases are sufficiently unlike as to be mistaken for different species, yet they belong to the same species. At first, there are many more solitary than gregarious individuals in the breeding area, but following a succession of dry, hot years favoring locust development and depletion of the food supply the locusts multiply chiefly as gregarious forms. The relatively few solitary individuals remain apart, so may be dismissed except for their role as residual breeding stock. However, the gregarious individuals have a strong mutual attraction. They concentrate in places with food which soon fill with agitated nymphal locusts jostling one another with every movement. Their movements are initially spontaneous and independent until suddenly massed individuals begin concerted marching on foot. Then winged adults appear, and they substitute flight for marching. They continue dispersing *en mass*, sometimes thousands of kilometers into the invasion area, stopping here and there and defoliating natural vegetation and crops. However, the invasion area reached is often far removed from the breeding area, so it provides inferior conditions for oviposition. The displaced locusts cannot perpetuate themselves here, and the population eventually dies back to the residual breeding area. Population pressure thus alleviated, the hatchlings again consist largely of solitary individuals, and the cycle begins anew.

Tent caterpillars and processionary caterpillars also have a strong mutual attraction. Tent caterpillars hatch from eggs laid on the host tree and migrate to a major fork where they construct a silken communal nest (*Fig. 17-1 b*). Their larvae feed independently on foliage during the day but at dusk retrace their path to the communal nest via the silken thread that each individual leaves in its wake. Processionary caterpillars are likewise communal and active on a daily basis, but they have the opposite periodicity. Their larvae rest in the nest during the day and parade at night for food. Their nightly columns consist of communally marching caterpillars, each with its head to the tail of the preceding animal.

Insect Aggregations

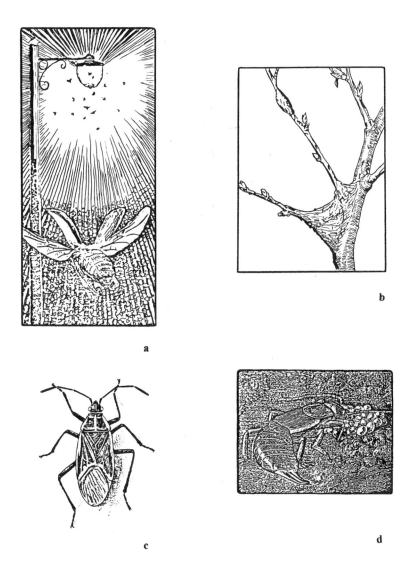

Fig. 17-1. (*a*) May beetles aggregating around an artificial light; (*b*) a tent caterpillar nest; (*c*) a boxelder bug; and (*d*) a female earwig tending her eggs within the darkness of the brood chamber (*a* is modified and redrawn from USDA, 1952; *b* is redrawn from Snodgrass, 1967; *c* is after USDA, 1952; *d* is after Fulton, 1924).

Subsocial Insects

A female sometimes remains with and cares for her eggs and occasionally her hatched young. This temporary "family" is an assemblage with neither division of labor nor sharing of tasks and with only one or two adults, the female and sometimes the male. The European earwig (*Fig. 17-1 d*) is an example. This insect, like many of its order, is gregarious. Its nymphs hatch in summer and stay together until autumn when the young adults appear, mate, and overwinter in crevices in the soil. In spring,

Chapter 17 / **INSECT AGGREGATIONS AND SOCIETIES**

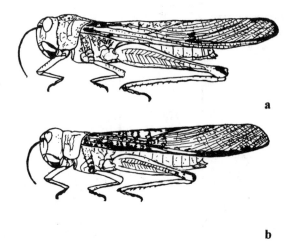

Fig. 17-2. An individual of the (*a*) solitary phase compared with one of the (*b*) gregarious phase of the African migratory locust (Modified and redrawn from Uvarov).

the female enlarges her crevice into a brood chamber and drives away the male. She then lays numerous eggs in a pile, grooms them of mold or debris, and protects them from intruders by her forceps and threatening gestures. She stays with her young up through two nymphal stages.

Burying beetles are subsocial insects renowned for their ability to move and bury the carcasses of mice and other small vertebrates found in nature. The female accomplishes this formidable task by crawling beneath the carrion, belly upward, and pushes from below with the legs. She pushes first on one side and then on the other until the carrion is moved to a patch of soft soil. She then buries the carrion by undermining it and piling soil atop it. The burial is sometimes a cooperative exercise involving other individuals that leave after eating their fill or after being dispersed by the prospective parents. The female then uses her mouthparts to work the dead flesh into a pasty ball into or near which she lays her eggs. The larvae hatch in the midst of plenty, and either eat the carrion for themselves or are fed by the parents.

Social Insects

Society is defined above as a usually permanent relationship among individuals of the same species that are variably dependent on one another and exhibit a division of labor. In insects, this relationship requires that the female remain with and care for her eggs and hatched young until they, in turn, take over the non-reproductive tasks and cooperate with her in rearing additional broods. Therefore, an insect society is an enlarged "family" consisting of one or more fertile females devoted to egg laying, progeny that do most of the nest building, defending, nursing, and foraging, and one or more males that mate with the fertile females but do not necessarily remain permanently within the nest. This *division of labor*

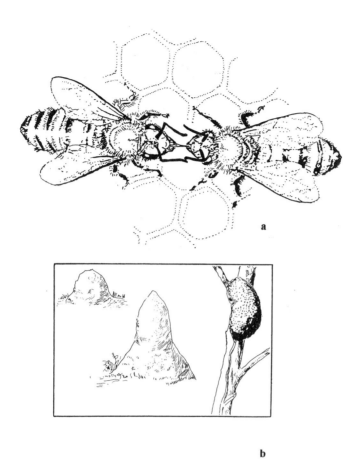

Fig. 17-3. (*a*) Two honey bee workers exchanging food; (*b*) above-ground nests of selected tropical termites, with the tree nest (*at right*) showing one of the covered runways leading to the ground (*a* is modified and redrawn from Ribbands, 1953; *b* is modified and redrawn from Snodgrass, 1967).

is often accompanied by *polymorphism*, the existence of different castes or different body forms, each devoted to specialized tasks. All society needs are transmitted through sounds, gestures, and, above all, *trophallaxis* (*Fig. 17-3 a*) which is the exchange of food and/or body secretions that obligates the individuals to act in a given manner, promptly, and in proper sequence.

Termites, ants, certain wasps, and various bees are among the thousands of known social insects. They exhibit remarkable behavioral parallels though all belong to different taxonomic groups.

Termites

Termites are specialized orthopteroid insects often called "white ants" though unrelated to ants and other social Hymenoptera. Most termites live within nests in sound or decayed wood or in the ground. Many tropical termites, however, construct conspicuous above-ground nests (*Fig. 17-3 b*) made of earth and masticated wood. These nests, whatever the type, are invariably closed to the outside which

Chapter 17 / INSECT AGGREGATIONS AND SOCIETIES

is adaptive to termites' delicate-skinned, weakly pigmented condition. Being unable to come out into the open to forage, termites construct tunnels in the soil or covered, above-ground runways through which they move from place to place.

Termites are wood-feeders though they lack the enzyme, cellulase, needed to digest wood. This enzyme is present, however, within the myriads of mutualistic flagellates or bacteria that live within the termite gut. These microorganisms digest the cellulose providing sustenance both for the termite host and for themselves. A termite from which the gut fauna has been excluded, either experimentally or by molting, continues to eat wood but starves owing to its inability to digest the wood that it eats. A recently molted termite loses its gut fauna owing to the casting of its internal gut lining, so it must reinfect itself either by exchange of anal liquids or by consumption of the excreta of other termites that contain the needed microorganisms.

Termites have three principal castes, each with males and females. Primary reproductives or kings and queens (*Fig. 17-4 d*) are comparatively large-bodied, winged, pigmented termites with well-developed compound eyes. They are usually the only individuals that mate. They break off the wings (*Fig. 17-4 e*) upon conclusion of the nuptial flight, establish a new colony, and thereafter devote themselves to repeated reproduction. The queen soon becomes distended with eggs (*Fig. 17-4 g*) and dwarfs her consort. So-called secondary reproductives (*Fig. 17-4 a*) are nymphal termites that may replace the kings and queens in the event of death of either of the royals. Workers (*Fig. 17-4 c*) are soft-bodied, usually wingless, pale, sterile males and females. They carry out all household duties including nest, gallery, and tunnel building, nursing, foraging, and constructing and tending fungus gardens, if any. These functions are apportioned according to age, the younger workers being concerned largely with brood care and the older ones with foraging and nest repair. Soldiers (*Fig. 17-4 b*) resemble worker termites in their sterility and their soft, non-pigmented body but differ in their sclerotized, pigmented head, often armed with massive jaws or sometimes, in the case of nasutes (*Fig. 17-4 f*), a snout that ejects a defensive fluid. Soldiers develop from workers of either sex and apparently serve a protective function, congregating where the nest has been invaded. They cannot use their over-sized jaws to eat, so they must be fed by workers.

Termite caste determination is based on trophallaxis. The interaction begins with the primary reproductives whose royal secretions are ingested by workers and nymphs and then passed, from termite to termite, through oral and anal feeding and licking of body surfaces. The royal exudates inhibit the development of secondary reproductives. Only when released from the royal influence by death or removal of the primary reproductives do the secondary reproductives or, in a few species, selected workers molt into fertile individuals that take over the reproductive chores. Soldiers likewise secrete an exudate that inhibits soldier development beyond a certain proportion of the nest population. This suggests that, according to the dictates of trophallaxis, all first-stage nymphs can become either fertile reproductives or sterile workers or soldiers.

A common termite life cycle starts with winged, sexually mature termites at the onset of the favorable season. They await the precise combination of moist atmospheric conditions required for the nuptial flight. Suddenly, they issue into the sky in dense clouds from nests throughout the area. Those termites not snapped up by insectivorous birds, insects, or other predators drop onto the vegetation or

Social Insects

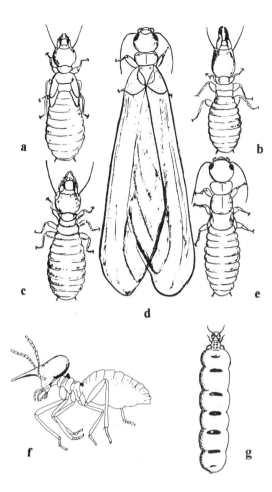

Fig. 17-4. Castes in a representative termite including (*a*) a secondary reproductive, (*b*) a soldier, (*c*) a worker, (*d*) a winged primary reproductive, (*e*) a primary reproductive after having shed its wings; (*f*) a nasute; and (*g*) a tropical termite queen visibly swollen with eggs (*a–e* are modified and redrawn from Ross, 1965, after Duncan & Pickwell; *f* is modified and redrawn from Grasse, 1949; *g* is modified and redrawn from Skaife, 1956).

ground, break off their wings, and await arrival of a member of the opposite sex. When a pair forms, the two scurry away in tandem, male following female, locate a crevice in the ground or in wood, enlarge it into a chamber, and seal themselves in. Then they mate. The young queen lays relatively few eggs at first, for the young that hatch are initially helpless juveniles that require parental care and regurgitated food to stay alive. Soon, however, a dozen or so workers begin chewing wood, and the first soldier appears. Thereafter, the colony increases in size as workers take over all nest maintenance, nursing, and foraging, soldiers assume their posts, and the king and queen devote themselves to repeated reproduction. The continued presence of the king in the nest, necessary for periodic mating with the queen, separates termites from most other social insects.

Chapter 17 / INSECT AGGREGATIONS AND SOCIETIES

Wasps

The hymenopterous superfamilies Sphecoidea and Vespoidea include wasps that differ from one another chiefly in pronotal structure and in their ability/inability to fold the wings lengthwise at rest. Sphecoids are largely solitary but also include a few subsocial and social species. Vespoids include solitary, subsocial, and social species. The social vespoids of temperate zones form an impermanent society consisting of queen and worker female castes and, of course, males. They eat nectar, sap, meat juices, honeydew, or similar fluids and dwell within characteristic paper-like nests of masticated wood molded into a circular plate or sac. The larval wasps, ensconced within the nest, eat masticated insects or spiders that the adults feed them.

The temperate-zone vespoids belong to subfamilies Polistinae and Vespinae. The Polistinae are often called paper wasps because of their plate-like, subcircular, horizontal paper nest supported by a slender stalk (*Fig. 17-5 a*). The comb is composed of a small number of cells that open below except during pupation, when they are temporarily closed by a silken cocoon. A young polistine queen upon breaking hibernation first fashions the stalk and attaches to it a single, shallow, six-sided cell in which she places an egg. She then construct additional cells at the periphery of the first, stocks each with an egg, and lengthens them with additional paper as required by larval growth. She feeds the larvae chewed insects and eagerly accepts the saliva that they offer in return. She does all the initial work including the progressive feeding and nest construction and is not relieved of these duties until the first offspring mature into workers, whereupon they take over all housekeeping and foraging. New queens and males emerge toward season's end, but only the inseminated queens overwinter to renew the life cycle.

The Vespinae, often called hornets and yellow jackets (*Fig. 17-6 a*), build large, bag-like nests (*Fig. 17-5 b*) suspended from tree branches or from beneath projecting objects or within underground cavities. Each mature nest consists of several tiers of combs wrapped within an oval envelope that opens below through a tubular entry. The inseminated queen breaks hibernation and begins the nest in spring by constructing a few cells enclosed within thin paper envelopes. She places an egg within each and cares for and feeds the young that emerge. They soon mature into winged workers that take over all duties except oviposition. Younger workers usually tend larvae, construct, repair, and guard the nest and fan the entry, while older workers forage out-of-doors for food and wood pulp. They begin building large queen cells toward the end of the season, and some workers graduate into additional egg producers. Inasmuch as their eggs are unfertilized, they give rise only to the males required for impregnating the new queens. The old queen is dead or moribund by this time, and the parental nest is near its end. Soon the males and young queens exit the nest and mate without swarming. The males and workers die within a few weeks leaving only the newly inseminated queens which hibernate within the ground or under logs or debris to begin the cycle anew the following spring.

Ants

The hymenopterous family Formicidae includes many thousands of entirely social insects called ants. They are readily separated from termites by possession of a constricted *pedicel*, or "wasp waist," and

Social Insects

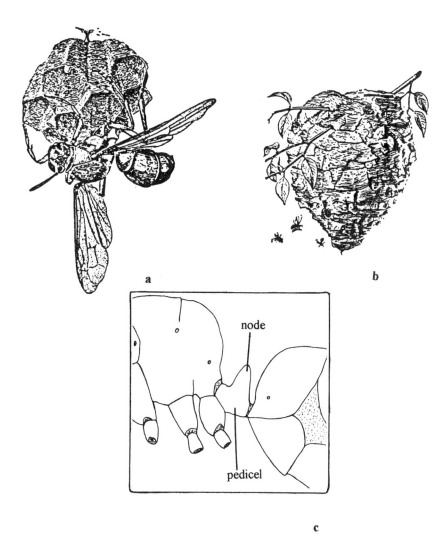

Fig. 17-5. (*a*) A young paper wasp queen tending her nest; (*b*) a nest of the bald-faced hornet, with the entry below; and (*c*) the node atop the pedicel of an ant; note how the pedicel separates the fused thorax and first abdominal segment from the remaining abdominal segments that constitute the gaster (*a* is from USDA; *b* is from USDA, 1952; *c* is modified and redrawn from Belkin, 1972).

from wasps by one or two dorsal protuberances called *nodes* atop the pedicel (*Fig. 17-5 c*). Ants are among the most abundant animals known. Most of their nests contain thousands of individuals, and army ant colonies may consist of millions of massed individuals but no actual nest. These figures pertain, of course, to individual colonies of which there are thousands in most localities. Most ants are soil inhabitants beneath logs or stones or sometimes within tunnels excavated within the earth itself. Others live in the lower vegetation or in trees. A few inhabit the dwellings of man or are parasitic in the nests of other ants.

247

Chapter 17 / INSECT AGGREGATIONS AND SOCIETIES

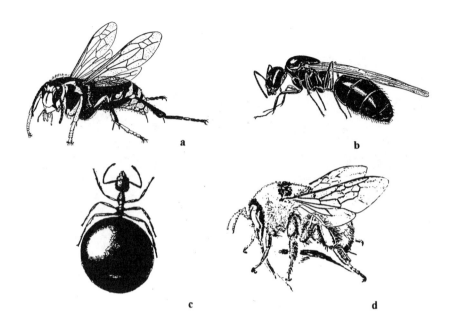

Fig. 17-6. (*a*) The bald-faced hornet; (*b*) a winged female carpenter ant; (*c*) the replete of a honey ant; and (*d*) a bumble bee (*a* is by courtesy of Illinois Natural History Survey; *b* is from U. S. Bureau of Entomology; *c* is modified and redrawn from Wheeler, 1910; and *d* is from Little, 1972, after Mohr).

Ant nests have single or multiple entries that are either left open or are closed with leaves, twigs, or debris. These entries may be guarded by soldiers or left unguarded. The dirt, pebbles, and debris accumulated during excavation may be heaped unceremoniously atop the nest making a mound, or it may be carried away and discarded at random, disguising the nest location. The tunnel system within the nest varies from simple to complex, the latter including specialized chambers for food storage, brood development, *etc.* Some ants dwell permanently within their nests, but most move readily whenever a new nest is needed. To do so, they find a suitable spot and bodily drag each egg, larva, and pupa, one by one, as well as the queen herself, to the new site until the move is completed.

Ants include the usual female castes, queens and workers, as well as males. Queens (*Fig. 17-6 b*) are fertile, usually winged ants, but they shed the wings after the nuptial flight. Males are winged, short-lived, and, unlike king termites, they are not involved in founding new nests. Workers are sterile females distinguishable from queens by their smaller size and winglessness. Workers are present in all species except for a few parasitic ones, and there are sometimes several distinct worker types. In this event, the smaller workers are largely housekeepers or foragers, and the larger ones, or soldiers, have powerful jaws used to protect the nest and crush hard food. The *repletes* (*Fig. 17-6 c*) that hang from the ceiling of the honey chamber in the honey ant genus are a worker subtype specialized to serve as veritable honey casks. Their distended gaster takes on the appearance of a spherical sac as returning foragers feed them enormous quantities of honeydew.

Social Insects

A common ant life cycle in temperate zones involves favorable weather or some unknown internal control that triggers the mass emergence and nuptial flight of queens and males from all nests in the local area assuring cross-fertilization. Queens often mate with a number of males which die soon afterward. Each queen sheds her wings, locates or digs a hollow, seals herself in, and waits for her eggs to hatch. She cares for the first larvae that appear, feeding them saliva until they mature into workers. These small-bodied individuals soon break out of the brood chamber and assume all household, nursing, and foraging duties leaving the queen free for egg production which she continues to do without benefit of additional fertilization. It may be some years before the sexual forms necessary for swarming are produced anew.

Most ants are omnivorous, eating whatever animal or vegetable food, living or dead, that comes to hand. Other ants are more specialized in their food-habits. Some ingest honeydew from aphids (*Fig. 12-5 a*); some eat fungi that they culture through a mutualistic kind of agriculture; some prey on insects, spiders, mice, lizards, and other small animals; and some even practice *social parasitism*, a kind of slavery involving the capture of the young of other ant species that they use as domesticated workers to carry out their own household chores. Some ants are blind, and others rely only in part on vision, light direction, and light intensity to find food. Their chemical sense is usually acute. Some foragers explore fortuitously and independently and neither return to a good source of food nor communicate its presence, once found, to others of the nest. Foragers of other species communicate a good source by laying down a chemical trail as clear to them as lines on our highways are to us. In this case, an ant returning with food touches the tip of her abdomen to the ground from time to time depositing pheromone. She then communicates with other foragers at the nest by antennal play and gesturing. They set out immediately, follow her trail, and soon a steady stream of them may be seen moving toward the food or returning from it, heavily laden. The trail dissipates as the supply dwindles.

Bees

The members of the superfamily Apoidea, known as bees, differ from their wasp relatives in possession of plumose body setae and in their use of nectar and pollen as food. Most bees are solitary, not social. They make cells similar to those of solitary wasps within burrows or cavities. They mass provision these cells, oviposit within them, and then seal them. The hatched larvae fend for themselves, using the honey and pollen provision. In contrast are the subsocial and social families, including the Apidae, especially bumble bees and honey bees, on which I concentrate below.

Bumble bees (*Fig. 17-6 d*) are robust, usually yellow and black insects densely covered by fur-like setae. Their life cycle is annual in temperate zones. Worker and male bumble bees die in autumn, but the young, newly impregnated queens overwinter in the ground, often near the parental nest. They break dormancy the following spring after appearance of the first flowers. Each new queen spends some days looking for a nest, usually an abandoned bird or mouse nest or an opening within a grass clump. She moves in, constructs a brood cell using wax from her abdominal pores, introduces into it a ball of pollen and perhaps eight eggs, seals the cell with wax, and, like a hen, settles atop it to brood.

Chapter 17 / INSECT AGGREGATIONS AND SOCIETIES

She also constructs a honey pot and fills it with nectar gathered during her foraging trips. Her incubated eggs hatch within a few days, whereupon she alternates between foraging out-of-doors and brooding over and caring for the young larvae. She either squirts regurgitated pollen and nectar directly at them, or she places dry pollen in pockets to the side from which they help themselves. The young spin separate cocoons about a week later. The queen continues brooding over them but also forages, constructs, and oviposits within additional brood cells near the pupae. The first workers emerge from their cocoons some weeks later and crawl to the honey pot for their initial adult meal. They subsequently take over all household duties and foraging, freeing the queen for egg laying. They increase the number of cells to accommodate population growth, and the nest gradually assumes the form of a comb. Spent cells are not necessarily abandoned because the workers trim off the cocoons within, rim them with wax, and convert them into additional honey pots. Toward autumn, bumble bees cease producing workers and give rise only to males and queens which, unlike the sexual forms of honey bees, do not swarm. They emerge individually and mate on foliage, flowers, or on the ground. Soon the workers and males die leaving the young queens to repair to their private hibernacula until the following year.

The family Apidae also includes the honey bees of which there is an introduced, highly specialized species, *Apis mellifera* (*Fig. 17-7*), in the United States. This domesticated insect is largely restricted to man-made hives. Its individuals are incapable of living alone, and its queens cannot even forage. However, honey bees nested in hollow trees long before mankind learned how to keep them, and even today a domesticated swarm may escape from the apiary and locate in the woods. The nest consists of several vertical waxen combs arranged parallel to one another, each comb two cells thick. The two cells at a given level attach base to base and open at opposite ends. They are uniformly hexagonal in cross-section whether in an artificial hive or in a hollow tree in nature though in the latter the overall comb necessarily takes on an irregular shape. Cells for rearing workers and storing pollen and honey are smaller than those for rearing drones. Those for rearing queens are larger still, being peanut-shaped structures suspended from near the bottom of the comb.

Honey bee sex and caste determination are by combined genetic and environmental means. The queen's spermathecal sphincter may be constricted preventing egress of the sperm with which she has been impregnated. In this case, her oocytes are not fertilized and accordingly develop parthenogenetically into haploid drones (males). Most of the time, however, the queen lays fertilized eggs which develop into diploid offspring destined to become either queens or workers, according to diet. If initially given royal jelly (the salivary secretion of nurse bees), and then switched to bee bread (a mixture of honey and pollen), they become workers. However, if fed royal jelly exclusively and reared within a special, large queen cell, they become queens. The controlling pheromone is produced by the queen's mandibular glands and spread over her body by grooming. This royal exudate is attractive to, and exhibits control over, the other bees in the nest as she is constantly attended by workers that lick her and pass her pheromone throughout the nest in the course of food exchange with others. All are apprized of her presence thereby. If she is removed from the hive, workers sense her absence and soon replace her by constructing a number of queen cells over eggs or young larvae that are not yet caste-determined.

Social Insects

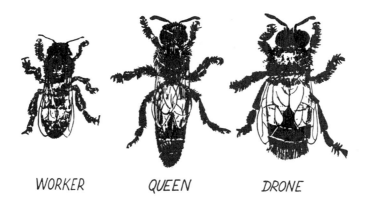

Fig. 17-7. Worker, queen, and drone castes of the honey bee (Modified and redrawn from USDA).

Honey bee eggs hatch into larvae in about three days, and the workers scurry about to feed them progressively. The nurse bees seal each larva within its private cell about six days later and leave it to spin its cocoon in seclusion. The transformation into the adult stage takes about a dozen days. The young animal then breaks its pupal skin, steps forth, inflates and hardens its wings, and becomes a new worker. Workers produced in autumn have a longevity of some months, but those that emerge in spring or summer live only a few weeks to a month or so. Upon emergence, they begin a sequence of activities apportioned largely according to age. They do not work during the first day or two. Thereafter, they function first as nurse bees, then as comb builders and soldiers, and finally as foragers.

Foraging is not a fortuitous activity carried out independently by each bee. It is a communal effort coordinated by a complex exchange of information between individuals using two main types of dance language according to the distance between the food and the nest. Returning foragers use the round dance (*Fig. 17-8 a*) to indicate a nearby source, and they communicate the kind of nectar discovered by food odors on their body and by food exchange. Other workers follow them as they dance about, their heads and antennae touching. The observers perceive the dance's round form indicating a nearby source; they apprize themselves of the kind of food by scent; and they learn something of its richness by the vigor and duration of the dance. Thus informed of the kind, quantity, and maximal distance of the food, they fly off to find it though they do not know precisely where it is. To communicate greater distances, returning foragers use the waggle dance (*Fig. 17-8 b*) conveying not only the kind of food but also its distance, quantity, and direction. For example, the worker may dance a clockwise half circle, make a straight vertical run while wagging its abdomen sideways, dance a counterclockwise half circle, make another tail-wagging straight run, and repeat the message again and again. The straight run indicates the food's actual direction under experimental conditions on a horizontal surface when the dancer can see the sun. However, bees normally dance in the darkness of the hive in which the vertical combs do not permit horizontal dancing. They obviate this difficulty by orienting the straight run to gravity rather than to the sun. They perform the straight run vertically and upward when the food is directly toward the sun, vertically and downward when directly away from

Chapter 17 / INSECT AGGREGATIONS AND SOCIETIES

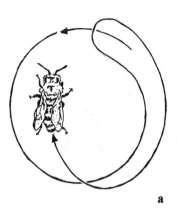
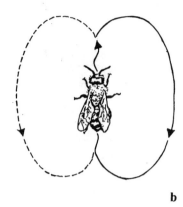

Fig. 17-8. (*a*) A returning forager of a honey bee executing a round dance; (*b*) a returning forager executing a waggle dance (Modified and redrawn from Eisner & Wilson, 1977, after von Frisch).

the sun, along a line 45 degrees to the right of vertical when it is 45 degrees to the right of the sun, *etc*. Thus, the orientation of the straight run indicates food direction with respect to the sun, but this is only part of the message. The frequency of tail wagging indicates food distance from the hive, and the vigor of dancing indicates its richness. Finally, as before, the bees convey information about the specific kind of food by smell and trophallaxis.

A strong hive of honey bees consists of thousands of workers and, at most, about a dozen drones and one queen. From time to time, however, the workers construct one or more large cells in which to rear additional queens. The dowager queen either kills these rivals before they emerge, or she departs the hive in a swarm of bees leaving the existing nest to one of the newcomers. The new queen is then fecundated in one or more nuptial flights with a few drones. She returns to the hive with a supply of sperm sufficient for some years' productivity. Honey bee dispersal is, as suggested, by swarming, initiated, at least in part, by overcrowding. Swarming generally involves the dowager queen leaving the nest with thousands of workers, finding a suitable site, and establishing a new hive there. The event is predictable. Foragers cease bringing in nectar and pollen, scouts explore the countryside, and all seem to take on an "excitement" as they gorge themselves on honey some days prior to the event. Then the dowager and thousands of workers exit in a cloud when conditions are suitable, buzzing in high-pitched fashion. They alight in an undulating cluster in some nearby bivouac and await the return of reconnoitering scouts that, from time to time, return with "messages" danced on the surface of the swarm. Other scouts visit the suggested place until the entire swarm departs for it.

Suggested Additional Reading

Von Frisch (1967); Holldobler & Wilson (1990); Michener (1974); Michener & Michener (1951); Ribbands (1953); O. W. Richards (1953);Wilson (1971, 1975).

Chapter 18

Nomenclature, Taxonomy, and Systematics

*"What's the use in their having names," the Gnat said,
"if they won't answer to them?" "No use to them," said Alice,
"but it's useful to the people who name them, I suppose."*
— Lewis Carroll

The world's total of living insects staggers the imagination. Entomologists have already described over 900,000 species, yet untold numbers remain to be recognized and named. To understand them and use them in biological research requires ordering them into a rational system based on their unifying features and giving them names. Such a system is the key to all published information. It involves principles and procedures that I now discuss.

Binomial Nomenclature

Taxonomy is the scientific study of organisms' nomenclature, classification, and evolutionary relationships. *Nomenclature*, in turn, is the naming of organisms so that they may be recognized and designated with accuracy throughout the scientific world. The current system, *binomial nomenclature*, was used for the first time by the naturalist Linnaeus. It uses two names to designate species: (1) a *generic name* (the *genus*), always capitalized, indicative of a general kind of animal, and (2) a *specific name* (the *species*), not capitalized, indicative of a particular kind of animal. The generic name is capitalized because it is a proper noun; the specific name (sometimes called the trivial name) is not capitalized because it is adjectival in the broad sense. Both names are italicized because they are Latin or Latinized.

To illustrate the binomial system this chapter will use the American cockroach *Periplaneta americana* (*Fig. 18-1 a*), the Australian cockroach *Periplaneta australasiae* (*Fig. 18-1 b*), the Oriental cockroach *Blatta orientalis* (*Fig. 18-1 c*), and the German cockroach *Blattella germanica* (*Fig. 18-1 d*). *Periplaneta*, *Blatta*, and *Blattella* are generic names each of which describes a general kind of cockroach, and *americana*, *australasiae*, *orientalis*, and *germanica* are specific names. Each of the four combined generic and specific names (genus and species names) is a *binomen* (pl., *binomina*) that designates a distinct, unique species. The fact that the generic name *Periplaneta* is common to two of the four suggests a closer relationship between them than between either of them and the other species, so

Chapter 18 / **NOMENCLATURE, TAXONOMY, AND SYSTEMATICS**

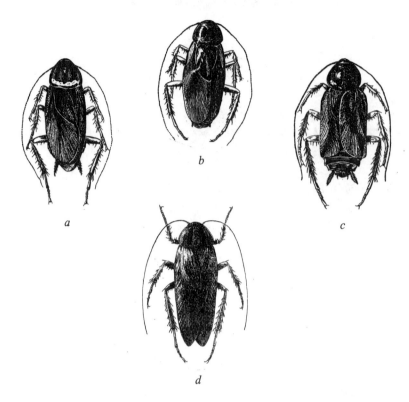

Fig. 18-1. Several common domestic cockroaches including (*a*) the American cockroach *Periplaneta americana*, (*b*) the Australian cockroach *Periplaneta australasiae*, (*c*) the oriental cockroach *Blatta orientalis*, and (*d*) the German cockroach *Blattella germanica* (*a*, *b*, and *c* are after Essig, 1926; *d* is after Chopard, 1951).

the binomial system lends itself to classification as well as naming. If the classification adopted is natural or biologically valid, it is consistent with phylogenetic or evolutionary relationships and, thus, reflects the genealogy of the species.

Species Concept

The *species* is the basic unit of taxonomy. It has been so for centuries, yet our present understanding of the term is different from that of the past. Linnaeus and those who followed in the Eighteenth and Nineteenth Centuries relied upon a *morphological species* concept. They felt that: (1) each species is separable structurally from all others; (2) species can be united with similar species into the next inclusive category (the *genus*), all species of which have properties in common; and (3) a species consists of its genus (the properties common to this inclusive group) and its differentia (the properties by which it is separable from others). Naturalists who used this system had merely to establish a number of *types* based on arbitrarily selected specimens and to give each a distinct name. Then, upon collecting additional specimens, they had only to compare the new specimens with the previously established types. If the new specimens agreed with one of these types, they were regarded as conspecific (*i. e.*, to belong to the same species) and were given the name of the type; if not, they were regarded as new species yet to be described. This approach emphasizing differences worked well to name and pigeon-hole specimens of small collections but soon led to chaos in groups with large collections. It simply could not accommodate the variability native to most populations, so the literature became mired in a swamp of species created on the basis of minute, superficial differences.

Today's *biological species concept* is a logical outgrowth. It emphasizes populations, not individuals, and uses structural, functional, distributional, behavioral, and other available characters to analyze both the similarities and the differences among populations. Inasmuch as this approach stresses variability, ideally it requires examination of the comparatively large samples that museum workers call *series*. It rejects the idea that the nomenclatural type is an idealized representative of the species though it is common for a species to be represented by one or only a few specimens.

Taxonomy vs. Systematics

Some research workers feel that taxonomy and systematics are synonymous. Others regard them as separate disciplines. I adopt an intermediate point-of-view that allows systematics to share the definition given earlier for taxonomy though they represent somewhat different approaches to the determination of species and evolutionary relationships. Taxonomy places the species in a central position, tends to neglect infraspecific categories, uses a structural approach involving small series, and emphasizes types and nomenclature. *Systematics* embraces a truly biological species concept that takes into consideration all available characters, not just the external structure of a few dead pinned specimens, and it uses comparatively large series adequate for detailed study of individual variation. There is, however, no fixed demarcation between the two. They are always, to a degree, mixed in today's research. Some studies involving small numbers of poorly known insects are necessarily more taxonomic, but others based on larger series of better-known insects are more systematic. Systematics remains an ideal that seldom is practiced in full.

Let us see how the two approaches worked in one case, that of the black field cricket genus *Gryllus*. By the turn of the Twentieth Century, some 47 species were known from the New World and 17 from the United States (*Fig. 18-2*). However, they posed an obvious taxonomic problem because the same collection of crickets sent to different authorities was likely to be returned with different names attached to them. One specialist who analyzed the genus statistically in 1908 found no specific entities within it saying essentially that they are all the same species. Two well-known authorities carried out a more detailed taxonomic analysis of *Gryllus* in 1915 with a result that was hardly more satisfactory. They concluded that the large series of crickets they examined constituted one great polytypic species that they arbitrarily called *Gryllus assimilis* which henceforth became the valid name for the previously recognized 17 United States species. However, variability in so-called *G. assimilis* continued to puzzle investigators, one of whom, B. B. Fulton, detected differences in the songs of field crickets in his home state of North Carolina. This led him to undertake an intensive systematic study lasting several decades as a result of which he concluded that there are four species of *Gryllus* just in North Carolina. He noted that one of them, a triller, is adult in fields from April to May and that the remaining three

Chapter 18 / NOMENCLATURE, TAXONOMY, AND SYSTEMATICS

Fig. 18-2. The common black field cricket of northeastern United States (Courtesy of R. D. Alexander).

species, all chirpers, live in other, different habitats during alternative months of the year. He found that the four species are structurally inseparable but can be distinguished on the basis of their song, geographic distribution, habitat selection, seasonal periodism, and, above all, reproductive isolation.

Conventional taxonomy proved incapable of accurate analysis of relationships within genus *Gryllus*. Only when investigators turned to a systematic approach using detailed behavioral, distributional, and rearing studies was the solution forthcoming. It is in instances such as this that the advantages of systematics become apparent. This is not to say that standard, structurally based taxonomy is archaic. Far from it. More often than not, taxonomy elicits the right answer with minimal time and effort, and often it is the only feasible approach because of the lack of adequate series, insufficient time or opportunity for detailed field observation, or the need for special equipment. It is only when taxonomy is confronted with highly varied populations such as those of genus *Gryllus* that is founders and must be supplemented by systematic research.

Species Definition

A species is a particular kind of organism. The idea is simple enough, but recognition of such is sometimes difficult. Some entomologists emphasize structural attributes, but others feel a genetic or biological definition is preferable. I am comfortable with the following definition: a *species* is a natural population whose individuals tend to be similar in structure, function, and behavior and, in nature, freely interbreed to produce fertile, living offspring.

Let us analyze this definition: (1) A species is an interbreeding, interrelated population of individuals of greater or lesser variability. The *type specimen* after which this assemblage is named is merely a single individual arbitrarily selected by the describer to bear the species' name. (2) Individuals of this population are structurally similar, so they tend to look alike and are usually distinguishable from individuals of other species. They are variable (*Fig. 18-3*), however, especially in cases of geographic or ecological separation into subpopulations, and they may vary remarkably with age and sex. (3) Given their similar structure, it is logical to expect a similar functional pattern or physiology subject to natural variability, sexual dimorphism (males and females being different), and developmental polymorphism (appearance changing with growth). (4) It is also logical to expect a similar habitat occupancy, periodicity, general deportment, feeding, and song, if any. As before, these

Nomenclature, Taxonomy, and Systematics

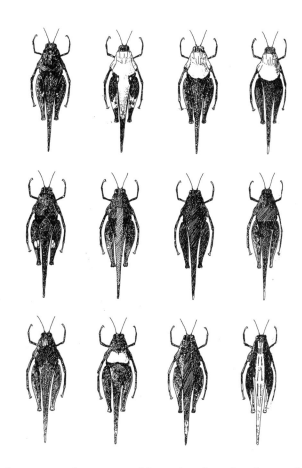

Fig. 18-3. Some color variants commonly encountered in a selected species of grouse locust (From Chopard, 1938, after Nabours).

attributes are subject to greater or lesser natural variability and to sexual and developmental disparities. Finally (5), species are usually *reproductively isolated* stemming from genetic, physiologic, structural, or other differences which prompt individuals to mate only within their own species, not with individuals of other species, and upon doing so they produce fertile living offspring. Cross mating between species occurs rarely in nature, and, when it does, hybrid offspring are either not produced or those few that arise tend to be sterile.

Thus, the *fertility* test is often a good criterion of speciation. Individuals caged together may be assumed conspecific if they produce fertile, living offspring but to belong to different species if they do not. However, the fertility test involves complicated rearing, cross breeding, and other often difficult, time-consuming procedures. Even so, it is more applicable to small insects with a brief life cycle than to large, long-lived vertebrates such as elephants! It fails only in certain laboratory situations and in the uncommon or non-existent self-fertilizing hermaphrodites, obligatory parthenogenetic animals, and asexual forms among insects.

Chapter 18 / NOMENCLATURE, TAXONOMY, AND SYSTEMATICS

Infraspecific Categories

From Linnaeus onward, naturalists compared everything with a type. If the specimen seemed similar, they judged it *conspecific* with the species represented by the type; if not, they assumed it to belong to a different species. Given this static morphological approach, it was inevitable that they would find specimens that seemed atypical of the species yet insufficiently divergent to be regarded as a new species. They termed this assemblage of ill-fitting individuals a *variety*. More recently, with our improved appreciation of natural variability, we expect a wide range of variants from within any population but no longer accord them formal taxonomic recognition. We no longer use the term *variety* for animals, at least in the formal sense.

No two places on earth are identical. Even places adjacent to one another may differ somewhat in soil, topography, climate, and biota and, hence, in the selective pressure that molds their inhabitants. We now accept that species consist of subpopulations more or less closely adapted to their particular place but somewhat different from neighboring subpopulations of the same species. It follows that the same or neighboring subpopulations of a species (which are the ones most likely to interbreed) tend to be more similar than are widely separated ones. This tendency is enhanced when the species' geographic range is broken into isolated segments by mountains, rivers, or other barriers, in which case the separated subpopulations may be recognizable as distinct *subspecies* that vary taxonomically and genetically from others of their species. However disparate they may be, these subspecies cannot be considered different species because their individuals are able to interbreed and intergrade with one another, at least where their ranges overlap.

Subspecies are designated by *trinomial nomenclature*. *Barytettix humphreysii* (*Fig. 18-4*), for example, is a grasshopper found in parts of southwestern United States down into Mexico. It consists of two subspecies (*Barytettix humphreysii humphreysii* and *Barytettix humphyreysii cochisei*) that differ in their genitalia and other physical characters, yet interbreed and intergrade with one another in certain mountains where their subpopulations overlap. This indicates they are subspecies rather than the separate species that their dissimilar appearance suggests they might be.

The preceding is not to suggest that only geographic barriers result in broken ranges. Food, climate, soil, and other factors may be operative. For example, the spring and fall field crickets of eastern United States have a similar geographic distribution, a similar appearance, similar food habits, and even similar calling songs, but one of these crickets overwinters in the egg stage and matures in the fall, while the other overwinters in the nymphal stage and matures in spring. Their disparate temporal distribution results in negligible overlap between them.

Higher Categories

Confronted by over 900,000 insect species, the entomologist must do more than name if he is to work effectively. He must also classify which means to arrange these creatures in some logical order. The standard means to do so is a nested system of natural *taxa* (sing., *taxon*) fitting within a hierarchy of formal categories. Thus, species are united into taxa identifiable by a particular combination of characters; these into other, greater taxa of a more general similarity; these into still others; and so on. The

Nomenclature, Taxonomy, and Systematics

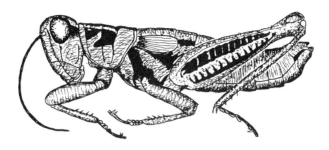

Fig. 18-4. *Barytettix humphreysii*, a grasshopper of southwestern United States and northern Mexico where it occurs as two subspecies (Modified and redrawn from Helfer, 1953).

universally recognized system of higher categories of insects includes genera < tribes < subfamilies < families < superfamilies < orders < classes < phyla. The reader's knowledge of these categories and their use is assumed.

Endings used are often indicative of taxonomic level. An insect name ending in *-ptera* such as Orthoptera, Hemiptera, or Lepidoptera is almost always an order; *-oidea* such as Scarabaeoidea or Ichneumonoidea indicates superfamily status; *-idae* such as Cimicidae or Acrididae family status; *-inae* such as Gryllinae or Papilioninae subfamily status; and *-ini* such as Melanoplini or Crabronini tribal status.

Assignment of an insect to a given taxon characterizes it as possessing that group's attributes. For example, we may know nothing of the screwworm fly *Cochliomyia hominivorax* (*Fig. 19-18 i*) having never seen it in nature or studied it in the laboratory, but we know much about it indirectly because it belongs to family Calliphoridae, order Diptera, *etc.*, with whose attributes we are familiar.

Nomenclature

Nomenclature, as noted, is the naming of organisms. It is an essential though purely legalistic scientific activity. No precision in naming is possible without it, yet its importance is ill-understood even by many biologists.

A species is a population, meaning a discrete, natural, non-arbitrary category, one of the fundamental units of biological sciences. Much past physiological research has been carried out on, for example, the "cricket" meaning the North American black field cricket which really includes six different species. Much research has been done on the American cockroach whose laboratory cultures have sometimes proved to consist of the southern brown cockroach rather than the American cockroach. These are but two examples from among many that can be cited from the literature. Inasmuch as different species are involved, is it not logical to expect a different physiology from the test animals? How valid is physiologic, genetic, or other research that ignores the test animals' true identity?

Naturalists subsequent to the Renaissance paid increased attention to describing and naming organisms. Their work was in Latin, then the language of scholarship, and it contributed to the knowledge of some insects. However, there was little advance in formal insect study until near the end of

Chapter 18 / NOMENCLATURE, TAXONOMY, AND SYSTEMATICS

the Seventeenth Century when John Ray introduced a species concept roughly consistent with that in use today though his system of naming bore little resemblance to ours. He used the common Latin name of an animal as the first word of its compound name to which he then added a long Latin phrase adjectival to the first word to describe the animal in question. This unwieldy pre-Linnaean system became applied so inconsistently that eventually it broke down of its own weight. It is in this context that the Swedish naturalist Linnaeus appeared on the scene. He used the system of the time, just described, for his early work but, in the tenth edition of his *Systema Naturae*, of 1758, he turned entirely to his newly devised binomial system. The advantages of this shortened system were immediately apparent, and soon it was adopted universally. It has remained essentially unaltered since then except for incorporation into it of the subspecific name expressing population differences not anticipated until modern times.

International Code of Zoological Nomenclature

The current naming of animals follows guidelines established in the *International Code of Zoological Nomenclature*. Much of the code is irrelevant here, but selected articles are important, so I shall paraphrase them, comment on them briefly, and provide examples.

The publication date of Linnaeus' *Systema Naturae*, 1758, is the starting date of zoological nomenclature. Zoological nomenclature is independent of botanical nomenclature. For example, *Conocephalus* and *Bacillus* are each generic names, both used for an insect and for a plant/bacterium.

Family names are formed by adding *-idae* and subfamily names by adding *-inae* to the stem of the type genus. For example, the water striders of family Gerridae and subfamily Gerrinae take their name from the type genus *Gerris*.

Species sometimes include subspecies in which case the subspecific name is added to the *binomen* to form a *trinomen*. The subspecific name, if any, begins with a lowercase letter and is written immediately after the specific name without interposition of punctuation marks. One subspecies always bears the specific name, for example, the carpenter ant subspecies *Camponotus abdominalis abdominalis* and *Camponotus abdominalis floridanus*. The first is the typical subspecies that repeats the specific name and contains the type, and the second uses a new subspecific name, *floridanus*, and has a new subspecific type.

The name of the person who first describes a species or subspecies is often added to the binomen or trinomen. This author name is neither italicized nor underlined when written because it is not part of the formal name, and its citation is optional. When cited, it appears after the scientific name without interposition of punctuation marks, sometimes along with the year of description. *Myodocha serripes* Olivier, 1811, is an example. Olivier described this bug, *M. serripes*, in 1811. The author name is retained even if the species is later transferred to another genus, but in this case the author name is placed within parentheses. The coulee cricket *Peranabrus scabricollis* (Thomas) illustrates. Its transfer from the genus in which it was originally placed by Thomas is shown by parentheses around Thomas' name.

The same species is sometimes mistakenly described under different names by scientists working independently of one another. These names must be united under the oldest valid name which

Nomenclature, Taxonomy, and Systematics

thereafter has priority over the other names. For example, *Locusta migratoria* Linnaeus, 1758, has priority over the synonyms *Pachytylus cinerascens* Finot, 1890, and *P. danicus* Azam, 1901. All three names apply to the same species, but only the first, oldest name is valid.

The designation of animals is uninomial and plural above the species, binomial for species, and trinomial for subspecies. Binomina and trinomina but not the names of higher taxa are italicized or underlined when written, and the names used are either Latin words or words from any other language Latinized. Their use follows the Law of Priority which holds that, with certain exceptions, the oldest available name must be used in its original spelling. For example, the dragonfly generic name *Aeshna* has priority over and must be used in place of *Aeschna*, a misspelling that appears in the literature.

A generic name consists of a noun in nominative singular or is treated as such, and it begins with a capital letter. Once published, it cannot be rejected in favor of another name because of inappropriateness even by its author. It is unique, being applicable only to one general kind of animal. If the same genus has been incorrectly described under two or more different generic names, the additional names must be rejected in favor of the oldest valid generic name. The katydid genus *Neobarrettia*, for example, consists of the former separate genus *Neobarrettia*, described in 1901, and the more familiar genus *Rehnia*, described in 1907.

A species sometimes mistakenly includes one or more additional species. These confused species, all under one name, must be divided so that each has its own name and its own type. The existing name is assigned to the one that contains the original type, and new names must be provided for the additional species. Thus, when it became necessary to separate the new species of the mosquito group Maculipennis from the old *Anopheles maculipennis* within which they were formerly concealed, the name *A. maculipennis* was assigned to the restricted species containing the original type, and new names were given to the additional species.

These are just a few of the many articles of the International Code of Zoological Nomenclature. The articles are merely guidelines or recommended procedures, but more is required to assure an accurate, uniformly employed system. Toward this end, the *International Commission on Zoological Nomenclature* (a permanent body empowered by the International Congresses of Zoology) recommends necessary changes in the code, renders opinions on formally submitted questions, compiles a list of official scientific names, and exercises plenary powers to suspend the rules under certain compelling situations.

The system is essentially self-policing. Comparatively few entomologists carry on taxonomic research, and there is little overlap in the taxa or faunas with which they work. Each specialist is free to apply the code's nomenclatural rules within the dictates of his conscience. It is natural, however, for disagreements to arise, in which event one of the interested researchers may petition the international commission for an official opinion. He publishes his petition in the *Bulletin of the International Commission of Zoological Nomenclature*, time is given others to comment on his proposal, and the commission eventually renders a decision. The matter decided, the petitioner and all others are obliged to abide by the opinion.

Chapter 18 / NOMENCLATURE, TAXONOMY, AND SYSTEMATICS

Binomial System vs. Common Names

Laymen often ask why it is necessary to use the long, complicated scientific names that entomologists have devised. Why not use the common names that everyone understands and remembers?

The binomial system is based on Latin, a dead, unchanging, precise language that is the basis of a number of present-day languages and is understood worldwide. Therein lies its utility. A scientific name such as that of the familiar domestic cockroach *Blattella germanica* (*Fig. 18-1 d*) means the same to workers throughout the world, notwithstanding their native tongue.

Common names are simply vernacular names that tend to be inappropriate and to lack consistency and universality and are not useful, except locally. *Blattella germanica* is called by hundreds of names depending on geographic locality including the Prussian cockroach in Russia, the Russian cockroach in Germany, and the German cockroach in much of the United States except New York, where it is called the Croton bug; June bugs are not bugs but beetles and are better termed May beetles; ladybugs or ladybirds, as they are variously called, are really beetles, and only about half are females, the rest being males; mayflies are not flies and do not necessarily emerge in May; the silkworm is not a worm but a moth larva; the kissing bug is hardly an osculator though it is a bug; and so on.

A more compelling argument against the use of common names is based on insects' tremendous numbers, their comparatively small size, and the layman's inability to distinguish them. Those common names in general use usually apply to larger insect groups such as subfamilies, families, orders, *etc.*, rather than to individual species. The term *grasshopper*, for example, is descriptive of individuals of the orthopteroid group Acridoidea; a *bug* is a member of order Hemiptera; a *beetle* a member of order Coleoptera; a *louse* a member of order Phthiraptera; a *flea* a representative of order Siphonaptera; and a *fly* a member of order Diptera.

The common names *bug*, *fly*, and *louse* may be modified by an appropriate second name and written in a conventionalized manner, as follows. The common name of a true fly (order Diptera) is written as two words, for example, black flies and crane flies, but the common names of non-dipterans called flies are written as a single word, as illustrated by Mayflies, stoneflies, and whiteflies. Likewise, bed bugs and assassin bugs, both of which are true bugs (members of order Hemiptera), are written as two words, but tumblebugs and Junebugs are written as a single word because they are not bugs but beetles. Similarly, body lice are true lice (members of order Phthiraptera), as opposed to plantlice, which only look like lice.

If a common name applies to a particular insect species, it is often a large, showy, readily identifiable creature such as the monarch butterfly. It may also be an economically important insect such as the Colorado potato beetle, the honey bee, the silkworm, or the house fly.

Pest control operators, extension workers, and other applied entomologists who discuss insect pests with the general public find it convenient to use common names. Hence, the national entomological society or other established organizations of those countries with a well-developed applied entomology attempt standardization of common names. In the United States, for example, the Entomological Society of America has a procedure for approving common insect names demonstrated to be appropriate and consistent. It publishes lists of these approved names at intervals correlated, of course, with the scientific names.

Nomenclature, Taxonomy, and Systematics

Types

Type specimens are one or more individuals or their parts set aside to bear the name of a described species, subspecies, or other taxon. These descriptions are sometimes ambiguous ignoring what later proves to be necessary diagnostic characters, so doubt may arise concerning which insect an author intended with his description. Availability of the type often resolves this question fixing the species' name on a particular specimen and leaving disparate specimens free for other names. Types were long regarded as idealized individuals representative of their species. Now, however, entomologists have an improved understanding of natural population variation, and they acknowledge the desirability of studying large series. They realize that they never have before them in their collection a complete representation of the species but only a small, inadequate sample. This realization has forever relegated types to the less glamorous but still important role of name bearer.

Investigators of past years commonly labeled as *cotypes* all specimens upon which they based a species description. This unfortunate practice led to problems when subsequent research revealed the presence of more than one species within the cotypic series. Inasmuch as only one specimen can be the intended name bearer, the others require new names. The quandary is resolved when a later entomologist arbitrarily selects one of the cotypes as *lectotype* to serve in place of the holotype (see below) and leaves the disparate specimens free for any required new names. Modern authors now select only a single type specimen, the *holotype*, which they regard as the best-available representative of the species or subspecies, and they designate all other specimens of the original series as *paratypes*. They also fix a type locality upon selection of the holotype. Several other classes of type are recognized today, but those listed above are sufficient for present purposes.

Today's authors are expected to make their type material as accessible as possible to the scientific community. They increasingly donate holotypes and paratypes to selected museums and list these institutions as repositories to which interested scientists may turn for direct information or specimen loans.

Museums

Museums are private or public repositories of collections. They contain specimens that are permanently maintained and curated and may be seen by visiting specialists. They also loan specimens, preferably ones other than types, to qualified investigators or institutions for more detailed study. Holotypes are often not loaned because they are irreplaceable and require especial protection from fire, flood, breakage, and pest damage.

Unfortunately, today's research museums with extensive insect collections are few in number. Most natural history museums are devoted largely to collections of vertebrate animals and have only small collections of insects and correspondingly reduced facilities for insect study. Moreover, most governmental, university, or private museums are associated with public exhibits. Their community function is often the one from which they derive their financial support. Few are independently endowed. Their staff is expected to prepare exhibits, make species determinations for the general public, advise on insect problems, give lectures, hold nature classes, and perform other

Chapter 18 / **NOMENCLATURE, TAXONOMY, AND SYSTEMATICS**

valuable but time-consuming services. Insect research is but one aspect of their overall role, and all too often it is neglected.

Suggested Additional Reading

Anderson (1975); Blackwelder (1967); Borror, Triplehorn, & Johnson (1989); Kevan (1973); Mayr (1969); Ross (1974); Simpson (1961); Steyskal (1978).

Chapter 19

Classification and Major Insect Groups

*To a person uninstructed in natural history,
his country or seaside stroll is a walk through
a gallery filled with wonderful works of art,
nine-tenths of which have their faces turned to the wall.*
— T. H. Huxley

Subclass Apterygota

Order Protura
(Telsontails)

Protura (*Fig. 19-1 a*) are an obscure group of minute, pale, soft-bodied, wingless, eyeless terrestrial arthropods consisting of about 260 species. They live a secretive existence in moist soil, under stones or leaf litter, in humus or decaying vegetation, or within decomposing logs where they eat fungi. Their primitive piercing-sucking mouthparts, differentiated thorax, three pairs of legs, and *styli* (vestigial legs) on the first three abdominal segments are insect-like attributes. However, their development is direct, by *anamorphosis* (an additional abdominal segment added per molt), as in myriapods; their genae and labium form a pocket surrounding the *entognathous* (withdrawn into head) mouthparts as in springtails and diplurans; their atrophied antennae are unique, functionally replaced by the upwardly-directed fore legs; they lack a tentorium (internal head support); their adult abdomen consists of 12 segments counting a well-developed *telson* (a greater number than in true insects except embryonically); and they lack cerci. Telsontails are non-insects whose peculiar mixture of characters places them either within an arthropod class of their own or perhaps as a sister group to Collembola (*Tables 1, 19-1*).

Order Collembola
(Springtails)

Springtails are small, soft-bodied, wingless, usually eyeless, insect-like hexapods found in places as divergent as within soil or decomposing logs, on rank vegetation or fleshy fungi, beneath bark, along

Chapter 19 / CLASSIFICATION AND MAJOR INSECT GROUPS

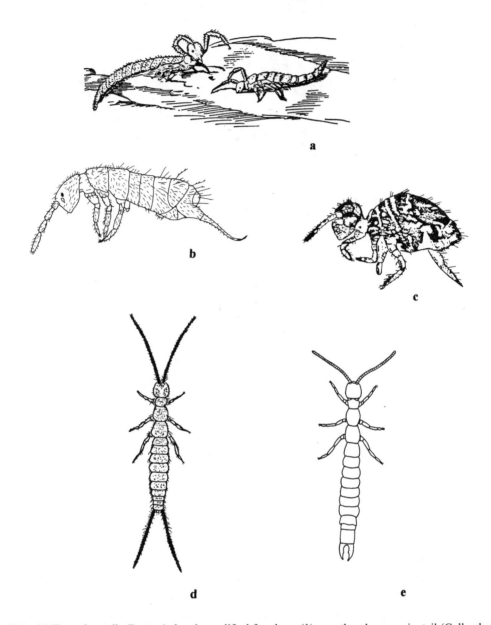

Fig. 19-1. (*a*) Two telsontails (Protura) showing uplifted fore legs; (*b*) an arthropleone springtail (Collembola) showing distinct abdominal segments; (*c*) a symphypleone springtail (Collembola) showing fused abdominal segments; (*d*) a campodeid (Diplura) with long, segmented cerci; (*e*) a japygid (Diplura) with caudal forceps (*a* is from Essig, 1942, after Berlese; *b* is modified and redrawn from Little, 1972, after Mills; *c* is modified and redrawn from Ross, 1965, after Mills; *d* and *e* are modified and redrawn from Essig, 1942).

the banks of ponds or streams, along the seashore, on the surfaces of quiet pools, on snow fields, within houses, cellars, greenhouses, or caverns, or within termite or ant nests. There are about 6,000 species, most of which are geographically widespread, and a few are cosmopolitan. They either scavenge on

Apterygota

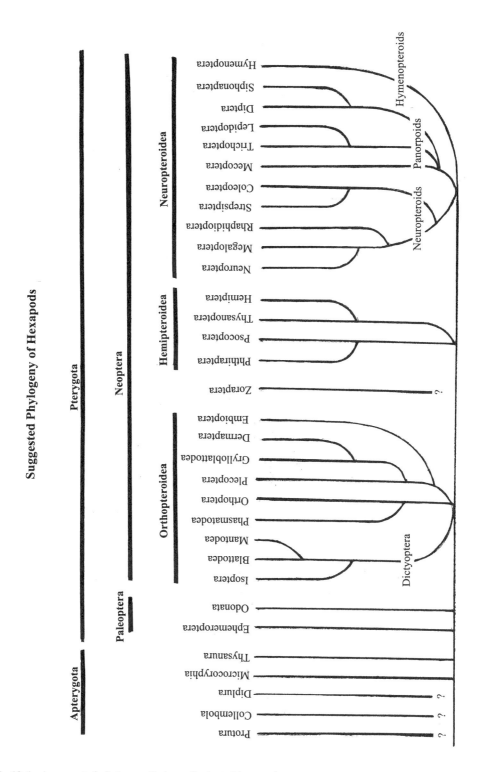

Table 19-1. A suggested phylogenetic tree of selected hexapods.

Chapter 19 / CLASSIFICATION AND MAJOR INSECT GROUPS

decaying organic material or eat leaves, flowers, or fungi with chewing, rarely sucking mouthparts. Some are pests of cultivated plants, and a few are predacious.

Springtails have an *ametabolous* (myriapod-like) development, a forked springing organ (*Fig. 2-19 b*) that allows them to jump, *segmental-type antennae* (shared with Diplura), an entognathous head (shared with Protura and Diplura), and a unique abdomen consisting of, at most, six segments either distinct in suborder *Arthropleona* (*Fig. 19-1 b*) or variably fused in suborder *Symphypleona* (*Fig. 19-1 c*). They are, however, more specialized than are those other apterygotes removing them from the immediate line of insect evolution.

Order Diplura
(Japygids and Campodeids)

Diplura (also called Entotrophi) are an obscure group of over 650 species of small, usually pale, soft-bodied, eyeless hexapods that frequent damp, sheltered places beneath logs, stones, or debris or live within caves. They scavenge and also take small prey with their entognathous biting mouthparts. They lack a tentorium and ocelli but have long bead-like segmental antennae. Their development is ametabolous. Their ten-segmented abdomen has paired styli like those of Symphyla on many segments and either a pair of long, segmented cerci in *Campodeidae* (*Fig. 19-1 d*) or unjointed forceps in *Japygidae* (*Fig. 19-1 e*). Their genitalia are vestigial. Diplura appear primitive, yet are specialized in their entognathous head and lack of eyes removing them from the immediate line of insect evolution. They were formerly placed within the combined order Thysanura *sens. lat.* but are readily distinguished inasmuch as they have two caudal abdominal processes as opposed to three in the combined order (now further separated into Microcoryphia and Thysanura).

Order Microcoryphia
(Bristletails)

Microcoryphia (also called Archeognatha) consist of about 350 species of small or moderate-sized, wingless, jumping hexapods with pigmented scales (*Fig. 19-2 a*). They have an *ectognathous head* (exposed mouthparts) with large compound eyes, ocelli, segmental-type antennae, primitive *unicondylic mandibles*, and conspicuous maxillary palpi. The abdomen bears numerous paired styli and three caudal appendages (one pair of short cerci and a relatively long median caudal filament). Bristletails are found mostly under logs, stones, bark, debris, or rock crevices from which they emerge nightly to eat algae, lichens, and plant debris. Their development is ametabolous, with molting continuing indefinitely even after sexual maturity. Previously grouped with Thysanura, bristletails are now placed within an order of their own based on their distinctive mandibles, well-developed eyes, and numerous abdominal styli.

Apterygota

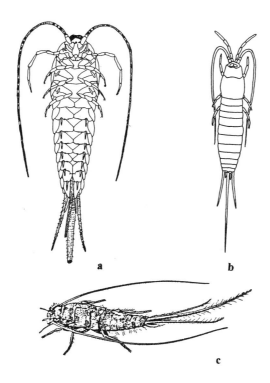

Fig. 19-2. (*a*) A jumping bristletail (Microcoryphia) showing long maxillary palpi between the fore legs and antennae and numerous abdominal styli; (*b*) a silverfish (Thysanura); and (*c*) the firebrat (Thysanura) (*a* is modified and redrawn from Comstock, 1948; *b* is modified and redrawn from Belkin,1972; *c* is by courtesy of Illinois Natural History Survey).

Thysanura
(Silverfish and Firebrats)

The approximate 370 species of order Thysanura (also called Zygentoma), the most advanced of the wingless orders of hexapods, probably qualify as primitive insects. They are small or moderate-sized, wingless creatures (*Fig. 19-2 b, c*) that superficially resemble Microcoryphia differing in possession of *dicondylic mandibles*, a tentorium, reduced compound eyes and no ocelli, *annulate antennae*, reduced numbers of abdominal styli, cerci sometimes nearly similar in length to the median caudal filament, and external genitalia. Their development is ametabolous, with molting continuing indefinitely even after sexual maturity. Many Thysanura hide under leaf litter, bark, or plant debris, often in moist environments. Others live in caves, in termite or ant nests, or are domiciliary. Their feeding varies from fungal-feeding to omnivorous, and the domestic species are prone to consume starchy materials.

269

Chapter 19 / CLASSIFICATION AND MAJOR INSECT GROUPS

Subclass Pterygota
Infraclass Paleoptera

Order Ephemeroptera
(Mayflies)

Mayflies are paleopterous insects of great antiquity, some from as far back as the Pennsylvanian. Their net-veined, triangular wings include a complete set of the primitive main veins and branches together with numerous intercalary veins. Modern mayflies consist of about 2,000 species of delicate, soft-bodied creatures associated with fresh, unpolluted lakes, ponds, and streams. Adults (*Fig. 19-3 a, b*) are non-feeding, large-eyed terrestrial insects with vestigial biting-type mouthparts. They are found flying in the vicinity of water. They usually have two unequal pairs of triangular-shaped, membranous wings held vertically at rest. The smaller hind wings are sometimes absent. Their ten-segmented abdomen has a pair of elongate, jointed cerci, usually a median caudal filament, and primitive external genitalia. The naiads (*Fig. 19-4 b*) are aquatic plant-feeders with abdominal tracheal gills. They undergo a hemimetabolous development that lasts up to three or four years. This development is unusual in that the final juvenile, or *subimago*, has functional wings and is able to fly to shore where it molts again into the winged, sexually mature adult which dies within days. This is only known case of a juvenile insect capable of flight.

Order Odonata
(Dragonflies and Damselflies)

Odonata include about 4,900 species of distinctive, medium- or large-sized Paleoptera with two pairs of elongate, membranous, net-veined wings having a costal *nodus* (incision) and a costal *stigma* (pigmented spot). The aquatic naiads live in lakes, ponds, and streams near which the strong-flying terrestrial adults patrol though sometimes far inland. All life stages are predacious. Adults have an ectognathous, maneuverable head with large compound eyes, three ocelli, short setaceous antennae, and strongly toothed biting-type mouthparts. Their enlarged meso- and metathoracic pleura displace the terga and wings backward and the sterna forward, bringing the legs close to the hypognathous mouthparts, facilitating predation. They consume Diptera, Lepidoptera, and other small flying insects found near water. Naiads use the modified, prehensile labium (*Fig. 19-4 c*) to catch worms, crustaceans, various insect juveniles, tadpoles, and even small fishes. They undergo a hemimetabolous development lasting for months to a year or more before crawling onto rocks or vegetation to molt again into sexually mature adults.

There are two main evolutionary lines of present-day Odonata: dragonflies of suborder *Anisoptera* (*Figs. 1-4 b, 19-3 c*) with unequal wings held horizontally at rest and stout, often flattened naiads with rectal gills (*Fig. 19-4 d*) and damselflies of suborder *Zygoptera* (*Fig. 19-3 d*) with subequal wings held

Pterygota / Paleoptera

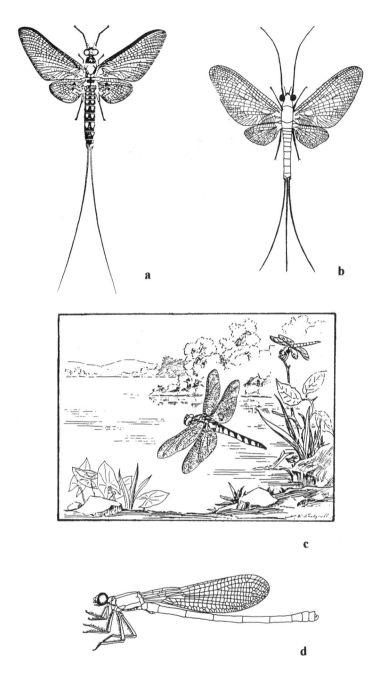

Fig. 19-3. (*a*) A mayfly adult (Ephemeroptera) lacking a median caudal filament; (*b*) a mayfly adult with a median caudal filament; (*c*) two dragonfly adults (Anisoptera) along a lake shore; and (*d*) a damselfly adult male (Zygoptera) (*a* is modified and redrawn from Borror *et al.*, 1989, after Needham & U. S. Bureau of Fisheries; *b* is modified and redrawn from Needham and other sources; *c* is after Snodgrass, 1967; *d* is modified and redrawn from Turtox Key Cards).

Chapter 19 / CLASSIFICATION AND MAJOR INSECT GROUPS

vertically at rest and slender naiads (*Figs. 13-1 c, 19-4 a*) with three caudal gills. The sex ducts are caudal as in other adult insects, but the male secondary genitalia are located forward on the third sternite in dragonflies and on the second in damselflies. This placement is responsible for the unusual copulatory behavior of odonates, discussed in Chapter 16.

Odonata are ancient and primitive. Some fossil damselflies date back to the Permian. Odonates' elongate, net-veined wings share many characters with the wings of other Paleoptera. They are probably closest to, and presumably evolved from, the extinct order Protodonata.

Subclass Pterygota
Infraclass Neoptera

The so-called "modern insects" of infraclass Neoptera have largely supplanted the archaic ones of infraclass Paleoptera. Neoptera are able to rotate the fore wings backward and fold them at specific articulations concealing the hind wings when at rest. This refinement allows them to hide within otherwise inaccessible microhabitats and makes way for subtle wing movements that result in a more efficient flight. They include distinct orthopteroid, hemipteroid, and neuropteroid lineages, to be discussed in the order listed.

Superorder Orthopteroidea
(Grasshoppers, Cockroaches, Katydids, Crickets, and Their Allies)

Orthopteroidea include several generalized insect orders characterized by hemimetabolous development, usually numerous Malpighian tubules, generalized biting-type mouthparts, cerci, and wings with many longitudinal veins and a well-developed jugal area. Wings first appear in orthopteroid development as external pads in the nymphal stage before becoming functional wings in the adult stage. The fore wings are sometimes leathery wing covers called *tegmina*, and the hind wings are usually membranous organs of flight (*Fig. 1-1 b*).

Orthopteroids have an ectognathous head with dicondylic, biting mandibles, paired compound eyes, usually two or three ocelli, annulate, multisegmented antennae, a well-developed thorax (often partly shielded beneath the prominent pronotum), legs with a several-segmented tarsus, paired claws, and a terminal arolium, wings as noted above, a ten-segmented abdomen with remnants of an eleventh, and well-developed cerci and external genitalia.

Pterygota / Orthopteroidea

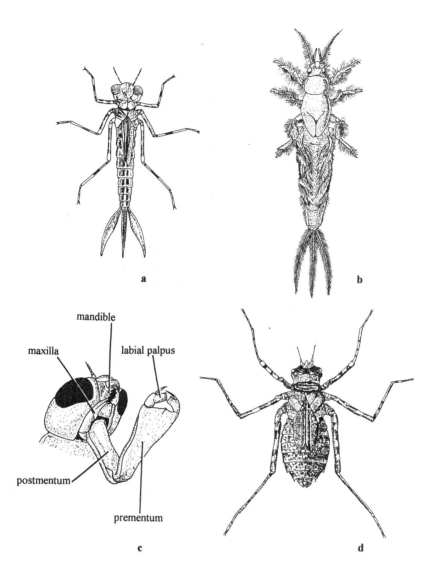

Fig. 19-4. (*a*) A damselfly naiad (Zygoptera); (*b*) a mayfly naiad (Ephemeroptera); (*c*) the head of a dragonfly naiad (Anisoptera) showing its prehensile mask; (*d*) a sprawling-type dragonfly naiad (Anisoptera) (*a* is modified and redrawn from Kennedy, 1962; *b* is modified and redrawn from Burks, 1953; *c* is modified and redrawn from Weber, 1933; *d* is modified and redrawn from Kennedy, 1916).

Order Blattodea or Blattaria (Cockroaches)

This familiar order includes cockroaches (*Fig. 19-5 a*). They are small to large, flattened, rapid-running, nocturnally active, omnivorous orthopteroids of largely tropical or subtropical distribution though some extend throughout the world. They have a leathery body, a shield-like pronotum obscuring an

Chapter 19 / CLASSIFICATION AND MAJOR INSECT GROUPS

opisthognathous (mouthparts directed backward) *head*, usually two ocelli, elongate, multisegmented antennae, short, several-segmented cerci, and an inconspicuous ovipositor. They place their eggs in packets, or *oothecae*, that are deposited in debris or carried about by the female, sometimes internally. Some species are short-winged or wingless. Cockroaches number about 4,000 species, most of which are wild, living under vegetation or within leaf litter or debris, but others are domestic, being carried around the world by commerce.

Cockroaches are often aggregated with mantises and sometimes with termites into the combined order *Dictyoptera* characterized by their mutual lack of jumping hind legs as opposed to the saltatorial orthopteroids discussed below.

Order Mantodea
(Mantises)

The approximately 1,500 species of praying mantises (*Fig. 19-5 b*) are generally elongate creatures of concealing form and coloration. They have a freely mobile, hypognathous head and are either winged or wingless. Like cockroaches, they lay their eggs in oothecae. Mantises are striking in appearance owing to their raptorial, prey-seizing fore legs usually held in an uplifted position enabling them to strike at and catch suitable prey near enough for capture. They inhabit shrubs and other low vegetation where, usually motionlessly, they await the approach of suitable prey. They are voracious feeders on prey insects of appropriate size, not too large, not too heavily armored. Like cockroaches, mantises are largely tropical or subtropical in distribution though some occur in temperate zones.

Order Isoptera
(Termites)

Isoptera include about 1,900 species of generalized social insects found chiefly in tropical or subtropical regions where, usually by the thousands, they hollow out and live within communal galleries in buried or exposed logs, living trees, or in underground and above-ground nests. Termites are *polymorphic*, usually having two soft-bodied, pale, usually sterile, eyeless castes, soldiers and workers (*Fig. 17-4 b, c, f*), as well as sclerotized, pigmented, primary reproductives with well-developed eyes (*Figs. 17-4 d, e, 19-7 e*). Their antennae are bead-like and their head prognathous with biting-type mouthparts, the latter enlarged in the soldier caste of many species (*Fig. 17-4 b*) but atrophied in nasute species (*Fig. 17-4 f*) that use chemical defenses. The primary reproductives' two pairs of wings are similar, elongate, membranous, held flat over the body at rest, and, following mating, are shed at *basal sutures*. They have a reduced number of forked longitudinal veins and no cross veins. Soldiers and workers are wingless. Development is hemimetabolous and slow, lasting over a year. The primary reproductives tend to be long-lived once they pair-bond as king and queen.

The earliest termites may have been solitary but more likely were either subsocial with habits similar to those of certain present-day cockroaches or primitively social with habits like those of one

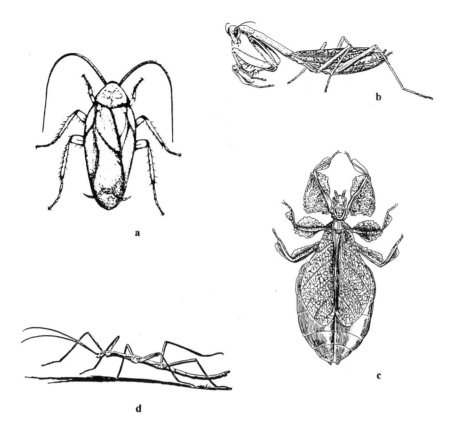

Fig. 19-5. (*a*) The American cockroach (Blattodea); (*b*) the Carolina mantis (Mantodea); (*c*) a leaf insect (Phasmatodea); (*d*) the common walkingstick (Phasmatodea) of northeastern United States (*a* is by courtesy of M. Dababneh; *b* and *d* are by courtesy of Illinois Natural History Survey; *c* is modified and redrawn from Sharp, 1895).

living Australian termite genus. There is even less certainty about when termites arose. Their fossil record goes back only to the Cretaceous though the order is assuredly older.

Order Phasmatodea
(Walkingsticks and Leaf Insects)

Phasmatodea are slow-moving, often large insects with both winged and wingless species. Our native species are almost entirely wingless. They have evolved into two major body forms: walkingsticks (*Fig. 19-5 d*) which are elongate, stick-like creatures represented by a number of North American species and leaf insects (*Fig. 19-5 c*) which are flattened, leaf-like forms chiefly of the Oriental tropics. Both are arboreal or woody plant foliage-feeders, especially at night. They are green or brown in color and of concealing form and behavior. The distinctly shaped eggs that the females drop resemble seeds. The insects themselves often fall to the ground and feign death when disturbed.

Chapter 19 / CLASSIFICATION AND MAJOR INSECT GROUPS

Order Orthoptera
(Grasshoppers, Locusts, Katydids, and Crickets)

Orthoptera may be classified either as short-horned species (suborder *Caelifera*) or as long-horned species (suborder *Ensifera*), each of which includes a number of superfamilies, as follows.

The Caeliferan superfamily *Acridoidea* includes the familiar grasshoppers, plague locusts, and their allies (*Fig. 19-6 a, b*). Grasshoppers are plant-feeders, chiefly of grassland and scrub, and some are highly destructive to agriculture. The term *locust* refers to perhaps 20 of their species (*Fig. 17-2 a, b*) that exist in two phases, solitary and gregarious, as well as intermediates, differing in behavior, color, shape of pronotum, wing length, *etc.*, as discussed in Chapter 17. Locusts can aggregate in immense swarms, migrate great distances, and inflict massive damage on native and cultivated vegetation.

The Caeliferan superfamily *Tetrigoidea* includes the pygmy grasshoppers or grouse locusts (*Fig. 19-6 c*), as they are popularly known. They may be recognized by their enlarged pronotum projecting backward over the abdomen. Many of them frequent moist environments where they may be found on the luxuriant mosses and algae that constitute their main food.

The Ensiferan superfamily *Tettigonioidea* consist of katydids (*Fig. 19-6 f*). They are often green in color, tree- or shrub-dwelling (accounting for the name *bush cricket* sometimes given them), and plant-feeding in habit, with some exceptions. Their ovipositor is a sword-like, laterally flattened structure, either straight or curved, and sometimes as long as or longer than the body itself.

Superfamily *Gryllacridoidea* are relatively primitive, cricket-like Ensifera whose ovipositor is laterally compressed rather than needle-like, as in true crickets. Camel crickets or cave crickets (*Fig. 19-6 d*), Jerusalem crickets or sand crickets (*Fig. 19-6 e*), the wetas of Australasia, and some other insects are included within the superfamily. They tend to be large-bodied insects found beneath rotting logs, under stones, in basements, or in other humid situations, and some are subterranean. Their food-habits are varied, but many are predacious or omnivorous.

The Ensiferan superfamily *Grylloidea* includes field crickets and tree crickets. Field crickets (*Fig. 19-7 a*) and certain other nocturnally active ground crickets frequent open fields, thickets, or woodland, but several species, including the familiar house cricket, are domestic. They are mostly omnivorous. The so-called tree crickets (*Fig. 19-7 b*) perch above ground on herbs, shrubs, or trees where they are either plant-feeders or predators.

The Ensiferan superfamily *Gryllotalpoidea*, or mole crickets (*Fig. 19-7 c*), are subterranean burrowers with shovel-like fore legs used to dig in loose soil.

Order Grylloblattodea
(Rock Crawlers)

Rock crawlers (*Fig. 19-7 d*) constitute about 20 species of wingless, nocturnally active scavengers that share a number of primitive blattodean and orthopteroid attributes as well as features peculiar to themselves. They are elongate creatures with long filiform antennae, reduced compound eyes, and no

Pterygota / Orthopteroidea

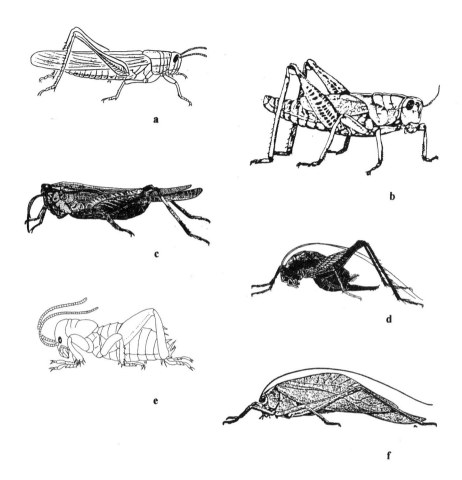

Fig. 19-6. (*a*) A long-winged North American grasshopper (Acridoidea) related to certain Old World plague locusts; (*b*) a short-winged lubber grasshopper (Acridoidea); (*c*) a grouse locust or pygmy grasshopper (Tetrigoidea); (*d*) a camel cricket (Gryllacridoidea); (*e*) a Jerusalem cricket or sand cricket (Gryllacridoidea); and (*f*) a katydid (Tettigonioidea) (*a* is modified and redrawn from Comstock, 1948, after Riley; *b* is by courtesy of M. Dababneh; *c*, *d*, and *f* are by courtesy of Illinois Natural History Survey; and *e* is modified and redrawn from Essig, 1942).

ocelli. They are restricted to the temperate Holarctic Realm where they inhabit cold, rocky situations such as talus slopes and the edges of glaciers, often at high elevations. Their origin is controversial. Some authorities regard them as survivors of a primitive protorthopteran stock, but others feel their affinities are with such specialized orthopteroid groups as Dermaptera.

Order Dermaptera
(Earwigs)

Earwigs include about 1,100 species of specialized, nocturnally active insects that derive their common name from the erroneous belief that they enter the human ear. They are, in fact, harmless creatures with

277

Chapter 19 / CLASSIFICATION AND MAJOR INSECT GROUPS

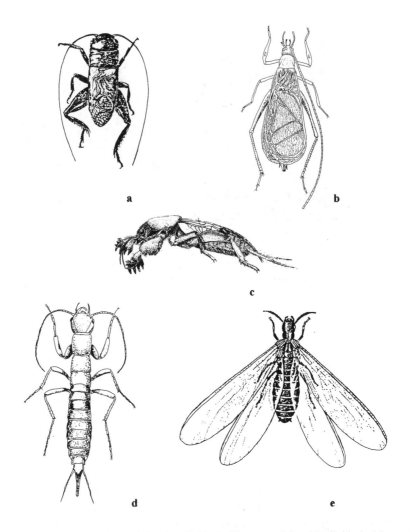

Fig. 19-7. (*a*) A common Old World field cricket (Grylloidea); (*b*) a tree cricket (Grylloidea); (*c*) a mole cricket (Gryllotalpoidea); (*d*) a rock crawler (Grylloblattodea); (*e*) a winged primary reproductive termite (Isoptera) (*a* is after Chopard, 1951; *b* is modified and redrawn from Essig, 1942, after Smith; *c* is by courtesy of Illinois Natural History Survey; *d* is modified and redrawn from CSIRO, 1970; *e* is modified and redrawn from Elzinga, 1981)

little tendency to interact with humans. They are small- to medium-sized, elongate, hard-bodied, dark colored, either winged or wingless, terrestrial insects (*Fig. 19-8 a, b*) with long, multisegmented antennae, prognathous biting mouthparts, and stout, opposable caudal *forceps* (modified cerci). Wings (when present) include truncate, veinless tegmina and large, semicircular flight wings capable of being pleated like a fan, folded twice on themselves, and stowed beneath the abbreviated tegmina. Earwigs superficially resemble the unrelated rove beetles (*Fig. 19-14 e*) except for their caudal forceps, found in only one other group, the dipluran family Japygidae. Earwigs occur throughout much of the world but are particularly common in the tropics. They frequent damp habitats, and many of them hide during the day under logs, bark, leaf litter, or debris, in crannies in the ground, or within plants but

Pterygota / Orthopteroidea

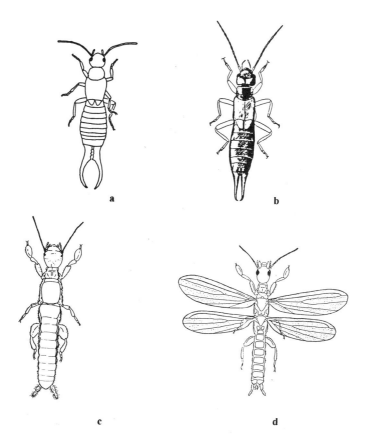

Fig. 19-8. (*a*) A male earwig (Dermaptera); (*b*) a female earwig; (*c*) a female webspinner (Embioptera); (*d*) a male webspinner (*a* is modified and redrawn from Belkin, 1972; *b* is after Fulton, 1924; *c* and *d* are modified and redrawn from Essig, 1942, after Ross).

emerge at night to forage. They are omnivorous except for a few plant-feeders or predators. A few aberrant species belonging to two suborders are wingless, viviparous, blind or mostly blind mammalian ectoparasites. Earwig development is hemimetabolous. Females of some species deposit their eggs in a brood chamber usually within the ground and give maternal care to the newly hatched young (*Fig. 17-1 d*).

Dermaptera are more specialized than are the other representatives of superorder Orthopteroidea in terms of their forceps, their shortened, veinless tegmina, their expanded, radially pleated, folding flight wings, their subsocial behavior, and their inclusion of a few mammalian ectoparasites, all of which bespeaks advanced status. They probably evolved separately and long ago from the main orthopteroid stem. Their earliest fossils are Jurassic.

Chapter 19 / **CLASSIFICATION AND MAJOR INSECT GROUPS**

Order Plecoptera
(Stoneflies)

The approximately 1,550 species of Plecoptera are mostly medium-sized, elongate, somewhat flattened insects with long multisegmented antennae and cerci (*Fig. 19-9 b*). Their development is hemimetabolous, lasting a year or more, and involves numerous molts in aquatic habitats, with the final molt to adult on land, often during the cold winter months. They differ from most other insects in their predilection for cool temperate or arctic zones where they live in association with rapidly flowing, stony-bottomed, fresh-water streams and lakes. Like mayflies, stoneflies are good ecological indicators of water condition, their presence indicating relatively unpolluted conditions. Stonefly naiads (*Figs. 13-1 b, 19-9 a*) are superficially similar to mayfly naiads but differ in having thoracic gills at the leg bases rather than on the abdomen and their lack of the median caudal filament of most mayflies. They are found beneath or between rocks or under debris in cool, rapid streams or wave-washed lakes where they use biting-type mouthparts to scavenge on organic material, prey on various aquatic invertebrates, or eat algae and other vegetation. Stonefly adults (*Fig. 19-9 b*) have two pairs of generalized, often net-veined, membranous wings used to fly feebly near water. These wings fold flat over the abdomen at rest. The anal lobe of the hind wing is greatly expanded and folded. Adults perch on exposed rocks, logs, tree limbs, bridges, and nearby objects. They range from plant-feeders that eat algae to non-feeders with reduced mouthparts.

Though superficially similar to mayflies, stoneflies are unrelated to them. They differ chiefly in their more advanced, folding wings. They are, in fact, closest to Orthoptera from which they differ in venation, lack of tegmina, lack of expanded hind femora, undeveloped ovipositor, and the aquatic habit of their juveniles.

Embioptera or Embiidina
(Webspinners)

The approximately 150 species of webspinners (*Fig. 19-8 c, d*) are curious, small-bodied, pale, secretive tropical or subtropical terrestrial insects that occupy silken tunnels woven under loose bark or stones, within moss or debris, or in the ground. They are seldom seen because of their small size, reduced numbers, and nocturnal habits. They live a casteless, subsocial existence in groups of a dozen or more individuals within interconnected silken tunnels that they construct using glands located in the swollen first tarsal segment. All stages spin, crossing and recrossing the fore tarsi, and all move about busily, backward and forward, within the tunnel system. Both nymphal and adult female webspinners eat dried herbs, mosses, lichens, and other vegetation, but adult males apparently do not eat. They die soon after mating.

Among webspinners' primitive characteristics are their elongate body, subequal thoracic segments, subequal wings, and generalized internal organs. Among their advanced features are their weakly sclerotized integument, female flightlessness, unique tarsal spinning organs, enlarged leaping-type hind

Pterygota / Orthopteroidea

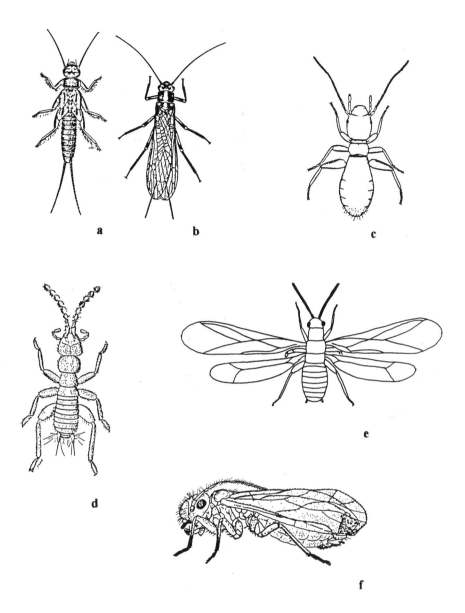

Fig. 19-9. (*a*) A stonefly naiad (Plecoptera); (*b*) a stonefly adult; (*c*) a book louse (Psocoptera); (*d*) a wingless zorapteran (Zoraptera); (*e*) a winged zorapteran; (*f*) a bark louse (Psocoptera) (*a* and *b* are by courtesy of Illinois Natural History Survey; *c* is modified and redrawn from Essig, 1942; *d* is modified and redrawn from Essig, 1942, after Silvestri; *e* is modified and redrawn from Belkin, 1972; *f* is modified and redrawn from Ross, 1982, after Sommerman).

femora, maternal care, and internal wing development until the last juvenile stage (as in holometabolous pupae). These attributes, together with webspinners' widespread pantropical distribution, suggest that they are an isolated branch from the main line of superorder Orthopteroidea. Their fossils are recorded from the Permian Period.

281

Chapter 19 / CLASSIFICATION AND MAJOR INSECT GROUPS

Having dealt with selected major taxa of superorder Orthopteroidea, one can now turn to their phylogeny which is notable because of the antiquity of many of their representatives. Orthopteroids were among the first abundant Neoptera to have arisen, having appeared as long ago as the Paleozoic. They could have stemmed from archaic ancestral forms, one line of which was characterized by adaptations for concealment within plant debris, branches, bark, *etc*. That dictyopterine line probably gave rise to cockroaches and later to mantises and termites. A second line of jumping forms (Saltatoria) adapted for life in more open habitats probably gave rise to cricket-like, long-horned insects and then to short-horned, grasshopper-like forms and phasmids. A final line probably evolved into Grylloblattodea, Plecoptera, Dermaptera, and their relatives.

Order Zoraptera
(Zorapterans)

Several dozen mostly tropical species within seven genera constitute the order Zoraptera, so-called because, when originally discovered, these minute, soft-bodied, usually pale, secretive insects were incorrectly assumed to be completely wingless. They occur in aggregations of a few to perhaps a hundred individuals under bark, within rotting sawdust or wood chips, or in humus or debris. Little is known of their biology, but presumably they eat fungal spores and mycelia and scavenge on small, dead insects and mites. They include both winged (*Fig. 19-9 e*) and wingless (*Fig. 19-9 d*) individuals, of which the former, like termites, shed their wings at a basal suture. This polymorphism, their gregarious habits, and their superficial resemblance to termites suggest that they are social insects. In fact, they exhibit neither division of labor nor care of the young, so their assemblages cannot be social.

Zorapterans have generalized hypognathous biting mouthparts, nine-segmented, bead-like antennae, subequal thoracic segments, and short, one-segmented cerci. They exhibit a mixture of blattoid and hemipteroid features but are generally considered to show blattoid affinities.

Subclass Pterygota
Infraclass Neoptera
Superorder Hemipteroidea

Representatives of the neopterous superorder Hemipteroidea were presumably next to arise following the orthopteroids. Hemipteroids, like orthopteroids, fold the wings over the abdomen at rest but differ in their reduced or lost jugal area, their reduced venation, their few Malpighian tubules, and their lack of cerci. Their development is generally hemimetabolous though a few of their taxa have evolved a rudimentary holometaboly. Above all, they tend toward development of suctorial mouthparts. The orders can be arranged sequentially from Psocoptera which chew solid food using modified biting mouthparts, through Thysanoptera with an intermediate mouthpart adaptation, and to Hemiptera with a well-developed suctorial proboscis for imbibing liquids.

Pterygota / Hemipteroidea

Order Psocoptera
(Booklice and Barklice)

The approximately 4,500 species of Psocoptera are small-sized, pale or protectively colored, soft-bodied, stout or flattened terrestrial hemipteroids with modified biting-type mouthparts featuring modified laciniae and reduced labial palpi. They are either winged or wingless. The winged forms have two pairs of membranous wings held roof-like over the abdomen at rest, the fore pair usually with a costal stigma. Venation is reduced to a few main veins and a few cross veins. The hind femora are sometimes enlarged. Cerci are lacking. Most of these psocids are solitary, but some gregarious individuals form small groups under silken tents or out in the open. Those psocids called barklice (*Fig. 19-9 f*) live out-of-doors on or beneath bark or vegetation, rocks, walls, or fences, or within mammal or bird nests. They usually eat molds, fungi, algae, and lichens though some scavenge on dead animals. Those psocids called booklice (*Fig. 19-9 c*) live indoors in barns, houses, libraries, and stores where they scavenge on organic debris or dead insects and eat cereals, glue, paste, and other starchy materials. Their gnawing habits sometimes pose a minor economic problem where books are kept or grain is stored.

Psocids are primitive general-feeding hemipteroids with known Cenozoic and Cretaceous amber fossils (some once-attributed Permian fossils are no longer considered psocopteran). Their slender, chisel-like laciniae anticipate the suctorial mouthparts of other hemipteroids which suggests that a gnawing group ancestral to Psocoptera could have given rise to the chewing lice and then to the sucking lice.

Order Phthiraptera
(Lice)

Phthiraptera include about 4,000 small, wingless, ectoparasitic hemipteroids commonly known as lice. Among them are the chewing, biting, or bird lice of suborder *Mallophaga* (*Fig. 19-10 c*) consisting of three major groups and the sucking lice of suborder *Anoplura* (*Fig. 19-10 a, b*) consisting of a single group. Mallophaga have a prominent head larger than the thorax. The thorax consists of a small prothorax and imperfectly separated meso- and metathoracic segments fused with the abdomen lending to the body an ovoid or slender, dorsoventrally flattened overall appearance. Their legs are short and stout, the fore pair kept close to the head for feeding and the mid and hind pairs extended behind for walking and clinging. Visual organs are reduced or absent, and the mouthparts are generally strongly modified chewing-type. Development is hemimetabolous, the ectoparasitic juveniles hatching within days from eggs glued onto the host's feather or hair bases. Some Mallophaga are host-specific, some genus-specific, some parasitize a given host family, and a few parasitize an entire avian order. Most are organic debris-feeders that gnaw feathers, scales, skin, dried blood, or hair (in the case of those few that parasitize mammals). Being wingless and with reduced powers of locomotion, biting lice die if unable to reestablish contact with the proper host. They rely for dispersal mostly on body contact between host individuals but also on the common use of bird nests. One genus found on ducks consists of lice that swim freely from one aquatic bird to another.

Chapter 19 / CLASSIFICATION AND MAJOR INSECT GROUPS

Mallophaga likely evolved from a psocid-like ancestor that continued chewing as it took on an ectoparasitic mode of life. Once this feeding relationship was established, the lineage became influenced by the hosts' evolution. No fossils are known. Certain elephant and wart hog parasites belonging to one group run like free-living insects and have reduced mouthparts at the tip of a weevil-like snout and a pharyngeal pump for aspirating blood. They seem to bridge the gap between Mallophaga and Anoplura and were responsible for prompting most authorities to combine the two formerly separate orders into the single order Phthiraptera recognized herein.

The approximately 500 ectoparasitic hemipteroids of suborder Anoplura have a more intimate host relationship, blood-sucking, than that of most Mallophaga and so are known as sucking lice. Anoplura are much like Mallophaga, differing chiefly in their suctorial mouthparts, relatively narrow head, and intimately fused thoracic segments. They are pale, wingless creatures with a small head narrower than the thorax, a depressed, elongate (*Fig. 19-10 a*) or crab-like (*Fig. 19-10 b*) body, and a tough, leathery integument. Their eyes are reduced or absent, and their segmented antennae are short. Development is hemimetabolous, with the nits, or eggs, being attached individually to the hair base from which they hatch within a week or so into ectoparasitic nymphs. The retractable mouthparts are strongly modified for piercing and sucking, the maxillae being fused into a dorsal food channel, the hypopharynx into an intermediate salivary channel, and the trough-like labium a ventral piercing stylet with recurved hooks for grasping the host's skin; mandibles are lacking. The intimately fused thoracic segments are provided with short, stout legs adapted for clinging to hair. The mostly host-specific Anoplura consist exclusively of mammalian parasites that die within a short time if removed from the host or if it dies.

Sucking lice are vectors of numerous diseases including relapsing fever, trench fever, and epidemic typhus, all transmitted by the body louse (*Fig. 19-10 a*). Through the ages, millions of people exposed to crowded conditions and poor sanitation have become sickened and died thereby. The related head louse poses a continuing problem today even in developed countries where it is spread by direct contact among individuals. The intense itching and irritation caused by head lice may result in serious dermatological problems and anemia as well as other diseases.

Suborder Anoplura seem to be a recent taxon that evolved in unison with their host mammals. They probably arose, along with suborder Mallophaga, from a psocopteran-like ancestral form.

Order Thysanoptera (Thrips)

The roughly 5,000 species of Thysanoptera are curious terrestrial hemipteroids known as thrips (same word, sing. or pl.). They are small, slender, soft-bodied, usually dark insects that are either winged or wingless. The former have long, strap-like wings with few or no veins but long marginal hairs (*Fig. 19-10 d*). Thrips' short legs bear a diagnostic, eversible bladder-like vesicle that enables them to walk on most surfaces. Female thrips lay eggs with either a saw-like or a chute-like ovipositor from the base of the tube-like tenth segment. Thrips occur throughout the world, often on flowers and grasses but also on leaves of trees, with many species breeding under bark or on dead leaves. They use their

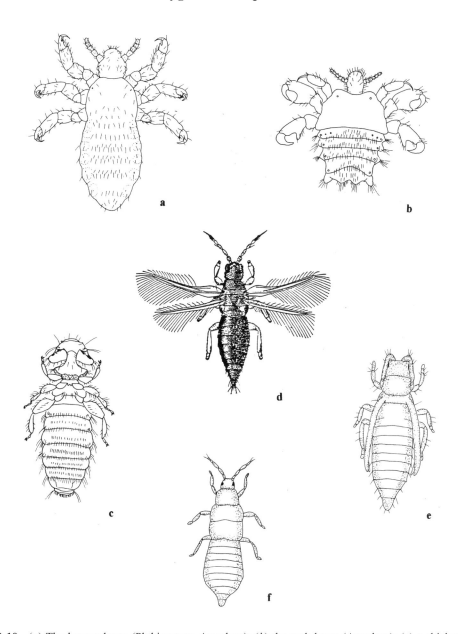

Fig. 19-10. (*a*) The human louse (Phthiraptera: Anoplura); (*b*) the crab louse (Anoplura); (*c*) a chicken body louse (Phthiraptera: Mallophaga); (*d*) a common greenhouse thrips (Thysanoptera); (*e*) a pupa and (*f*) a second larva of the citrus thrips (Thysanoptera) (*a* is modified and redrawn from Essig, 1942; *b* and *c* are modified and redrawn from USDA; *d* is modified and redrawn from Essig, 1942, after Russell; *e* and *f* are modified and redrawn from Comstock, 1948).

asymmetrical sucking mouthparts to imbibe sap from living plants except for the few that prey on small insects. About half of the species eat only fungal hyphae or spores. Some species damage cereals, vegetables, fruit trees, and other crops by their feeding and/or oviposition habits.

Chapter 19 / CLASSIFICATION AND MAJOR INSECT GROUPS

Thrips were long regarded as hemimetabolous. Now, however, it is recognized that they undergo a rudimentary holometabolism. The first two instars are active, feeding larvae (*Fig. 19-10 f*) that resemble adult thrips except for the reduced number of antennal articles and lack of wings. The next two or three instars are resting, non-feeding pupae (*Fig. 19-10 e*) with external wing buds. A few species produce a pupal cocoon.

Thrips have long been in existence based on their Permian fossils. Their body features are distinctive, yet unmistakably hemipteroid and suggestive of an hemipteroid derivation.

Order Hemiptera
(Cicadas, Hoppers, Aphids, Scale Insects, Bugs, and Their Allies)

The approximately 90,000 species of bugs and their allies were included within two distinct orders, Homoptera and Hemiptera, until recently. Most present-day workers classify these insects within the combined order Hemiptera now including at least the following suborders: *Sternorrhyncha, Auchenorrhyncha, Coleorrhyncha,* and *Heteroptera.*

Hemiptera are named after the "half" character of the fore wings of Heteroptera. They are hardened at the basal half but membranous and provided with a characteristic venation at the apical half (*Fig. 19-12 g*). The hind wings, when present, are membranous and folded flat beneath the fore pair at rest.

Hemiptera range from small to large. They have suctorial mouthparts consisting of two pairs of piercing stylets (mandibles and maxillae) enclosed within a flexible, segmented labial sheath (*Figs. 2-12 a, 2-13 a*). They have two pairs of wings, the fore pair of which is usually harder and less membranous than the hind pair, and cerci that are either reduced or absent. Development is usually hemimetabolous. All Homoptera and many Heteroptera are terrestrial except for certain Heteroptera adapted for an aquatic existence. All Homoptera suck plant juices, as do many Heteroptera except for certain predacious bugs.

Despite their similarities, the two formerly recognized suborders of Hemiptera are readily separated. Homoptera have uniformly textured fore wings, an opisthognathous head, a well-developed tentorium, and, at most, a membranous gula. Heteroptera, in contrast, generally have hemelytra, a hypognathous head with a sclerotized gula bridging the underside behind the beak, and they lack a tentorium.

The homopteran suborder Auchenorrhyncha includes such diverse terrestrial hemipteroids as cicadas or Cicadidae (*Fig. 19-11 a*), treehoppers or Membracidae, planthoppers or Fulgoridae (*Fig. 19-11 b*), leafhoppers or Cicadellidae (*Fig. 19-11 c*), and spittlebugs or Cercopidae. The homopteran suborder Sternorrhyncha includes plantlice, aphids, or Aphididae (*Fig. 19-11 d, e*), whiteflies or Aleyrodidae, and scale insects or Coccoidea (*Fig. 19-11 f, g*). So heterogeneous are the Auchenorrhyncha and Sternorrhyncha that it is difficult to characterize them. The best that can be said is that they are often small, relatively soft-bodied insects with opisthognathous piercing-sucking mouthparts that arise behind the head and extend backward between the fore and mid coxae. They usually have two pairs of wings capable of being folded, roof-like, over the abdomen, the fore pair of which is either leathery or membranous and uniformly textured, and the hind pair is membranous.

Pterygota / Hemipteroidea

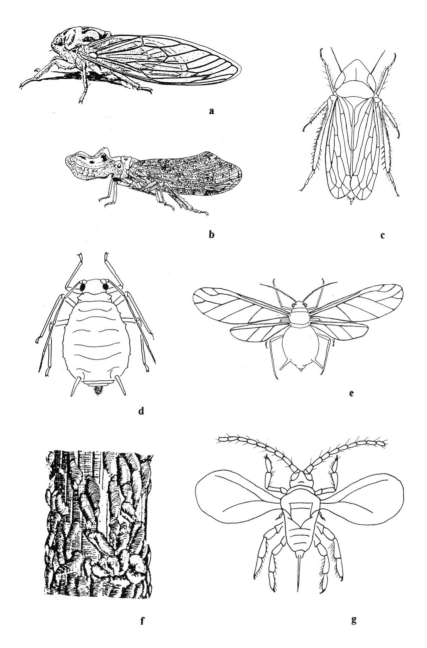

Fig. 19-11. (*a*) A common cicada (Cicadidae); (*b*) a lanternfly (Fulgoridae); (*c*) a leafhopper (Cicadellidae); (*d*) a wingless and (*e*) a winged aphid (Aphididae); (*f*) the oystershell scale insect (Coccoidea); (*g*) a male San Jose scale insect (Coccoidea) (*a* is by courtesy of Illinois Natural History Survey; *b* is modified and redrawn from Comstock, 1940; *c* is modified and redrawn from Essig, 1942; *d* and *g* are modified and redrawn from *Turtox Key Cards*; *e* is modified and redrawn from Belkin, 1972; *f* is after USDA, 1952).

Chapter 19 / CLASSIFICATION AND MAJOR INSECT GROUPS

Venation varies from complex, with many closed cells, to simple, with only a few sclerotized veins. Color ranges from pale green, yellow, or beige, to brightly colored and strikingly patterned. Certain cicadas and the bizarre tropical lanternflies (*Fig. 19-11 b*) are large. Female scale insects (*Fig. 19-11 f*) are wingless, legless creatures often without antennae. They live an attached existence beneath a resinous or mealy secretion or under a scale of cast skins and wax. Their male counterparts (*Fig. 19-11 g*) are delicate, gnat-like insects without functional mouthparts but with four "eyes," long antennae, an abdomen terminating in a unpaired style-like process, and a single mesothoracic pair of wings. Development is hemimetabolous except in certain groups with a rudimentary holometaboly.

Adult male scale insects lack functional mouthparts. They obviously do not eat. The remaining Homoptera are exclusively plant-feeders using their elongate beak to suck sap from the foliage, stems, or sometimes roots of herbs, shrubs, or trees. Many are serious agricultural pests because of the damage they inflict by feeding, by ovipositing, or by transmitting viral and other plant diseases.

The suborder Heteroptera, often referred to as *true bugs*, are distinguished from other insects for which the layman uses the word *bug* inappropriately. The pronotum is enlarged, and a triangular-shaped scutellum overlaps the anterior part of the abdomen between the wing bases. The scent glands of adults commonly vent through paired apertures near the hind coxae and those of nymphs through dorsal abdominal apertures. Development is hemimetabolous. The nymphs resemble adults except for their scent gland apertures and lack of functional wings.

Heteroptera consist of about 38,000 species widely distributed throughout the world's terrestrial and fresh-water environments except polar zones, being abundant in both tropical and temperate regions. A few are marine. Among the fresh-water Heteroptera are giant water bugs or Belostomatidae (*Fig. 19-12 c*), water boatmen or Corixidae (*Fig. 19-12 a*), and backswimmers or Notonectidae (*Fig. 19-12 b*). Among the terrestrial Heteroptera are plant bugs or Miridae (*Fig. 19-12 d*), assassin bugs or Reduviidae (*Fig. 19-12 e*), leaf-footed bugs or Coreidae (*Fig. 19-12 f*), stink bugs or Pentatomidae (*Fig. 19-12 g*), and the ectoparasitic bed bugs or Cimicidae (*Fig. 19-12 h*).

Most bugs use their hypognathous piercing-sucking mouthparts to take sap from wild or cultivated vegetation, and some are highly injurious to crops. Other bugs are predacious, for example, the giant water bugs and certain other bugs that use raptorial fore legs to capture and hold small insects and other appropriate-sized prey. A few bugs, including bed bugs and bat bugs, are parasitic on warm-blooded vertebrates.

As noted, an ancestral "homopteran" line presumably gave rise to modern Sternorrhyncha, Auchenorrhyncha, Coleorrhyncha, and Heteroptera. These four suborders are so closely related as to merit combination into the single order Hemiptera, as adopted herein. The Coleorrhyncha have a primitive austral disjunct distribution in New Zealand, Australia, and the southern tip of South America. Their primitive structure combines features of both suborders and even includes paranota with wing-like tracheation.

Pterygota / Hemipteroidea

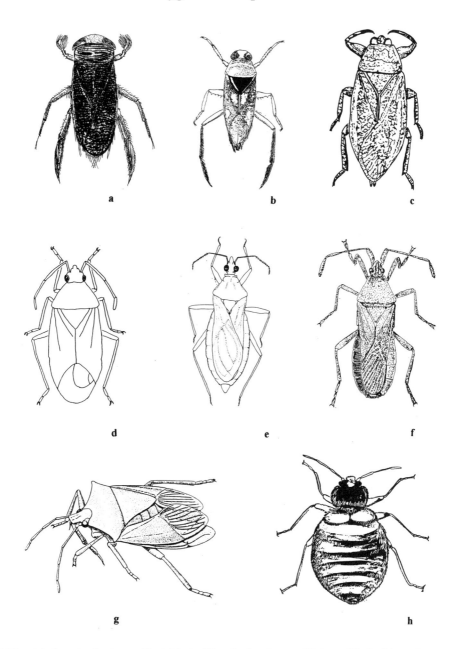

Fig. 19-12. (*a*) A water boatman (Corixidae); (*b*) a backswimmer (Notonectidae); (*c*) a giant water bug (Belostomatidae); (*d*) a leaf bug or plant bug (Miridae); (*e*) an assassin bug (Reduviidae); (*f*) the squash bug (Coreidae); (*g*) a stink bug (Pentatomidae) with the left flight wing partly exposed from beneath the left hemelytron; (*h*) the common bed bug (Cimicidae) (*a* is modified and redrawn from Usinger, 1974; *b* is modified and redrawn from Viedma *et al.*, 1984; *c* is after Comstock, 1940; *d* is modified and redrawn from Romoser, 1981; *e* is modified and redrawn from Kettle, 1995, after Schofield & Polling; *f* is redrawn from USDA, 1952; *g* is by courtesy of Illinois Natural History Survey; and *h* is redrawn from various sources).

Chapter 19 / CLASSIFICATION AND MAJOR INSECT GROUPS

Subclass Pterygota
Infraclass Neoptera
Superorder Neuropteroidea or Endopterygota

The final stage in insect phylogeny may have involved replacement of hemimetaboly by holometaboly. Insects with this more advanced development proceed through distinct egg, larval, pupal, and adult stages. The larva is essentially a feeding and growing stage often characterized by reduced eyes, antennae, and body sclerotization. Its suppression of adult characteristics imparts to it a complete dissimilarity from the structure, physiology, and behavior of the adult. It may even occupy a different environmental niche and use different kinds of mouthparts and a different digestive tract to take advantage of foods other than those that adults eat. The adult, in contrast, is a dispersal and reproductive stage that may consume foods different from those eaten by the larva and occasionally does not even eat. So unlike are the two stages that an apparently quiescent pupa which, however, is physiologically quite active intervenes toward accomplishment of the necessitated internal and external body reorganization. Once accomplished, this changeover gives the species access to two different ways of life and eliminates any juvenile-adult competition for food and space.

Those advanced insects that have availed themselves of holometaboly constitute the final superorder, Neuropteroidea. Neuropteroids are a great polyphyletic assemblage with holometabolous development, usually four or fewer Malpighian tubules, and wings with a reduced jugal area. They include three separate branches: neuropteroid *sens. str.*, panorpoid, and hymenopteroid, each of which is sometimes regarded as a superorder of its own.

Order Neuroptera
(Lacewings and Their Relatives)

Neuroptera are a diverse, primitive order of about 4,000 species represented in all regions of the world and especially common in warm zones. Most adult Neuroptera are small-bodied, weakly flying insects with well-developed compound eyes, long, multisegmented antennae, and biting mouthparts. They have paired, usually subequal, net-veined wings held roof-like over the abdomen at rest. The longitudinal wing veins often have a branched, comb-like radial sector and numerous costal and other cross veins. The ten-segmented abdomen lacks cerci. Development is holometabolous resulting in larvae with thoracic but no abdominal legs.

Neuroptera are terrestrial in distribution except for sisyrids, osmilids, and a few other aquatic families. They are almost invariably predacious in the larval stage and sometimes also in the adult stage. Their choice of prey varies with the environment. Adults use biting mouthparts to eat, and larvae use sickle-like suctorial mouthparts.

Among common Neuroptera are the lacewings of family Chrysopidae (*Fig. 19-13 g*) whose greenish, delicate-bodied adults are common on herb, shrub, and tree foliage. Adult Mantispidae (*Fig. 19-13 f*) are mantis-like insects with an elongate prothorax and large, raptorial fore legs used to capture and hold prey. Their larvae undergo hypermetamorphosis and are parasitic on spider egg masses and the larvae of some subterranean insects. Adults of the antlion family Myrmeleontidae (*Fig. 19-13 h*) are delicate, feeble-flying, damselfly-like Neuroptera with net-wings, a long, slender abdomen, and clubbed antennae. The odd-looking, ovoid larvae of certain genera, so-called doodlebugs (*Fig. 14-3 a*), occupy conical depressions (*Fig. 14-3 b*) dug within loose sandy soil from the bottom of which they lie in wait with their jaws exposed to seize ants that stumble downward onto them. However, most myrmeleontid genera occupy soil surface debris, tree holes, tree trunks, and rock faces from which they hunt freely.

Order Megaloptera
(Dobsonflies and Alderflies)

Megaloptera are a small cosmopolitan group of about 300 species. Adults of the group are relatively large, soft-bodied insects with chewing mouthparts, large compound eyes, elongate antennae, and two pairs of membranous wings with a primitive venation and a large number of veins and cross-veins. Their ten-segmented abdomen is without either ovipositor or cerci. The elongate larvae are aquatic creatures provided with biting mouthparts and seven or eight pairs of abdominal gills. Among common Megaloptera are the large, soft-bodied dobsonflies of family Corydalidae usually found flying awkwardly near water. Male dobsonflies (*Fig. 19-13 a*) have long, prominent mandibles that distinguish them from the female sex. The aquatic juveniles of Corydalidae, called hellgrammites (*Fig. 19-13 b*), are found under stones in streams. The alderflies of family Sialidae (*Fig. 19-13 d*) are a sister group. Alderfly adults are smaller, darker-colored insects than dobsonflies. They frequent shore plants near the slowly moving water inhabited by their larvae (*Fig. 19-13 c*) which differ from hellgrammites chiefly in their lack of hooked anal prolegs having instead an elongate terminal filament. Both larvae and adults are predacious.

Order Rhaphidioptera
(Snakeflies)

Adult snakeflies (*Fig. 19-13 e*) include over 200 species of peculiar Neuroptera characterized by a head with biting mouthparts, well-developed compound eyes, an elongate prothorax, two pairs of membranous, identical wings with many cross-veins, and a ten-segmented abdomen without cerci but with an elongate ovipositor. Snakeflies superficially resemble mantispids (*Fig. 19-13 f*) but lack the latter's raptorial fore legs. They frequent flowers, foliage, *etc.*, and oviposit in cracks beneath tree bark or within the tunnels of wood-boring insects. They consume wood-boring insects, caterpillars, and similar prey.

The neuropteroid branch of Holometabola differs from the two other branches, the panorpoids and the hymenopteroids, primarily in their more complete wing venation. Certain Megaloptera with

Chapter 19 / CLASSIFICATION AND MAJOR INSECT GROUPS

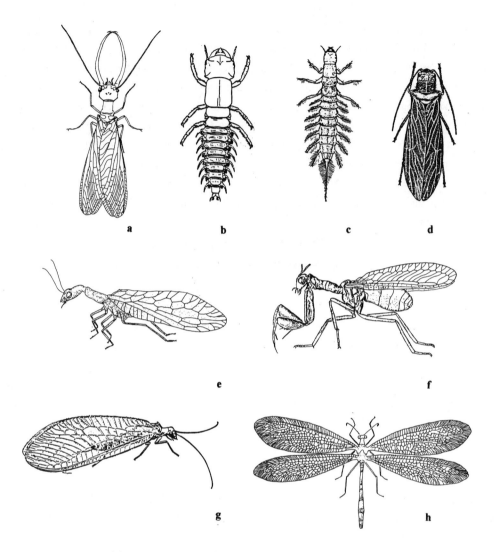

Fig. 19-13. (*a*) A male dobsonfly (Corydalidae); (*b*) a hellgrammite or dobsonfly larva; (*c*) a larval alderfly (Sialidae); (*d*) an adult alderfly; (*e*) a snakefly (Raphidioptera); (*f*) a mantispid (Mantispidae); (*g*) the common green lacewing (Chrysopidae), (*h*) an adult antlion (Myrmeleontidae) (*a* is modified and redrawn from USDA, 1952; *b* and *f* are modified and redrawn from Essig, 1942; *c*, *d*, and *g* are by courtesy of Illinois Natural History Survey; *e* and h are modified and redrawn from Borror *et al.*, 1989).

a generalized venation, primitive larvae, and archaic forms qualify as "living fossils." Their immediate precursors are recorded from the Jurassic and Triassic and more distant ones from the Permian. Living groups, some quite primitive, are known from the Tertiary. Orders Megaloptera and Rhaphidioptera probably diverged first, followed by order Neuroptera.

Pterygota / Neuropteroidea

Order Coleoptera
(Beetles and Weevils)

The vast insect order Coleoptera, with over 370,000 known species, is the largest in the Animal Kingdom. Included within it are 200 or so families of strongly sclerotized insects called beetles. Their species range from some of the smallest insects known to some of the largest, but most are intermediate in size. They occupy such diverse terrestrial situations as the soil surface, crannies under rocks, logs, or debris, within dung, carrion, or soil, on or within moss and other "lower plants," or on or within herbs, shrubs, or trees. Some dwell on or within dry cereals, animal products, timber, or fabrics. Many are aquatic.

The mouthparts of Coleoptera are biting-chewing type associated with a developed gula except in weevils in which the head is drawn into a beak at the apex of which are located the jaws (*Fig. 19-15 n*). There are usually two pairs of wings, the fore pair horny or leathery, often striated veinless elytra, or wing covers, that meet in a straight line down the back to conceal the long, membranous, complexly folded hind wings (*Fig. 19-14 d*). Venation is reduced, and cross veins are few or absent altogether. Only segments five to eight of the ten-segmented abdomen are visible, and there are no cerci. Development is holometabolous except in blister beetles with their modified hypermetamorphosis (having two or more sequential larval types). Beetle larvae vary from elongate, active types with well-developed legs and antennae to legless types. Beetle pupae lack functional mandibles but usually have free appendages.

Several suborders are recognized, of which two, Adephaga and Polyphaga, are the most important. *Adephaga* include those comparatively primitive beetles with thread-like antennae, hind coxae that divide the first abdominal sternum into two (*Fig. 19-14 a*), and elongate, active larvae that usually have tarsi with paired claws. The more advanced, larger suborder *Polyphaga* includes beetles with varied, often apically enlarged antennae, hind coxae that do not subdivide the first abdominal sternum (*Fig. 19-14 b*), and larvae that are either legless or have legs that terminate in a tibiotarsus and a single claw.

Among important terrestrial Adephaga are the colorful, diurnally active, predacious tiger beetles of family Cicindelidae that occupy sandy, sunny open places and the secretive, nocturnally active, predacious ground beetles of family Carabidae (*Fig. 19-14 c*) that live under stones, leaves, or debris but are often seen running over the ground. Among important aquatic Adephaga are the hard-bodied, streamlined, predacious diving beetles of family Dytiscidae (*Fig. 9-8 a*) known to submerge for long periods and the hard-bodied, ovoid whirligigs of family Gyrinidae that swim in endless gyrations on the surface of ponds and quiet streams.

Among important aquatic *Polyphag*a are the water scavenger beetles of family Hydrophilidae (*Fig. 9-8 b*). These ovoid insects resemble predacious diving beetles except for their clubbed antennae, long maxillary palpi, and their usual ventral keel as opposed to the diving beetles' thread-like antennae and short palpi. Among important terrestrial Polyphaga are the elongate, earwig-like rove beetles of family Staphylinidae (*Fig. 19-14 e*) with short elytra; the ovoid, powerfully built scarabs

Chapter 19 / CLASSIFICATION AND MAJOR INSECT GROUPS

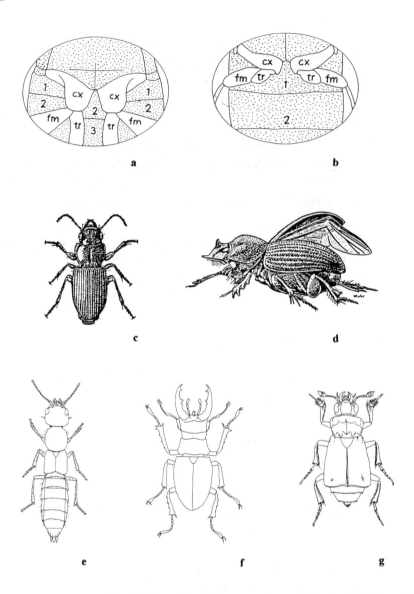

Fig. 19-14. (*a*) The metasternum and the first few abdominal sterna of a ground beetle (Adephaga: Carabidae) showing the hind coxae dividing the first abdominal sternum into two parts and (*b*) the same view of a click beetle (Polyphaga: Elateridae) showing an undivided first abdominal sternum; (*c*) a ground beetle (Carabidae); (*d*) a scarab (Scarabaeidae) with the right elytron lifted to reveal the partly folded right flight wing; (*e*) a rove beetle (Staphylinidae); (*f*) a male stag beetle (Lucanidae) showing its enlarged mandibles; (*g*) a carrion beetle (Silphidae) (*a* and *b* are modified and redrawn from Borror et al., 1989; *c* is by courtesy of Kansas State University; *d* is by courtesy of Illinois Natural History Survey; *e* is modified and redrawn from USDA, 1952; *f* is modified and redrawn from Imms, 1961; *g* is modified and redrawn from Essig, 1942).

or Scarabaeidae (*Fig. 19-14 d*) with leaf-like antennae; the stag beetles of family Lucanidae (*Fig. 19-14 f*) whose males have large, branched mandibles; the soft-bodied carrion beetles of family Silphidae (*Figs. 14-4 b, 19-14 g*) with clubbed antennae and often short elytra; the elongate click beetles of family Elateridae (*Fig. 19-15 h*) named after their ability to right themselves when upside-down, producing an audible "click;" the elliptically shaped, soft-bodied lightningbugs of family Lampyridae (*Figs. 16-3 a, 19-15 i*) known for their bioluminescence; the soft-bodied blister beetles of family Meloidae (*Fig. 19-15 j*) certain of which secrete the vesicating substance cantharidin; the small, often scale-covered, scavenging skin beetles of family Dermestidae (*Fig. 19-15 k*); the elongate, hard-bodied, often colorful, wood-boring long-horned beetles of family Cerambycidae (*Fig. 19-15 l*) named for their antennae; the often colorful, ovoid leaf beetles of family Chrysomelidae (*Fig. 19-15 m*) with shorter antennae than cerambycids; the weevils of family Curculionidae (*Fig. 19-15 n*) with elbowed antennae borne on a slender snout with small mouthparts at the apex; and the sap beetles of family Nitidulidae that infest putrifying vegetation and garbage (*Fig. 19-15 0*).

Adephaga are predacious. Polyphaga are largely plant-feeders though some of their representatives are predators, scavengers, or even parasites. Among the plant-feeding Polyphaga are fruit-, stem-, and root-feeders, many of which do serious damage both to cultivated and to wild vegetation.

Coleoptera constitute a relatively compact, homogeneous order not closely related to any other major neuropteroid group. Their ancestral stock probably arose from near the base of the neuropteroid line at an early, perhaps Permian date subsequent to which they diverged into Adephaga and Polyphaga.

Order Strepsiptera
(Stylopids)

Strepsiptera include less than 400 species of minute insects with a strong sexual dimorphism. The weakly sclerotized males (*Figs. 14-1 f, 19-16 a*) are short-lived, free-living animals with reduced biting-type mouthparts, well-developed legs, eyes, antennae, and a single metathoracic pair of membranous, longitudinally folding flight wings. The parasitic females (*Figs. 14-1 e, 19-16 b*) are sac-like, larva-like, eyeless, quiescent creatures without appendages. They remain enclosed within the persistent last larval skin but protrude their cephalothorax from between the host's abdominal segments. The first instar larvae, or triungulins (*Fig. 14-1 d*), are agile juveniles with eyes, legs, and caudal appendages. The remaining larvae are legless, maggot-like endoparasites.

A common life cycle begins with the male within his puparium (hardened last larval skin) between the host's abdominal sclerites. He breaks free, seeks out, and fertilizes the permanently endoparasitic female by inserting his aedeagus into her brood canal aperture or by rupturing her exposed cephalothorax. The resulting larvae hatch within the mother, pass through genital pores on her underside, and issue from her brood canal aperture. The active triungulins then move to the ground or onto nearby vegetation where they await contact with a suitable host insect that chances by. If that host is a social insect, the triungulins may "hitch-hike" to the visitor's nest to seek out and enter a host larva before molting into successive quiescent legless stages and eventually reaching maturity.

Chapter 19 / CLASSIFICATION AND MAJOR INSECT GROUPS

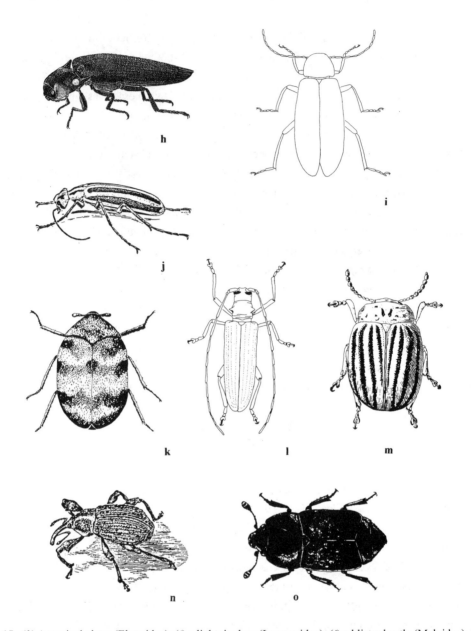

Fig. 19-15. (*h*) A tropical elater (Elateridae); (*i*) a lightningbug (Lampyridae); (*j*) a blister beetle (Meloidae); (*k*) the varied carpet beetle (Dermestidae); (*l*) a long-horned beetle (Cerambycidae); (*m*) the Colorado potato beetle (Chrysomelidae); (*n*) the cotton boll weevil (Curculionidae); (*o*) a sap beetle (Nitidulidae) (*h* is after Essig, 1942; *i* is modified and redrawn from Borror *et al.*, 1989; *j* is by courtesy of Illinois Natural History Survey; *k* is modified and redrawn from Arnett, 1985; *l* and *m* are modified and redrawn from USDA, 1952; *n* is after USDA. 1952; *o* is modified and redrawn from Borror *et al.*, 1989).

Pterygota / Neuropteroidea

Contrary to earlier belief, stylopid larvae have a functional mouth used to swallow host blood, and they feed in this manner until full grown. Each then protrudes its head from between the host's abdominal sclerites, molts, and, if destined to become female, remains permanently sequestered within the first larval skin, awaiting fertilization by a male. A male larva transforms into a pupa within the last larval skin and, when mature, escapes as above to repeat the cycle.

Various wasps, ants, and other Hymenoptera, cicadas and other Homoptera, and Heteroptera are among the hosts commonly chosen by stylopids and, therefore, are said to be stylopized.

Stylopids were long regarded as aberrant beetles based on their males, their triungulin larvae, their hypermetamorphosis, and their parasitic habits which are reminiscent of those of certain Coleoptera. Now most specialists acknowledge their separation from Coleoptera and accord them full ordinal status.

Order Mecoptera
(Scorpionflies, Hangingflies, and Their Allies)

The approximately 480 species of scorpionflies, hangingflies, and their relatives of order Mecoptera are distinctive, usually moderate-sized insects with an elongate, deflected, beaked head tipped by small, biting mouthparts. They have two pairs of long, subequal, primitively veined, membranous wings that are held apart and nearly flat over the abdomen or to the side during repose. Development is holometabolous, the larvae usually being caterpillar-like and the pupae having functional mandibles and free appendages.

Male genitalia in the scorpionfly family Panorpidae (*Fig. 19-16 c*) are bulbous, recurved structures reminiscent of a scorpion sting (hence the common name). Adult scorpionflies frequent damp, heavily wooded ravines and similar places and perch on rank vegetation from which they make short flights. Panorpids, like most Mecoptera, are terrestrial insects largely known from temperate zones. However, there are many tropical hangingflies of family Bittacidae (*Fig. 19-16 e*). Adult hangingflies suspend themselves from vegetation by the fore legs and grasp small flying prey by their curious raptorial hind legs. The short-winged snow scorpionflies of family Boreidae (*Fig. 19-16 d*) live on mosses in cold or temperate zones and are active during winter.

The literature mentions nectar, pollen, flowers, and fruit as items of mecopteran diet. This interpretation, based largely on caged scorpionflies, is not necessarily representative of the entire order. Mecoptera are commonly carnivorous in the wild, eating dead insects and occasionally snails, vertebrate carcasses broken open by predators, *etc.*, and they drink much water. However, the food-habits of particular taxonomic groups may depart from the above. Adult bittacids prey on living insects; larval bittacids and larval and adult panorpids scavenge on dead insects; and boreids eat mosses.

Mecoptera are ancient, with numerous attributed fossils from as far back as the Lower Permian, long before the other panorpoid orders (Trichoptera, Lepidoptera, Diptera, and Siphonaptera) appeared. They are correspondingly primitive, having an archaic venation that includes all of the principal veins and branches. This suggests that they arose early from the ancestral line and probably gave rise to the other panorpoid orders.

Chapter 19 / CLASSIFICATION AND MAJOR INSECT GROUPS

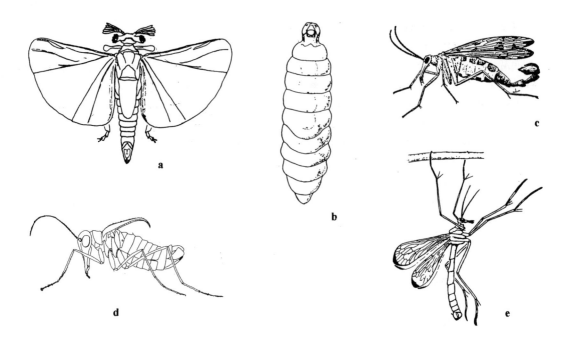

Fig. 19-16. (*a*) A male stylopid (Strepsiptera); (*b*) the sac-like, legless, wingless parasitic female of the preceding stylopid; (*c*) a male scorpionfly (Panorpidae); (*d*) a snow scorpionfly (Boreidae); (*e*) a hangingfly (Bittacidae) (*a* and *b* are modified and redrawn from CSIRO, 1970; *c* is by courtesy of Illinois Natural History Survey; *d* is modified and redrawn from Imms, 1961; *e* is modified and redrawn from Thornhill, 1980).

Order Trichoptera
(Caddisflies)

The approximately 10,000 species of Trichoptera (*Fig. 19-17 b, c*) are small- or medium-sized, brownish or streaked, moth-like insects with two pairs of often densely "hair"-covered, primitively veined wings. They are terrestrial, spending most of their time on vegetation, usually near water. They fly weakly, mostly at night, and are often attracted to artificial light. They use channeled mouthparts to lap fluids or take particulates but some do not eat. Development is holometabolous, and the larvae, called caddisworms, are exceptional in their preference for cold, unpolluted water. Caddisworms take two different forms: elongate, active, often predacious juveniles with legs and antennae and a prognathous head (*Fig. 19-17 a*) and the more common, largely vegetarian type with a hypognathous head (*Fig. 19-17 e*). The former are not case-builders but may construct aquatic nets to capture prey. If so, the net-makers live in a nearby recess and feed, spider-like, on trapped prey or drifting organic matter. The latter live within a distinctive case of their own construction made from pebbles, debris, *etc*. The pupae have functional mandibles and free appendages.

Trichoptera took their origin long ago from panorpoid stock. There are fossil records of Trichoptera from the late Triassic and of apparent precursors from the Permian. Structure in the order is primitive. It indicates a close relationship with Lepidoptera, and the two orders seem linked by the existence of certain mandibulate moths that closely resemble primitive caddisflies. Trichoptera resemble Lepidoptera in wing venation, genitalia, larval structure, and occasionally possession of wing scales. They differ chiefly in the aquatic habit of the larvae, longer antennae, haustellate mouthparts, and fore-wing venation though even these differences are not absolute. Their mouthparts consist of a protrusible haustellum, usually channeled, analogous to that of Diptera.

Order Lepidoptera
(Moths and Butterflies)

Lepidoptera include about 150,000 described species of familiar, often large-bodied, attractively colored and patterned, economically important insects known as moths and butterflies. Most moths (*Fig. 19-17 f*) are stout-bodied, nocturnal or crepuscular Lepidoptera that couple their fore and hind wings by various means including a spine-like frenulum (*Fig. 15-10 c*), fold them flat over the body at rest, and have antennae of thread-like, plumed, bristle-like, or other type. Butterflies (*Fig. 19-17 g, i*) are slender-bodied, diurnally active Lepidoptera with clubbed antennae that in one group, the so-called skippers (*Fig. 19-17 h*), are apically recurved. They couple their wings without a frenulum and fold them vertically at rest.

The fore and hind wings of Lepidoptera are dissimilar in shape, in size (the fore wings usually being larger), and usually in venation. Wings in certain families are deeply fissured, reduced, or absent altogether. The overall body is soft, cylindrical, and heavily clothed in scales and other articulated processes on all surfaces, especially the wings. These structures grade from hair-like to scale-like and come off like dust when the insects are handled. Scales take on a scattered distribution in primitive forms but assume a regularly arranged, imbricate pattern like the shingles of a roof in more advanced Lepidoptera. They are often brilliant owing to contained pigments and sometimes to physical interference phenomena.

Lepidoptera occur virtually throughout the terrestrial world but are not exclusively land-dwellers. There are a few aquatic Lepidoptera with gilled larvae that live on or mine within aquatic plants and adults that swim with facility.

Lepidopteran development is invariably holometabolous involving young known as caterpillars (*Fig. 19-17 d*). These soft-bodied, mobile larvae have a well-developed, sclerotized, hypognathous head and a soft, cylindrical body composed of three thoracic and ten abdominal segments. The head (*Fig. 2-6 a*) bears chewing-type mouthparts, a pair of short antennae, and a cluster of six stemmata (a type of simple eye) behind each antennal base. The limbs include a pair of walking legs on each thoracic segment and pairs of stump-like larvapods, or prolegs, on abdominal segments three-to-six and ten. These caterpillars resemble those of certain Hymenoptera but are easily separable on the basis of the six-to-eight pairs of prolegs and the single pair of stemmata (*Fig. 19-19 d*) of the latter. Lepidopteran pupae are hard-bodied, usually immobile creatures without functional mandibles and with appendages usually fused to the body surface.

Chapter 19 / CLASSIFICATION AND MAJOR INSECT GROUPS

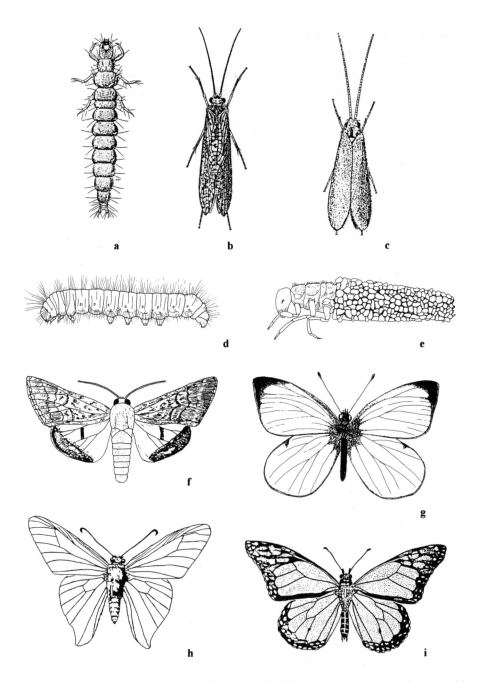

Fig. 19-17. (*a*) A larva and (*b*) an adult of a primitive caddisfly (Trichoptera) close to the micropterygid moths; (*c*) a caddisfly showing the characteristic "hairy" body of Trichoptera; (*d*) a larval tent caterpillar (Lasiocampidae); (*e*) a case-bearing caddisworm (Trichoptera); (*f*) the corn earworm moth (Noctuidae); (*g*) a cabbage butterfly (Pieridae); (*h*) a skipper butterfly (Hesperiidae); (*i*) the monarch butterfly (Danaidae) (*a, b,* and *c* are by courtesy of Illinois Natural History Survey; *d* is modified and redrawn from Snodgrass, 1967; *e* is modified and redrawn from Ceballos, 1962; *f, h,* and *i* are modified and redrawn from *Turtox Key Cards*; *g* is modified and redrawn from Viedma *et al.*, 1985).

Pterygota / Neuropteroidea

The protective shelters that most Lepidoptera make are composed at least partly of a secretion from the paired salivary glands that open by a median spinneret at the anterior labial margin. The silk produced therefrom forms a structure varying from a portable case, to a communal nest, or to a cocoon in which to pupate.

Caterpillars are commonly vegetarians that use their biting-type mouthparts to ingest large quantities of the host plant, whether flowers, fruit, leaves, stems, or roots. Their feeding latitude varies from that of the gypsy moth which eats the foliage of over 200 tree species to that of the silkworm moth which eats only mulberry leaves. The characteristic damage that they inflict on the host varies. Large caterpillars generally margin-feed (*Figs. 14-5 d, 14-9 b*), excising and swallowing strips from the edge of the leaf or flower being eaten. Small caterpillars are often restricted to gnawing holes in the leaf center, to leaf-skeletonization, or to mining within leaves or boring within roots, stems, fruits, or other plant parts. A few Lepidoptera depart from plant-feeding altogether, consuming woolens or other fabrics. Still others eat wax, leaf litter, and detritus, and a few are carnivorous or endoparasitic.

Adult Lepidoptera feed by means of their unique suctorial mouthparts (*Figs. 2-12 b, 2-13 b*) consisting of the paired, modified galeae appressed and interlocked to form an elongate, flexible proboscis thrust forward to eat but coiled beneath when not in use. Unlike their caterpillars, adults are necessarily fluid feeders that imbibe nectar, sap, other plant discharges, and liquids from overripe fruit and dung, and drink water from puddles on the ground. A few adults are short-lived creatures that eat little or nothing, and some of them have non-functional, atrophied mouthparts.

The primitive so-called "mandibulate moths" possess more or less developed jaws. For example, adults of the family Micropterygidae lack the proboscis of other Lepidoptera but have strong mandibles and orthopteroid-like maxillae used to ingest pollen particles. Other "mandibulate moths" have non-functional mandibles and lack laciniae making chewing impossible, for which they compensate by using an abbreviated galeal proboscis.

Lepidoptera may be divided into suborders based on their genitalic characters. Females of suborder *Monotrysia* have a combined copulatory opening/gonopore or cloaca on the ninth sternite (*Fig. 10-3 a*), while those of suborder *Ditrysia* have a copulatory opening on the eighth sternite and a separate gonopore, together with an anus, on the ninth (*Fig. 10-3 b*). This classification allows the micropterygids either to be set aside as an order of their own (order *Zeugloptera*) or to be kept within the Lepidoptera as a third suborder (suborder *Zeugloptera*). Several artificial classifications of past years are no longer accepted, for example, separation on the basis of the wing-coupling mechanism, *viz.*, Jugatae *vs.* Frenatae (*Fig. 15-10 b, c*, respectively); the similarity or dissimilarity of the fore and hind wings, *viz.*, Homoneura *vs.* Heteroneura; or separation based on body size, *viz.*, Microlepidoptera *vs.* Macrolepidoptera.

One micropterygid genus appears to be more primitive than even the primitive Trichoptera, on which basis one may conclude that the micropterygids arose first, followed by the mandibulate moths that lost functional mandibles. Further specialization then involved loss of the jugum, hind-wing reduction, and genital modification until finally an ancestral monotrysian evolved into the first ditrysian.

Chapter 19 / CLASSIFICATION AND MAJOR INSECT GROUPS

Order Diptera
(True Flies)

Diptera include approximately 120,000 species of highly specialized insects called flies, as distinguished from many other so-called *flies* recognized by laymen. Flies are minute- to moderate-sized insects with a single mesothoracic pair of membranous wings and a soft, elongate or compact body, often bristly, and usually darkish but sometimes metallically colored. The modified hind wings have become knobbed, vibratory halteres that function as a kind of gyroscope.

Few animals surpass Diptera in terms of their impact on the human population. Fly-borne pathogens and parasites such as malaria, leishmaniasis, and sleeping sickness transmitted by mosquitoes, black flies, tsetse flies, and other biting flies, and gastrointestinal diseases spread by house flies are responsible for untold human suffering and death. The crop losses inflicted annually by fruit flies such as the Mediterranean fruit fly and the apple maggot and the gall midge known as the Hessian fly exceed those of most other animal groups. On balance, however, flies rank with parasitic wasps in their value as agents useful in the biological control of pest insects, and their role in the degradation of dead animal and plant materials is exceeded only by that of bacteria and fungi. Some flies are useful also as insect biocontrol agents, and others are plant pollinators.

Diptera are common even in austere arctic zones from which most other insects are excluded. Larvae of many Nematocera and Brachycera are aquatic forms that inhabit streams, lakes, marshes, ponds, puddles within tree holes, and brackish water. Other fly larvae live in a variety of habitats including damp soil, fresh dung, decaying wood or fruit, moss, mushrooms, and living plant parts, or they occur as internal parasites of living animals. Adult flies are terrestrial, being found on the wing everywhere from dry desert to lush rain forest.

The higher classification of flies is controversial inasmuch as the phylogenetic relationships within some suborders are not resolved. In particular, some families of Nematocera may be more closely related to Brachycera, but different character data support competing hypotheses. Likewise, the lower Brachycera do not seem to be a natural group although the Cyclorrhapha probably are. The traditional system that recognizes suborders *Nematocera*, *Brachycera*, and *Cyclorrhapha* is artificial and likely to be supplanted by a more natural system. However, it suffices for want of better interpretation.

Adults of suborder Nematocera (*Fig. 19-18 c, d, e*) are generally slender, long-legged flies with multisegmented antennae usually longer than the combined head and thorax and without an arista (a bristle-like antennal style, *Fig. 2-8 i*); they have four-to-five-segmented palpi; and their wings tend to be elongate, narrow at the base and usually provided with a moderate number of longitudinal veins and few cross veins but no discal cell (*Fig. 19-18 b*). Most larval Nematocera have a sclerotized, generally protuberant head capsule and horizontally moving mandibles. Nematocerous pupae are obtect (*Fig. 11-2 e*), their appendages being glued to the general body surface.

Adults of suborder Brachycera (*Fig. 19-18 f*) are generally stout-bodied, often large flies usually

Pterygota / Neuropteroidea

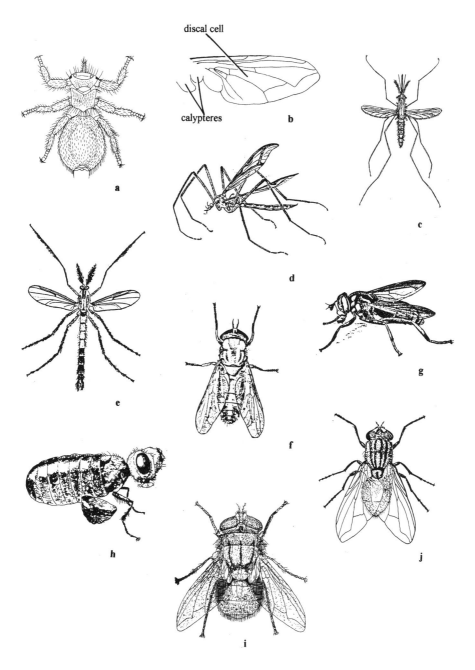

Fig. 19-18. (*a*) A louse fly (Hippoboscidae); (*b*) a calyptrate wing showing selected characters; (*c*) a mosquito (Culicidae); (*d*) a crane fly (Tipulidae); (*e*) a midge (Chironomidae); (*f*) a horse fly (Tabanidae); (*g*) a flower fly (Syrphidae); (*h*) a house fly (Muscidae) escaping from the puparium by means of its everted ptilinum, with the detached puparial cap discarded below; (*i*) the screwworm fly (Calliphoridae); (*j*) the house fly (Muscidae) (*a* is modified and redrawn from USPHS; *b* is modified and redrawn from Borror *et al.*, 1989; *c* is modified and redrawn from Essig, 1942, after Nutall & Shipley; *d*, *e*, and *g* are by courtesy of Illinois Natural History Survey; *f* and *i* are modified and redrawn from USDA, 1952; *h* is modified and redrawn from Greenberg, 1965; *j* is modified and redrawn from *Turtox Key Cards*).

Chapter 19 / CLASSIFICATION AND MAJOR INSECT GROUPS

with three-segmented antennae shorter than the combined head and thorax, one-or-two-segmented palpi, and wings with a discal cell. Larval Brachycera have a poorly developed head capsule partly retracted into the body and fitted with vertically moving mandibles, and their pupae are exarate (*Fig. 11-2 f*), the appendages being free of the body surface.

The remaining flies are included within suborder Cyclorrhapha. Adult Cyclorrhapha (*Fig. 19-18 g–j*) are typically short, stout flies with three-segmented, aristate antennae, single-segmented palpi or none at all, and wings with a discal cell (*Fig. 19-18 b*). Only four or five of their abdominal segments may be obvious. Larval Cyclorrhapha, called maggots (*Fig. 9-5 b*), lack an external head capsule but have a characteristic set of sclerotized, mostly internal head parts including vertically moving mandibles. Cyclorrhaphan pupae are coarctate, developing within the puparium, a hardened last larval skin from which the emerging adult escapes through a circular opening, hence the name *Cyclorrhapha* (*Fig. 19-18 h*).

Adult flies use their suctorial mouthparts for piercing, sponging, and occasionally tearing. Their feeding ranges from flower-, pollen-, nectar-, or sap-feeding, to secretion- or dung-feeding, and to blood-sucking. Some blood-sucking flies are ectoparasitic, flattened creatures with widely separated coxae and legs (*Fig. 19-18 a*). Various species including at least one whole family have vestigial mouthparts and do not eat as adults. Larval feeding ranges widely encompassing leaf-mining, stem- or root-boring, gall-making, fungus-, leaf-, fruit- or flower-feeding, scavenging of dead plant and animal materials, predation, and ecto- and endoparasitism.

Dipteran reproduction is bisexual except for some flies that are parthenogenetic. Development is holometabolous and either oviparous by eggs laid externally or ovoviviparous by eggs that hatch within the female parent. The legless vermiform larvae, or maggots (*Fig. 9-5 b*), vary in form as do pupae. The latter range from exarate, having free appendages, to obtect, having appendages glued to the external body surface, and to coarctate, enclosed within a hardened last larval skin (*Fig. 19-18 h*).

Diptera are highly specialized panorpoid insects whose divergence must have involved modification of the hind wings into halteres, reduction of fore wing venation, and mesothoracic expansion accompanied by pro- and metathoracic reduction and fusion of the three segments into a pterothorax. There are fossil flies from the Jurassic and upper Triassic, but flies probably did not become common until the Tertiary.

Nematocera are the most primitive representatives of the order based on the fact that some of them are structurally similar to Mecoptera in terms of the larval head, adult antennae, wing venation, *etc*. Brachycera and Cyclorrhapha are more derived and more recent. Evolution in the latter involved, among other modifications, development of the puparium to protect the pupa. In flower flies and other *Aschiza* (which lack a ptilinum), the anterior end of the puparium splits into two parts that pop open, so emergence takes place through a dorsal slit. In the remaining Cyclorrhapha, those flies called *Schizophora*, emergence is facilitated by a ptilinum that everts making a circular opening for adult emergence (*Fig. 19-18 h*). The schizophoran families are grouped into *calyptrates* (flies with calypteres, enlarged, lobular connecting membranes at the wing base) (*Fig. 19-18 b*) and *acalyptrates* (flies that lack calypteres).

Pterygota / Neuropteroidea

Order Siphonaptera
(Fleas)

Siphonaptera (*Fig. 19-19 a*) include some 2,250 species of small, strongly sclerotized, darkly pigmented, wingless, laterally flattened ectoparasites of mammals and birds. Their short, three-segmented, clubbed antennae are concealed within a groove on either side of the head. They are usually heavily bristled, having strong setae on the legs and head, single or double rows of them on the body segments, and special, thick bristles here and there. They may also have short, strong combs along the underside of the head (*Fig. 19-19 b*), the posterior margin of the pronotum, on the abdomen, and elsewhere. These bristles are backwardly directed, facilitating forward progression through the hosts' closely set hair or feathers. The flat, powerful legs are adapted for clinging, and the hind pair is enlarged for leaping. The fore pair is articulated forward, often appearing to arise from the head.

Fleas are a source of annoyance owing to their "bites." Their continued irritation may cause the host to develop a dermatitis and/or, in some particularly susceptible individuals, a severe allergic reaction. Fleas are also intermediate hosts of the dog tapeworm that occasionally infests humans. More importantly, however, fleas act as vectors of such serious diseases as epidemic typhus and bubonic plague. Both of these rodent diseases are transmitted to humans through flea "bites." The "*Black Plague*," a flea-transmitted form of bubonic plague, resulted in the death of over a quarter of the total population of Europe during the Fourteenth Century.

Fleas move with alacrity through the host's coat and readily transfer to other individuals upon contact between hosts. They spend much time on the host though some fleas that reside within the host's nest mount only to eat. Fleas feed by means of piercing-sucking mouthparts consisting of an unpaired labrum-epipharynx and paired, blade-like laciniae within a short labium; there are no mandibles. The maxillary palpi are positioned forward, reminiscent of the antennae of other insects. Whatever their habits, fleas tend to be long-lived, hardy, and able to subsist for weeks without eating. Most are weakly host-specific, being at least temporarily capable of infesting other warm-blooded animals with which they establish bodily contact. However, female "sticktights," or chigoes, are subcutaneous parasites that burrow into the host's feet or other body parts, remain there, develop, and cause a painful sore. Male chigoes differ in that they live much like other fleas, feeding ectoparasitically on the infested mammalian host.

Most fleas cannot reproduce until the females have had a blood meal. Following mating, they drop their minute, whitish eggs onto litter within the nest, glue them onto debris, or scatter them about on the host's fur or feathers from which the eggs eventually fall and begin to develop. They hatch into blind, legless larvae some days or weeks later, depending on temperature. The young usually do not live on the host and, being mandibulate, cannot suck blood. Instead, they scavenge on feculae, hair, feathers, scales, cast skins, or other organic debris. Not surprisingly, they can be reared on floor sweepings. After two molts, they spin a silken, debris-covered cocoon in which to pupate. Flea pupae are adecticous (having fixed, immovable mandibles) and exarate (having appendages free of the body surface). They give rise to adults which remain inactive pending floor vibration or other mechanical

Chapter 19 / CLASSIFICATION AND MAJOR INSECT GROUPS

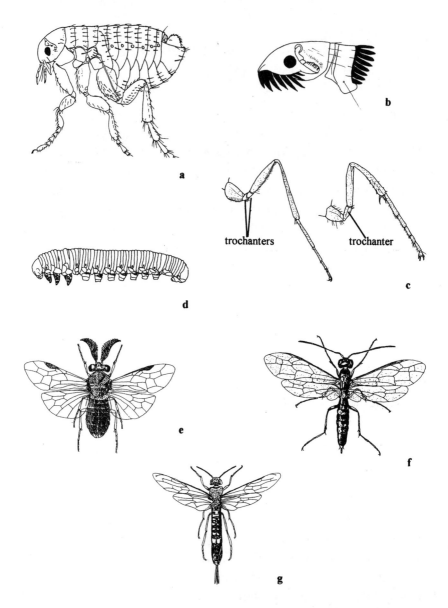

Fig. 19-19. (*a*) The human flea (Pulicidae); (*b*) the head and pronotum of the cat flea (Pulicidae) showing genal and pronotal combs; (*c*) the leg of a symphytan (left) compared to that of an apocrite (right), showing differences in the trochanter; (*d*) a hymenopteran caterpillar (Diprionidae); (*e*) a conifer sawfly (Diprionidae); (*f*) a wheat stem sawfly (Cephidae); (*g*) a horntail (Siricidae) (*a* is modified and redrawn from Essig, 1942; *b* is modified and redrawn from Kettle, 1995, after Patton & Evans; *c* is modified and redrawn from Borror *et al.*, 1989; *d* and *e* are modified and redrawn from Ceballos, 1962; *f* is modified and redrawn from USDA; *g* is modified and redrawn from Sharp, 1895).

disturbances, whereupon they issue in number to seek food.

Siphonaptera are a distinctive, homogeneous order of highly specialized ectoparasites. Their phylogeny is uncertain. They have neither wings nor ovipositor though some have pupal wing buds indicative of derivation from a winged ancestral stock. A likely dipteran relationship is suggested by the similarity of fleas' adult mouthparts and overall larval and pupal structure to that of Nematocera. Little is known of fleas' first geological appearance though it assuredly coincided with the evolution of their warm-blooded hosts. The few available fossil fleas are mostly from the Tertiary.

Order Hymenoptera
(Sawflies, Ants, Wasps, and Bees)

Hymenoptera consist of about 103,000 specialized species whose several common names (sawflies, horntails, ichneumonflies, chalcids, gall wasps, ants, hornets, bees, *etc.*) are indicative of their diversity. Their adults usually range from small-to-medium-sized, but some species are minute (1-2 mm) and others large (5-6 cm). They may be brilliantly colored or distinctively patterned and either naked or "hairy," and they are invariably well-sclerotized. They usually have two pairs of membranous wings supported by a distinctive venation. There are, however, species that belong to a few families having wings that are reduced or absent, as well as certain small-bodied parasites (*Fig. 19-20 h, j*) from which most venation has disappeared. The larger fore wings of many have a dark costal spot, or stigma, and couple with the hind wings by hooks from the hind costal margin. Development is holometabolous and reproduction either bisexual or unisexual (parthenogenetic). Bisexual reproduction results only in female offspring, whereas unisexual reproduction leads only to male offspring. Sometimes there is an alternation of bisexual and parthenogenetic generations and/or even polyembryony (identical embryos resulting from mitotic division of single eggs).

Most Hymenoptera are strong-flying, diurnally active, often flower-frequenting insects found throughout the terrestrial environment, even in arctic zones. Most are solitary, but a few (especially ants and some wasps and bees) are either subsocial or social.

The impact of Hymenoptera on our biotic world is significant. They are associated with plants either as pollinators or as pests. Bees alone account for more than 80 per cent of the pollination upon which depends most of our agricultural and horticultural crops except grasses. Many larval Hymenoptera eat broad-leaved or coniferous vegetation or are gall-makers, leaf-miners, or wood-borers that damage nursery and forest trees, but some are parasitic on other insects and, hence, may be useful in biological control. Many adult Hymenoptera are predacious on insects, using the caught prey to stock their cells with food for the developing young. Hymenoptera also play an important role as scavengers in the detritus cycle that returns organic materials to soil.

The order Hymenoptera is currently divided into two suborders, Symphyta and Apocrita. Adult *Symphyta* (*Fig. 19-19 e, f, g*) are primitive Hymenoptera that usually have long multisegmented antennae (generally nine-segmented or less), an abdomen that is broadly joined to the thorax, hind wings with three or more basal cells, an ovipositor adapted for sawing or boring, and two apparent

Chapter 19 / **CLASSIFICATION AND MAJOR INSECT GROUPS**

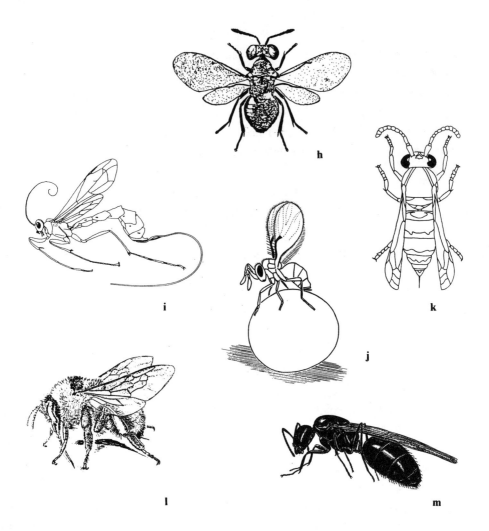

Fig. 19-20. (*h*) A chalcidoid; (*i*) an ichneumonfly (Ichneumonidae); (*j*) a trichogrammatid wasp (Trichogrammatidae); (*k*) a hornet or yellow jacket (Vespidae); (*l*) a bumble bee (Apidae); (*m*) a carpenter ant queen (Formicidae) (*h* is modified and redrawn from Imms, 1939; *i* and *j* are modified and redrawn from USDA, 1952; *k* is modified and redrawn from *Turtox Key Cards*; *l* is by courtesy of Illinois Natural History Survey; *m* is from Essig, 1942, after Back, U. S. Bureau of Entomology).

trochanters (*Fig. 19-19 c, left*). Most of their larvae (*Fig. 19-19 d*) are caterpillar-like creatures with a well-developed head capsule, a single stemma (simple eye) on each side of the head, and both thoracic and abdominal legs. Adult *Apocrita* (*Fig. 19-20 h–m*) usually have relatively short, sometimes elbowed antennae, a "wasp waist" (the second abdominal segment constricted into a narrow pedicel leading to the gaster, or apparent abdomen) (*Fig. 17-5 c*), hind wings having two or fewer basal cells each, an ovipositor adapted for piercing or stinging, and one (*Fig. 19-19 c*) or two apparent trochanters. Larval apocrites are grub-like creatures having neither eyes nor legs.

Most Symphyta are phytophagous, being destructive to forests, agricultural crops, and ornamental shrubs and trees, but Apocrita are more varied in their food-habits. Many primitive apocrites are parasitic, and some are gall-makers. The primitive ants are carnivorous, but the advanced ants may be omnivorous, herbivorous, or resort to specialized seed-eating, harvesting fungal gardens, *etc.* Sphecoid and most vespoid wasps prey on insects or spiders. They use the paralyzed bodies of these hosts to provision their cells. Bees and certain social wasps eat stored pollen and honey.

Hymenoptera are recent except for some Symphyta recorded from the Triassic. The advanced symphytans resemble "higher Diptera" and Lepidoptera in their distribution from the Tertiary onward. This suggests that these three orders of flower-pollinating insects evolved in close association with the flowering plants that burgeoned at about the same time.

Hymenoptera are among the most highly evolved of insects. They have a structural specialization surpassed only by Diptera and a diversity of habits and behavior unique among insects. This suggests that they are different from the other endopterygote insects, yet they have preserved the large number of Malpighian tubules and the ovipositor characteristic of generalized insects. I place them alongside the neuropteroids *sens. str.* and the panorpoids as a third major branch of the primeval neuropteroid *sens. lat.* stock based primarily on the overall similarity of their development to that of Lepidoptera and for want of better interpretation.

The generalized venation, other structural attributes, and the early fossils of some primitive Symphyta suggest they were first to evolve, followed by Apocrita, of which ichneumonoids (*Fig. 19-20 i*) are the most primitive. Next to arise must have been the chalcidoids (*Fig. 19-20 h, j*) and other Hymenoptera in which the ovipositor issues from the abdominal venter forward of the apex. Then the Hymenoptera in which the ovipositor issues from, and is withdrawn into, the abdominal apex must have evolved. These apocrites have a comparatively large number of antennal segments and usually two apparent trochanters (*Fig. 19-19 c, left*), but the remaining apocrites (which presumably arose next) modified the ovipositor into a sting, reduced the number of antennal segments, and returned to a single trochanter (*Fig. 19-19 c, right*). Most of these *aculeates*, or "stinging wasps," as they are called, are paralyzers that immobilize prey on which to lay eggs. This stock probably gave rise to a number of lesser wasp groups and to ants, vespoid and sphecoid wasps, and bees. Bees are provisioning wasps that are phylogenetically close to sphecoids and sometimes grouped together with them in the same superfamily. They differ from sphecoids in their use of pollen as food, development of branched "hairs," a pollen basket, *etc.*

Suggested Additional Reading

General Insects

Arnett (1985); Berenbaum (1995); Bland & Jaques (1978); Borror *et al.* (1989); Borror & White (1970); Chu & Cutkomp (1992); CSIRO, vol. 2 (1991); Davies (1988); Dunn (1996); Gilbert & Hamilton (1990); Gullan & Cranston (2000); Kettle (1995); Klots & Klots (1971); Lehmkuhl (1979); Merritt & Cummins (1996); Metcalf & Metcalf (1993); Richards & Davies, vol. 2 (1977); Walker (1994); Williams & Feltmate (1992).

Chapter 19 / CLASSIFICATION AND MAJOR INSECT GROUPS

Apterygota

Christiansen & Bellinger (1978); Conde & Pages (1991); Greenslade (1991); Hopkin (1997); Imadate (1991); Smith & Watson (1991); Tuxen (1964); Watson & Smith (1991); Wygodzinsky (1972); Wygodzinsky & Schmidt (1980).

Ephemeroptera

Edmunds *et al.* (1976); Needham *et al.* (1935); Peters & Campbell (1991).

Odonata

Corbet (1963); Needham & Westfall (1955); Watson & O'Farrell (1991).

Orthoptera *sens. lat.*

Balderson (1991); Bland (2003); Chapman & Joern (1990); Cornwell (1968); Gangwere *et al.* (1997); Helfer (1953); Key (1991); Otte (1981, 1984); Rentz (1991); Rentz & Kevan (1991); Roth (1991); Uvarov (1966, 1977).

Isoptera

Skaife (1956); Snyder (1948, 1954); Watson & Gay (1991).

Embioptera

E. S. Ross (1944, 1991).

Zoraptera

Smithers (1991).

Plecoptera

Frison (1935); Theischinger (1991).

Psocoptera

Mockford (1993).

Phthiraptera

Calaby & Murray (1991); Ferris (1951).

Thysanoptera

Lewis (1973); Mound & Heming (1991); Stannard (1968).

Hemiptera

Carver *et al.* (1991); Miller (1971); Slater & Baranowski (1978).

Megaloptera, Raphidioptera, & Neuroptera

Aspock & Aspock (1991); New (1991); Theischinger (1991).

Coleoptera

Arnett (1973); Crowson (1981); Dillon & Dillon (1972); Evans (1975); Headstrom (1977); Jaques (1951); Lawrence & Britton (1991); White (1983).

Strepsiptera

Kathirithamby (1991).

Mecoptera

Byers (1991).

Trichoptera

Neboiss (1991); H. H. Ross (1944).

Lepidoptera

Corell (1984); Douglas (1986); Ehrlich & Ehrlich (1961); Ford (1955, 1957); Holland (1968); Howe (1975); Klots (1951); Nielsen & Common (1991); Scott (1986).

Diptera

Carpenter & LaCasse (1974); Cole (1969); Oldroyd (1964).

Siphonaptera

Dunnet & Mardon (1991); Ewing & Fox (1943); Hubbard (1947).

Hymenoptera

Creighton (1950); Evans & Eberhard (1970); Holldobler & Wilson (1990); Michener (1974); Naumann (1991); Spradbery (1973); Wilson (1971, 1975).

Bibliography

*When I see the list of books which I read and abstracted,
I am surprised at my industry.*
— Darwin

Alexander, R. D. 1964. The evolution of mating behavior in arthropods. *Symp. Roy. Entomol. Soc. London,* 2: 78-94.

Anderson, D. M. (ed.). 1975. Common names of insects. *Entomol. Soc. Amer.*, Special Pub. 75-1, 37 pp.

Arnett, R. H. 1973. *The Beetles of the United States.* Amer. Entomol. Inst., xii + 1112 pp.

——. 1985. *American Insects.* Van Nostrand Reinhold, xiv + 850 pp.

Aspock, H., & U. Aspock. 1991. Raphidioptera, ch. 33, pp. 521-524. In: CSIRO *The Insects of Australia*, vol. 2, Melbourne Univ. Press & Cornell Univ. Press.

Balderson, J. 1991. Mantodea, ch. 21, pp. 348-356. In: CSIRO *The Insects of Australia*, vol. 1, Melbourne Univ. Press & Cornell Univ. Press.

Balduf, W. V. 1939. *The Bionomics of Entomophagous Insects*, pt. 2. Swift, 384 pp.

Berenbaum, M. R. 1995. *Bugs in the System. Insects and Their Impact on Human Affairs.* Addison-Wesley, Helix Books, xiii + 377 pp.

Bernays, E. A., & R. F. Chapman. 1994. *Host-Plant Selection by Phytophagous Insects.* Chapman & Hall Contemporary Topics Entomol., 2, xiii + 312 pp.

Berry, S. J. 1985. Reproductive systems, ch. 11, pp. 437-466. In: M. S. Blum (ed.) *Fundamentals of Insect Physiology.* John Wiley.

Blackwelder, R. E. 1967. *Taxonomy: A Text and Reference Book.* John Wiley, 698 pp.

Bland, R. G. 2003. *The Orthoptera of Michigan—Biology, Keys, and Descriptions of Grasshoppers, Katydids, and Crickets.* Mich. State Univ., Extension Bull. E-2815, 220 pp.

Bland, R. G., & H. E. Jaques. 1978. *How to Know the Insects,* 3rd ed. Wm. C. Brown Pictured Key Nature Ser., xi + 409 pp.

Blum, M. S. (ed.) 1985. *Fundamentals of Insect Physiology.* John Wiley, xiv + 598 pp.

Borror, D. J., C. A. Triplehorn, & N. F. Johnson. 1989. *Study of Insects,* 6th ed. Saunders, xiv + 875 pp.

Borror, D. J., & R. E. White. 1970. *A Field Guide to the Insects of America North of Mexico.* Houghton-Mifflin Peterson Fld. Guide Ser., 404 pp.

Brown, W. D., & D. T. Gwynne. 1997. Evolution of mating in crickets, katydids, and wetas, ch. 13, pp. 281-314. In: S. K. Gangwere, M. C. Muralirangan, & Meera Muralirangan (eds.) *The Bionomics of Grasshoppers, Katydids, and Their Kin.* CAB International.

Brues, C. T. 1946. *Insect Dietary.* Harvard Univ. Press, xxvi + 466 pp.

——. 1952. How insects choose their food plants. In: *USDA Yearbook of Agriculture*, pp. 37- 42.

Bullock, T. H., & G. A. Horridge. 1965. *Structure and Function in the Nervous System of Invertebrates.* W. H. Freeman, vol. 2, pp. 801-1719.

Bursell, E. 1970. *An Introduction to Insect Physiology.* Academic Press, xiv + 276 pp.

——. 1974. Environmental aspects—-temperature, ch. 1, pp. 1-41. In: M. Rockstein (ed.) *The Physiology of Insecta*, 2nd ed., vol. 2. Academic Press.

ENTOMOLOGY

——. 1974a. Environmental aspects—-humidity, ch. 2, pp. 43-84. *Ibid.*

Byers, G. W. 1991. Mecoptera, ch. 37, pp. 696-704. In: CSIRO *The Insects of Australia*, vol. 2, Melbourne Univ. Press & Cornell Univ. Press.

Calaby, J. H., & M. D. Murray. 1991. Phthiraptera, ch. 29, pp. 421-428. In: CSIRO *The Insects of Australia*, vol. 1, Melbourne Univ. Press & Cornell Univ. Press.

Carpenter, S. J., & W. J. LaCasse. 1974. *Mosquitoes of North America*. Univ. Calif. Press, vii + 360 pp.

Carver, M., G. F. Gross, & T. E. Woodward. 1991. Hemiptera, ch. 30, pp. 429-509. In: CSIRO *The Insects of Australia*, vol. 1, Melbourne Univ. Press & Cornell Univ. Press.

Chapman, R. F. 1974. Feeding in leaf-feeding insects. *Oxford Biol. Readers,* 69, 16 pp.

——. 1990. Food selection, ch. 2, pp. 39-72. In: R. F. Chapman & A. Joern (eds.) *Biology of Grasshoppers.* John Wiley.

——. 1991. General anatomy and function, ch. 2, pp. 33-67. In: CSIRO *The Insects of Australia*, vol. 1, Melbourne Univ. Press & Cornell Univ. Press.

——. 1998. *The Insects—Structure and Function*, 4th ed. Cambridge Univ. Press, xvii + 770 pp.

Chapman, R. F., & G. de Boer (eds.). 1995. *Regulatory Mechanisms in Insect Feeding*. Chapman & Hall, xviii + 398 pp.

Chapman, R. F., & A. Joern (eds.). 1990. *Biology of Grasshoppers*. John Wiley. x + 563 pp.

Cheng, L. (ed.). 1976. *Marine Insects*. American Elsevier, xii + 581 pp.

Christiansen, K., & P. F. Bellinger. 1978. *The Collembola of North America North of the Rio Grande*. Entomol. Reprint Specialists.

Chu, H. F., & L. K. Cutkomp. 1992. *How to Know the Immature Insects*. Wm. C. Brown Communications.

Clausen, C. P. 1940. *Entomophagous Insects*. McGraw-Hill, x + 688 pp.

Cochran, D. G. 1985. Excretory systems, ch. 3, pp. 91-138. In: M. S. Blum (ed.) *Fundamentals of Insect Physiology*. John Wiley.

Cole, F. R. 1969. *The Flies of Western North America*. Univ. Calif. Press, xi + 693 pp.

Colless, D. H., & D. K. McAlpine. 1991. Diptera, ch. 39, pp. 717-786. In: CSIRO *The Insects of Australia*, vol. 2, Melbourne Univ. Press & Cornell Univ. Press.

Conde, B., & J. Pages. 1991. Diplura, ch. 13, pp. 269-271. In: CSIRO *The Insects of Australia*, vol. 1, Melbourne Univ. Press & Cornell Univ. Press.

Corbet, P. S. 1963. *A Biology of Dragonflies*. Quadrangle, xvi + 247 pp.

Corell, C. V. 1984. *A Field Guide to the Moths of Eastern North America*. Houghton-Mifflin Peterson Fld. Guide Ser., xv + 496 pp.

Cornwell, P. B. 1968. *The Cockroach*, vol. 1. Hutchinson, 391 pp.

Cranston, P. S., P. J. Gullan, & R. W. Taylor. 1991. Principles and practice of systematics, ch. 4, pp. 109-124. In: CSIRO *The Insects of Australia*, vol. 1, Melbourne Univ. Press & Cornell Univ. Press.

Creighton, W. S. 1950. *The Ants of North America*. Bull. Mus. Comp. Zool., 104, 585 pp.

Crowson, R. A. 1981. *The Biology of the Coleoptera*. Academic Press, xii + 802 pp.

CSIRO. 1991. *The Insects of Australia*. Melbourne Univ. Press & Cornell Univ. Press. Vol. 1: xv + 542; vol. 2: vi + 1137 pp.

Davies, R. G. 1988. *Outlines of Entomology*, 7th ed. Chapman & Hall, viii + 408 pp.

Delcomyn, F. 1985. Factors regulating insect walking. *Annu. Rev. Entomol.,* 30: 239-256.

Dethier, V. G. 1963. *The Physiology of Insect Senses*. John Wiley, ix + 266 pp.

Dillon, E. S., & L. S. Dillon. 1972. *A Manual of Common Beetles of Eastern North America*. Dover, x + 894 pp.

Douglas, M. M. 1986. *The Lives of Butterflies*. Univ. Mich. Press, vii + 241 pp.

Bibliography

Dunn, G. A. 1996. *Insects of the Great Lakes Region*. Univ. Mich. Press, viii + 324 pp.

Dunnet, G. M., & D. K. Mardon. 1991. Siphonaptera, ch. 38, pp.705-716. In: CSIRO *The Insects of Australia*, vol. 2, Melbourne Univ. Press & Cornell Univ. Press.

Edmunds, G. F., S. L. Jensen, & L. Berner. 1976. *The Mayflies of North and Central America*. Univ. Minn. Press, x + 330 pp.

Ehrlich, P. R., & A. H. Ehrlich. 1961. *How to Know the Butterflies*. Wm. C. Brown Pictured Key Nature Ser., v + 262 pp.

Engelmann, F. 1970. *The Physiology of Insect Reproduction*. Pergamon, ix + 307 pp.

Evans, G. 1975. *The Life of Beetles*. Hafner Press, Macmillan. 232 pp.

Evans, H. E., & M. J. W. Eberhard. 1970. *The Wasps*. Univ. Mich. Press, vi + 265 pp.

Ewing, H. E., & I. Fox. 1943. *The Fleas of North America*. USDA Misc. Pub., 500, 142 pp.

Felt, E. P. 1965. *Plant Galls and Gall Makers*. Hafner Press, viii + 364 pp.

Ferris, G. F. 1951. *The Sucking Lice*, vol. 1. Mem. Pac. Coast Entomol. Soc., ix + 320 pp.

Ford, E. B. 1955. *Moths*. Macmillan, xix + 266 pp.

———. 1957. *Butterflies*, 3rd ed. Wm. Collins, xiv + 368 pp.

von Frisch, K. 1967. *The Dance Language and Orientation of Bees*. Belknap Press, Harvard Univ., xiv + 566 pp.

Frison, T. H. 1935. The stoneflies, or Plecoptera, of Illinois. *Bull. Ill. Nat. Hist. Surv.*, 20: 281- 471.

Gangwere, S. K. 1972. Host finding and feeding behavior in the Orthopteroidea, especially as modified by food availability: a review. *Rev. Univ. Madrid,* 21: 107-158.

———. 1991. Food habits and feeding behavior of locusts and grasshoppers. In: V. R. Vickery (ed.) *Field Guides to Locusts and Grasshoppers of the World*. Orth. Soc. Fld. Guides, 4BE, 56 pp.

Gangwere, S. K., M. C. Muralirangan, & Meera Muralirangan (eds.). 1997. *The Bionomics of Grasshoppers, Katydids, and Their Kin*. CAB International, xiii + 529 pp.

Gilbert, P., & C. J. Hamilton. 1990. *Entomology. A Guide to Information Services*, 2nd ed., Mansell.

Gordh, G., & D. H. Headrick. 2001. *A Dictionary of Entomology*. CAB International, ix + 1032 pp.

Gray, J. 1953. *How Animals Move*. Cambridge Univ. Press, xii + 114 pp.

Greenslade, P. J. 1991. Collembola, ch. 11, pp. 252-264. In: CSIRO *The Insects of Australia*, vol. 1, Melbourne Univ. Press & Cornell Univ. Press.

Gressitt, J. L. 1967. *Entomology of Antarctica*. Amer. Geophys. Union, Ant. Res. Ser., 10, xii + 395 pp.

Gullan, P. J., & P. S. Cranston. 2000. *The Insects: An Outline of Entomology*, 2nd ed. Blackwell Sci. Ltd., xvi + 470 pp. + appendix.

Headstrom, R. 1977. *The Beetles of America*. Barnes & Noble, 488 pp.

Helfer, J. R. 1953. *How to Know the Grasshoppers, Cockroaches, and Their Allies*. Wm. C. Brown Pictured Key Nature Ser., v + 353 pp.

Hepburn, H. R. (ed.). 1976. *The Insect Integument*. American Elsevier, xix + 571 pp.

———. 1985. The integument, ch. 4, pp. 139-183. In: M. S. Blum (ed.) *Fundamentals of Insect Physiology*. John Wiley.

Holland, W. J. 1968. *The Moth Book*. Dover, xxiv + 479 pp.

Holldobler, B., & E. O. Wilson. 1990. *The Ants*. Belknap Press, Harvard Univ., xii + 732 pp.

Hopkin, S. P. 1997. *Biology of the Springtails*. Oxford Univ. Press, x + 330 pp.

Howe, W. H. 1975. *The Butterflies of North America*. Doubleday, 633 pp.

Hubbard, C. A. 1947. *Fleas of Western North America*. Iowa State Coll. Press, ix + 533 pp.

Huffaker, C. B., & R. L. Rabb (eds.). 1984. *Ecological Entomology*. John Wiley, xiii + 844 pp.

ENTOMOLOGY

Hughes, G. M. 1952. The co-ordination of insect movements, pt. 1. The walking movements of insects. *J. Exp. Biol.*, 29: 267-284.

———. 1958. The co-ordination of insect movements, pt. 3. Swimming in *Dytiscus*, *Hydrophilus*, and a dragonfly nymph. *J. Exp. Biol.*, 35: 567-583.

Hughes, G. M., & P. J. Mill. 1974. Locomotion: terrestrial, ch. 5, pp. 335-379. In: M. Rockstein (ed.) *The Physiology of Insecta*, 2nd ed., vol. 3. Academic Press.

Imadate, G. 1991. Protura, ch. 12, pp. 265-268. In: CSIRO *The Insects of Australia*, vol. 1, Melbourne Univ. Press & Cornell Univ. Press.

Jaques, H. E. 1951. *How to Know the Beetles*. Wm. C. Brown Pictured Key Nature Ser., iii + 372 pp.

Jones, J. C. 1977. *The Circulatory System of Insects*. Charles C. Thomas, xvi + 255 pp.

Kathirithamby, J. 1991. Strepsiptera, ch. 36, pp. 684-695. In: CSIRO *The Insects of Australia*, vol. 2, Melbourne Univ. Press & Cornell Univ. Press.

Kettle, D. S. 1995. *Medical and Veterinary Entomology*, 2nd ed. ix + 725 pp. CAB International.

Kevan, D. K. McE. 1973. The place of classical taxonomy in modern systematic entomology, with particular reference to orthopteroid insects. *Can. Entomol.* 105: 1211-1222.

Key, K. H. L. 1991. Phasmatodea, ch. 25, pp. 394-404. In: CSIRO *The Insects of Australia*, vol. 1, Melbourne Univ. Press & Cornell Univ. Press.

Klots, A. B. 1951. *A Field Guide to the Butterflies of North America East of the Great Plains*. Peterson Fld. Guide Ser., Houghton-Mifflin, xvi + 349 pp.

Klots, A. B., & E. B. Klots. 1971. *Insects of North America*. Doubleday, 250 pp.

Lawrence, J. F., & E. B. Britton. 1991. Coleoptera, ch. 35, pp. 543-683. In: CSIRO *The Insects of Australia*, vol. 2, Melbourne Univ. Press & Cornell Univ. Press.

Lawrence, J. F., E. S. Nielsen, & I. M. Mackerras. 1991. Skeletal anatomy and key to orders, ch. 1, pp. 1-32. In: CSIRO *The Insects of Australia*, vol. 1, Melbourne Univ. Press & Cornell Univ. Press.

Leather, S. R., & J. Hardie (eds.). 1995. *Insect Reproduction*. CRC Press, 255 pp.

Lehmkuhl, D. M. 1979. *How to Know the Aquatic Insects*. Wm. C. Brown Pictured Key Nature Ser., xi + 168 pp.

Lewis, T. 1973. *Thrips: Their Biology, Ecology, and Economic Importance*. Academic Press, xv + 349 pp.

Locke, M. 1974. The structure and formation of the integument in insects, pp.123-213. In: M. Rockstein (ed.) *The Physiology of Insecta*, 2nd ed., vol. 6. Academic Press.

———. 2001. The Wigglesworth Lecture: insects for studying fundamental problems in biology. *J. Insect Physiol.*, 47: 495-507.

Maddrell, S. H. P. 1971. The mechanisms of insect excretory systems. *Advan. Insect Physiol.*, 8: 200-331. Academic Press.

Mani, M. S. 1962. *Introduction to High Altitude Entomology*. Methuen, xix + 302 pp.

———. 1968. *Ecology and Biogeography of High Altitude Insects*. W. Junk, xiv + 527 pp.

Marshall, A. G. 1981. *The Ecology of Ectoparasitic Insects*. Academic Press, xvi + 459 pp.

Matsuda, R. 1965. *Morphology and Evolution of the Insect Head*. Mem. Amer. Entomol. Inst., 4, 334 pp.

———. 1970. *Morphology and Evolution of the Insect Thorax*. Mem. Entomol. Soc. Can., 76, 431 pp.

———. 1976. *Morphology and Evolution of the Insect Abdomen*. Pergamon, viii + 534 pp.

Mayr, E. 1969. *Principles of Systematic Zoology*. McGraw-Hill, xi + 428 pp.

McFarlane, J. E. 1985. Nutrition and digestive organs, ch. 2, pp. 59-89. In: M. S. Blum (ed.) *Fundamentals of Insect Physiology*. John Wiley.

Merritt, R. W., & K. W. Cummins (eds.). 1996. *An Introduction to the Aquatic Insects of North America*, 3rd ed. Kendall-Hunt, xiii + 862 pp.

Bibliography

Metcalf, R. L., & R. A. Metcalf. 1993. *Destructive and Useful Insects*, 5th ed. McGraw-Hill.

Michener, C. D. 1974. *The Social Behavior of the Bees: A Comparative Study*. Belknap Press, Harvard Univ., xii + 404 pp.

Michener, C. D., & M. H. Michener. 1951. *American Social Insects*. Van Nostrand Reinhold, xiv + 267 pp.

Miller, N. C. E. 1971. *The Biology of the Heteroptera*. E. W. Classey, xiii + 206 pp.

Miller, T. A. 1975. Insect visceral muscle, ch. 10, pp. 545-606. In: P. N. R. Usherwood (ed.) *Insect Muscle.* Academic Press.

Mockford, E. L. 1993. *North American Psocoptera*. Sandhill Crane Press, xviii + 455 pp.

Mound, L. A., & B. S. Heming. 1991. Thysanoptera, ch. 31, pp. 510-515. In: CSIRO *The Insects of Australia*, vol. 1, Melbourne Univ. Press & Cornell Univ. Press.

Nachtigall, W. 1974. *Insects in Flight*. McGraw-Hill, 153 pp.

Nation, J. L. 1985. Respiratory systems, ch. 5, pp. 185-225. In: M. S. Blum (ed.) *Fundamentals of Insect Physiology*. John Wiley.

———. 2002. *Insect Physiology and Biochemistry*. CRC Press, 485 pp.

Naumann, I. D. 1991. Hymenoptera, ch. 42, pp. 916-1000. In: CSIRO *The Insects of Australia*, vol. 2, Melbourne Univ. Press & Cornell Univ. Press.

Neboiss, A. 1991. Trichoptera, ch. 40, pp. 787-816. In: CSIRO *The Insects of Australia*, vol. 2, Melbourne Univ. Press & Cornell Univ. Press.

Needham, J. G., J. R. Traver, & Y.-C. Hsu. 1935. *The Biology of Mayflies*. Comstock, xvi + 759 pp.

Needham, J. G., & M. J. Westfall. 1955. *A Manual of the Dragonflies of North America*. Univ. Calif. Press, xii + 615 pp.

New, T. R. 1991. Neuroptera, ch. 34, pp. 525-542. In: CSIRO *The Insects of Australia*, vol. 1, Melbourne Univ. Press & Cornell Univ. Press.

Nielsen, E. S., & I. F. B. Common. 1991. Lepidoptera, ch. 41, pp. 817-915. In: CSIRO *The Insects of Australia*, vol. 2, Melbourne Univ. Press & Cornell Univ. Press.

Nijhout, H. F. 1994. *Insect Hormones*. Princeton Univ. Press.

Norris, K. R. 1991. General biology, ch. 3, pp. 68-108. In: CSIRO *The Insects of Australia*, vol. 1, Melbourne Univ. Press & Cornell Univ. Press.

Oberlander, H. 1985. Hormone action during insect development, ch. 13, pp. 507-534. In: M. S. Blum (ed.) *Fundamentals of Insect Physiology*. John Wiley.

Odum, E. P. 1971. *Fundamentals of Ecology*, 3rd ed. Saunders, xiv + 574 pp.

Oldroyd, H. 1964. *The Natural History of Flies*. Weidenfeld & Nicolson, xiv + 324 pp.

Otte, D. 1981. *The North American Grasshoppers*. Harvard Univ. Press, vol. 1, ix + 275 pp.

———. 1984. *The North American Grasshoppers*. Harvard Univ. Press, vol. 2, x + 366 pp.

Patton, R. L. 1963. *Introductory Insect Physiology*. Saunders, vi + 245 pp.

Peters, W. L., & I. C. Campbell. 1991. Ephemeroptera, ch. 16, pp. 279-293. In: CSIRO *The Insects of Australia*, vol. 1, Melbourne Univ. Press & Cornell Univ. Press.

Peterson, A. 1948-1951. *Larvae of Insects*, pts. 1 & 2. Edwards Bros., pt. 1: 315 pp.; pt. 2: 416 pp.

Price, P. W. 1997. *Insect Ecology*, 3rd ed. John Wiley, xii + 874 pp.

Pringle, J. W. S. 1975. Insect flight. *Oxford Biol. Readers,* 52: 16 pp.

Rentz, D. C. F. 1991. Grylloblattodea, ch. 22, pp. 357-359. In: CSIRO *The Insects of Australia*, vol. 1, Melbourne Univ. Press & Cornell Univ. Press.

———. 1991. Orthoptera, ch. 24, pp. 369-393. In: CSIRO *The Insects of Australia*, vol. 1, Melbourne Univ. Press & Cornell Univ. Press.

Rentz, D. C. F., & D. K. McE. Kevan. 1991. Dermaptera, ch. 23, pp. 360-368. In: CSIRO *The Insects of*

Australia, vol. 1, Melbourne Univ. Press & Cornell Univ. Press.

Ribbands, C. R. 1953. *The Behavior and Social Life of Honeybees.* Dover, 352 pp.

Richards, A. G. 1951. *The Integument of Arthropods.* Univ. Minn. Press, xvi + 411 pp.

———. 1953. Structure and development of the integument, chs. 1-3, pp. 1-54. In: K. D. Roeder (ed.) *Insect Physiology.* John Wiley.

Richards, O. W. 1953. *The Social Insects.* Macdonald, xiii + 219 pp.

Richards, O. W., & R. G. Davies. 1977. *Imms' General Textbook of Entomology.* Halsted Press, Chapman & Hall, vol. 1: viii + 418 pp.; vol. 2: viii + 1354 pp.

Rodriguez, J. G. (ed.). 1972. *Insect and Mite Nutrition.* North-Holland Pub., xiii + 702 pp.

Ross, E. S. 1944. A revision of the Embioptera, or web-spinners, of the New World. *Proc. U. S. Nat. Mus.*, 94: 401-504.

———. 1991. Embioptera, ch. 26, pp. 405-409. In: CSIRO *The Insects of Australia*, vol. 1, Melbourne Univ. Press & Cornell Univ. Press.

Ross, H. H. 1944. *The Caddis Flies, or Trichoptera, of Illinois.* Bull. Ill. Nat. Hist. Surv., 23, 326 pp.

———. 1974. *Biological Systematics.* Addison-Wesley Pub. Co., 345 pp.

Roth, L. M. 1991. Blattodea, ch. 19, pp. 320-329. In: CSIRO *The Insects of Australia*, vol. 1, Melbourne Univ. Press & Cornell Univ. Press.

Salt, G. 1970. *The Cellular Defence of Insects.* Cambridge Univ. Press, vi + 118 pp.

Scott, J. A. 1986. *The Butterflies of North America.* Stanford Univ. Press, xiii + 583 pp.

Scudder, G. G. E. 1961. The comparative morphology of the insect ovipositor. *Trans. Roy. Entomol. Soc. London*, 113: 25-40.

———. 1971. Comparative morphology of insect genitalia. *Annu. Rev. Entomol.*, 16: 379-406.

Simpson, G. G. 1961. *Principles of Animal Taxonomy.* Columbia Univ. Press, xii + 247 pp.

Skaife, S. H. 1956. *Dwellers in Darkness.* Longmans & Green, x + 134 pp.

Slater, J. A., & R. M. Baranowski. 1978. *How to Know the True Bugs.* Wm. C. Brown Pictured Key Nature Ser., x + 256 pp.

Smith, G. B., & J. A. L. Watson. 1991. Thysanura, ch. 15, pp. 275-278. In: CSIRO *The Insects of Australia*, vol. 1, Melbourne Univ. Press & Cornell Univ. Press.

Smithers, C. N. 1991. Zoraptera, ch. 27, pp. 410-411. In: CSIRO *The Insects of Australia*, vol. 1, Melbourne Univ. Press & Cornell Univ. Press.

Smyth, T. 1985. Muscle systems, ch. 6, pp. 227-252. In: M. S. Blum (ed.) *Fundamentals of Insect Physiology.* John Wiley.

Snodgrass, R. E. 1935. *Principles of Insect Morphology.* McGraw-Hill, ix + 667 pp.

———. 1952. *A Textbook of Arthopod Anatomy.* Comstock, viii + 361 pp.

———. 1954. Insect metamorphosis. *Smithson. Misc. Colln.*, 122, iii + 124 pp.

———. 1967. *Insects: Their Ways and Means of Living.* Dover, iv + 362 pp.

Snyder, T. E. 1948. *Our Enemy, the Termite*, rev. ed. Comstock, 257 pp.

———. 1954. *Order Isoptera—the Termites of the United States and Canada.* Nat. Pest Control Ass., 64 pp.

Southwood, T. R. E. 1972. The insect/plant relationship—an evolutionary perspective. In: H. F. Van Emden (ed.) *Symp. Roy. Entomol. Soc. London*, 6, pp. 3-30.

Spradbery, J. P. 1973. *Wasps. An Account of the Biology and Natural History of Solitary and Social Wasps.* Univ. Wash. Press, xvi + 408 pp.

Stannard, L. J. 1968. *The thrips, or Thysanoptera, of Illinois.* Bull. Ill. Nat. Hist. Surv., 29: vi + 215-552.

Steinmann, C. Sc., & L. Zambori. 1985. *An Atlas of Insect Morphology*, 2nd ed. Akademiai Kiado, Budapest,

Bibliography

253 pp.

Steyskal, G. C. 1978. What is systematic entomology? *Proc. Entomol. Soc. Wash.* 80: 43-50.

Theischinger, G. 1991. Plecoptera, ch. 18, pp. 311-319. In: CSIRO *The Insects of Australia*, vol. 1, Melbourne Univ. Press & Cornell Univ. Press.

———. 1991. Megaloptera, ch. 32, pp. 516-520. In: CSIRO *The Insects of Australia*, vol. 1, Melbourne Univ. Press & Cornell Univ. Press.

Thornhill, R., & J. Alcock. 1983. *The Evolution of Insect Mating Systems.* Harvard Univ. Press, 547 pp.

Tuxen, S. L. 1964. *The Protura. A Revision of the Species of the World with Keys for Determination.* Hermann, 360 pp.

——— (ed.). 1970. *Taxonomist's Glossary of Genitalia in Insects*, 2nd ed. Munksgaard, 359 pp.

USDA. 1952. *Insects. The Yearbook of Agriculture.* U. S. Gov. Printing Office, Wash., D. C., xviii +780 pp. + pls.

Uvarov, B. P. 1966, 1977. *Grasshoppers and Locusts. A Handbook of General Acridology.* Vol. 1 (1966) Cambridge Univ. Press, xi + 481 pp.; vol. 2 (1977) Centre for Overseas Pest Res., ix + 613 pp.

Walker, A. 1994. *The Arthropods of Humans and Domestic Animals.* Chapman & Hall, xx + 213 pp.

Wallwork, J. A. 1976. *The Distribution and Diversity of Soil Fauna.* Academic Press, xiv + 355 pp.

Ward, J. V. 1992. *Aquatic Insect Ecology. 1. Biology and Habitat.* John Wiley, xi + 438 pp.

Watson, J. A. L., & F. J. Gay. 1991. Isoptera, ch. 20, pp. 330-347. In: CSIRO *The Insects of Australia*, vol. 1, Melbourne Univ. Press & Cornell Univ. Press.

Watson, J. A. L., & A. F. O'Farrell. 1991. Odonata, ch. 17, pp. 294-310. In: CSIRO *The Insects of Australia*, vol. 1, Melbourne Univ. Press & Cornell Univ. Press.

Watson, J. A. L., & G. B. Smith. 1991. Archeognatha (Microcoryphia), ch. 14, pp. 272-274. In: CSIRO *The Insects of Australia*, vol. 1, Melbourne Univ. Press & Cornell Univ. Press.

White, R. E. 1983. *A Field Guide to the Beetles of North America.* Houghton-Mifflin Peterson Fld. Guide Ser., xii + 368 pp.

Wigglesworth, V. B. 1954. *The Physiology of Insect Metamorphosis.* Cambridge Univ. Press, viii + 152 pp.

———. 1959. *The Control of Growth and Form.* Cornell Univ. Press, 140 pp.

———. 1970. *Insect Hormones.* W. H. Freeman, ix + 159 pp.

———. 1972. *The Principles of Insect Physiology*, 7th ed. John Wiley, viii + 827 pp.

———. 1974. Insect hormones. *Oxford Biol. Readers*, 70, 16 pp.

———. 1984. *Insect Physiology*, 8th ed. Chapman & Hall, x + 191 pp.

Williams, C. M. 1950. The metamorphosis of insects. *Sci. Amer. Reprint*, 49, 5 pp.

———. 1958. The juvenile hormone. *Sci. Amer.* (Feb.), 67-74.

———. 1967. Third-generation pesticides. *Sci. Amer.*, 217 (1): 13-17.

Williams, D. D., & B. W. Feltmate. 1992. *Aquatic Insects.* xiii + 358 pp. CAB International.

Wilson, E. O. 1971. *The Insect Societies.* Belknap Press, Harvard Univ., x + 548 pp.

———. 1975. *Sociobiology.* Belknap Press, Harvard Univ., ix + 697 pp.

Woodring, J. P. 1985. Circulatory systems, ch. 1, pp. 5-57. In: M. S. Blum (ed.) *Fundamentals of Insect Physiology.* John Wiley.

Wygodzinsky, P. 1972. A review of the silverfish of the United States and the Caribbean area. *Amer. Mus. Novit.*, 2481, 26 pp.

Wygodzinsky, P., & K. Schmidt. 1980. Survey of the Microcoryphia (Insecta) of the northeastern United States and adjacent provinces of Canada. *Amer. Mus. Novitat.*, 2071, 17 pp.

Glossary

Abdomen. The 3rd, most posterior region of the insect body usually consisting, in adults, of 9-10 legless segments. Compare with *head* and *thorax*.

Accessory glands. Glands of the female reproductive system that empty into the *copulatory pouch* and coat *eggs* in a protective covering, or glands of the male system that empty into the *vasa deferentia* or *seminal vesicles* and produce seminal fluid or *spermatophores*.

Accessory vessels. Paired, loosely valved segmental tubes of the circulatory system that exit from the heart, bifurcate around the tergosternal muscles, and end in fat body.

Acron. The unsegmented preoral part of the head immediately forward of the antennal, or 2nd segment, apparently comparable to the prostomium of other invertebrates.

Acrotrophic. See *ovarioles*.

Action. Contraction or shortening of a muscle bringing the insertion closer to the origin, thereby pulling on any membrane, appendage, or sclerite to which the muscle attaches causing movement of the limb or other part with respect to the body as a whole.

Action potential. The transient change in electrical potential within a nerve or muscle cell upon excitation. Compare with *resting potential*.

-ad. Suffix used to express direction, as with *dorsad* (upward), *ventrad* (downward), *cephalad* (forward), *caudad* (backward), *etc.*

Adecticous. A *pupa* with immovable, non-functional mandibles. Compare with *decticous*.

Adhesive organs. Soft, pad-like structures on the legs that seem instrumental in climbing surfaces. They include *arolia*, each a hoof-like, pretarsal structure at the apex of the tarsus between the tarsal claws; *plantulae*, pad-like structures on the underside of the tarsal segments; and, in Diptera, *pulvilli*, paired pads associated with the tarsal claws, and sometimes *empodia*, each either an unpaired pad-like or spinous structure between the tarsal claws.

Adult. See *imago*.

Aedeagus. The distal part of the phallus or male intromittent organ.

Air sacs. Pouch-like expansions of the *tracheae* generally lacking spiral supportive *taenidia* and, hence, capable of accommodating inflation or deflation, making possible breathing.

Air stores. Buoyant air bubbles or channels on the body surface of certain aquatic insects into which the *spiracles* open, acting as a physical gill for gas exchange under water.

Alar sclerites. Supportive thoracic structures (anterior *basalar* and posterior *subalar* sclerites) on either side of the wing process that are involved in wing articulation.

Alimentary canal. The digestive tract traversing the body from *mouth* to *anus*. See *intestine* or *gut*.

Alinotum. The dorsal thoracic sclerite; specifically the *tergum* or *notum* of the meso- or metathorax of a wing-bearing insect. See *pleuron* and *sternum*.

Ametabolous or *direct development.* A postembryonic development that lacks an obvious *metamorphosis* as in Collembola, Diplura, Microcoryphia, and Thysanura whose juveniles differ little from one another and from the adult, except in size and fertility.

Ampullae. Proximal swellings of the *Malpighian tubules* that, in some insects, intervene between the tubules and the gut into which they discharge.

Anal wing margin. See *wing margins*.

ENTOMOLOGY

Anamorphosis. The type of *ametabolous* or *direct development* that occurs in Protura in which an additional abdominal segment is added per molt up to the adult instar.

Antennae. The paired, movable, sensory appendages of the 2nd segment of the head between or below the compound eyes. They are either *segmented,* in which each *article* or subsegment is independently movable by intrinsic muscles from the preceding article, as in Diplura and Collembola, or *annulated,* in which intrinsic muscles occur only in the 1st article, as in Thysanura and Pterygota.

Anterior. Before or in front of. Compare with *posterior.*

Anterior intestine. The cephalic section of proctodaeum behind the entry of the *Malpighian tubules,* usually opening into the *posterior intestine* through a rectal valve.

Antiperistalsis. See *peristalsis.*

Anus. The posterior opening of the alimentary canal.

Aorta. The anterior, non-chambered portion of the dorsal blood vessel opening into the head. See *dorsal blood vessel* and *heart.*

Apical. At or near the tip of an appendage or other structure.

Apical wing margin. See *wing margins.*

Apneustic. A type of respiration carried out in absence of functional *spiracles,* as in those larval Diptera that use cutaneous respiration and various aquatic naiads that rely on *tracheal gills.* Compare with *oligopneustic* and *polypneustic.*

Apodemes. The "endoskeleton" of insects consisting of rigid body wall invaginations or inpocketings visible externally as a crease along the body surface. They support and serve in muscle attachment. Compare with *apophyses.*

Apolysis. Separation of the *epidermis* from the old *cuticle* during *molting.*

Apophyses (sing., *apophysis*). Slender, arm-like, rigid invaginations of the external body wall used for muscle attachment. Compare with *apodemes.*

Apposition. The image formed by the eyes of day-active insects adapted so that only light rays normal to the lens are perceived and all oblique rays are absorbed. Compare with *superposition.*

Apterous. Wingless. Compare with *brachypterous* and *macropterous.*

Apterygota. The arthropod subclass consisting of primitively wingless, insect-like creatures with a *direct development.* Compare with *Pterygota.*

Arista. A large, usually dorsal bristle located on the 3rd antennal article in "higher" Diptera.

Arolium. See *adhesive organs.*

Arthrodial membrane. The flexible connections by which the *articles,* or subsegments, of jointed appendages attach to one another. They are comparable to the *conjunctivae,* or intersegmental membranes, that articulate abdominal and other body segments.

Articles. The subsegments of insects' jointed appendages.

Asynchronous wing muscles. Rapidly contracting muscles of swift fliers whose individual contractions can merge into a sustained, rapid, vibratory contraction stemming from the inherent elasticity of cuticle, muscle, wings, and *resil*in pads at the wing base; they do not beat in direct response to individual contractions of the muscles that drive them. See *synchronous wing muscles.*

Athrocytes. Specialized phagocytes associated with the *Malpighian tubules* or gut. See *phagocytosis.*

Atrium. An integumentary depression through which a *spiracle* opens to the outside.

Atrophied or *vestigial.* Non-functional structures; ones that are rudimentary or reduced in size.

Attenuated. Slender and tapering distally.

Axillary sclerites. Sclerites at the wing base that permit wing flexion, hence, developed only in Neoptera, the winged insects.

Axillary wing margin. See *wing margins.*

Basal. Toward the base of an appendage or structure. See *apical.*

Basement membrane. The simple, homogeneous layer that overlies the *hemocoele,* muscles, and other body tissues and underlies the *epidermis.*

Beak. The protruding mouthparts, proboscis, or snout of a weevil, snakefly, or certain other insects.

Benthic. Pertaining to the *benthos,* or bottom-dwelling organisms, of lakes or sea.

Bifurcate. To divide or fork into two parts.

Binomen. See *species* and *scientific name.*

Glossary

Biotic potential. The capacity of organisms exposed to optimal conditions to increase in population numbers up to the carrying capacity of the environment.

Bisexual. Species having two sexes, males and females, being either *monoecious* (both male and female organs found in the same hermaphroditic individual) or *dioecious* (having separate sexes, either males or females).

Biting, chewing, or *mandibulate mouthparts.* See *mouthparts.*

Blood. See *hemolymph.*

Blood gills. See *gills.*

Boring. Pertaining to the habits of adult insects or their larvae that tunnel in woody or other tissues. Compare with *mining.*

Brachypterous. Having short wings that do not cover the abdomen. Compare with *apterous* and *macropterous.*

Brain. The *supraesophageal ganglion,* or brain, is the complex fusion product of 3 pairs of ganglia located in the head above the pharynx or esophagus and responsible for monitoring body activities.

Brain neurosecretory cells. See *PTTH.*

Breathing. The mechanical facilitation of the gas exchanges of respiration involving an enhanced, usually directed air flow through the major *tracheae* by the rhythmic expansion and contraction of the *air sacs,* chiefly those of the abdomen.

Bursa copulatrix. A diverticulum separate from the female's *copulatory pouch* that, where developed, receives the *aedeagus* and associated male parts during coitus.

Caecum (pl., *caeca*). A blind sac or pouch that opens into the mid gut and is responsible for increasing the effective internal secretory and/or absorptive surfaces.

Calypteres. One or two lobular membranous scales located above the *haltere*s at the wing base of calyptrate Diptera.

Campaniform organs. Dome-like, mechanoreceptive *sensilla* that are widely distributed over the body, especially at or near *conjunctivae* and *arthrodial membrane.*

Campodeiform larvae. Elongate, flattened, active larvae with well-developed legs and antennae.

Cannibalism. A type of *carnivory* involving the eating of one's own species.

Cardo (pl., *cardines*). The basal sclerite of a *maxilla.*

Carnivory. Eating the flesh of animals. Compare with *predation* and *phytophagy.*

Castes. The different body forms of social insects, each devoted to specialized tasks, as with workers, soldiers, queens, *etc.*

Caterpillars. See *eruciform larvae.*

Caudal. Belonging to, attached to, or toward the posterior end.

Cavernicoly. Occupancy of caves, caverns, or mines. Compare with *fossorial* and *interstitial.*

Cell. Any partly or completely closed area of wing membrane surrounded by *veins* and *cross veins.*

Central nervous division. That part of the nervous system consisting of the *brain* and *ventral nerve cord.* Compare with *peripheral* and *sympathetic divisions.*

Centrolecithal. An egg cell in which, as in insects, the yolk is centrally located.

Cephalic. Belonging to, attached to, or toward the head.

Cerci (sing., *cercus*). The paired feeler-like, mechanoreceptive appendages of the primitive 11th abdominal segment that often appear to arise from the 10th or 9th segments.

Cervix. See *neck.*

Chaetotaxy. The study and use in *taxonomy* of insects' setae, bristles, spines, knobs, and other external *integumentary processes.*

Chemoreceptors. Sensilla such as those of the antennae and palpi responsible for insects' perception of taste and smell.

Chitin. The colorless, nitrogenous polysaccharide within the protein matrix of the *endocuticle.*

Chorion. The outer protective shell of an insect *egg* secreted by the female's follicular cells.

Chorionic sculpturing. Impressions of the follicular cells, usually seen as hexagonal ridging on the external chorionic surface, and often usable taxonomically.

Chrysalis. The *pupa* of a butterfly.

Circumesophageal connectives. Nerves that fork around the gut and connect the *brain* with the *subesophageal ganglion* of the *ventral nerve cord.*

ENTOMOLOGY

Claws. Articulated, multicellular *integumentary processes* of the *pretarsus*.

Click mechanism. The tendency of insect wings to be stable in only two positions, either wings up or wings down, owing to thoracic resiliency.

Climate. The long-term, overall macroclimatic manifestations of *weather*. Compare with *microclimate*.

Clotting. Formation of a plug that seals a wound on the insect body, stopping bleeding, apparently involving *hyaline hemocytes* or *coagulocytes*.

Clypeus. The unpaired sclerite on the insect face to which the *labrum* attaches.

Coagulocytes. See *clotting*.

Coarctate pupa. A pupa enclosed within the hardened last larval skin.

Cocoon. The silken or other case within which a *pupa* develops.

"Cold bloodedness." See *poikilothermy*.

Common name. A vernacular name applied to a *species* or other taxonomic unit. Such designations tend to be inappropriate, to lack consistency and universality, and often are not useful.

Common oviduct. See *oviducts*.

Communal insects. Gregarious insects that share a nest but are not truly social.

Community. All the species populations, both plant and animal, that live together in a particular place, interact with one another, and have a degree of mutual adjustment. Compare with *ecosystem* and *population*.

Community metabolism. The energy flow that proceeds from sun, to producers, to insects and other consumers, and back to the physical environment.

Complete development. See *holometabolous development*.

Compound eyes. The large, paired, lateral eyes of most insects, each composed of individual visual elements, or *ommatidia*, seen externally as a *facet*. Compare with *simple eyes* including *ocelli* and *stemmata*.

Compressed. Laterally flattened, as with the body of fleas. Compare with *depressed*.

Condyles. Knob-like processes by which an appendage articulates with the body.

Congeners. Species belonging to the same genus. Compare with conspecific.

Conjunctivae or *intersegmental membrane*. The membranous articulations between *sclerites*. Compare with *arthrodial membrane* and *obsolete sutures*.

Conspecific. To belong to the same species. Compare with *congener*.

Coordinating nerve centers. See *reflex centers*.

Copulatory pouch or *vagina*. The terminal portion of the female reproductive system opening to the outside.

Cornea. The external, cuticular surface of a visual unit, whether of *simple* or of *compound eyes*.

Corpora allata (sing., *corpus allatum*). One or two small glandular bodies located behind the *brain* and *corpora cardiaca* and known to secrete *juvenile hormones*.

Corpora cardiaca (sing., *corpus cardiacum*). Usually paired neurohemal organs closely associated with the *dorsal aorta* behind the *brain* that store, mix, and release *PTTH* and other endocrine secretions.

Costal wing margin. See *wing margins*.

Coxa (pl., *coxae*). The basal segment of a leg that joins it to the thorax.

Coxal cavity. The lateral opening into which the coxa fits to articulate with the thorax.

Coxal processes. The two *condyles,* or points of articulation, of a leg to the thorax.

Creeping. The telescopic, syringe movements of certain often legless insects that crawl slowly over the substrate.

Cremaster. A hold-fast spine or hook at the posterior end of certain *pupae*.

Crepuscular. Active at dusk or dawn. Compare with *diurnal* and *nocturnal*.

Crochets. Hooked spines at the tip of the abdominal *larvapods,* or prolegs, of lepidopterous caterpillars.

Crop. The fore gut's dilated portion which holds food received from the *esophagus*.

Cross veins. Short crosswise veins that connect adjacent longitudinal wing veins and/or their branches. See *veins*.

Cryptonephridial. Malpighian tubules whose tips are imbedded within the wall of the *hind gut*, thereby increasing water retention.

Cuticle. The body wall's non-cellular outer layer secreted by the epidermal cells of the *epidermis*.

Decticous. A *pupa* with movable, functional mandibles. Compare with *adecticous*.

Density-dependent factors. Environmental factors, mostly biotic, responsible for a level of mortality that varies with population size.

Density-independent factors. Environmental factors, mostly physical, that cause a given level of mortality independent of fluctuations in population size.

Glossary

Depress. To lower a leg, wing, or other appendage with respect to the horizon. Compare with *elevate.*

Depressed. Dorsoventrally flattened, as with the body of bird lice and sucking lice. Compare with *compressed.*

Dermal gland ducts. Canals responsible for spreading cement from the dermal glands of the *epidermis* to the *epicuticle.* Contrast with *pore canals.*

Deutocerebrum. See *mid brain.* Compare with *fore brain* and *hind brain.*

Diapause. A period of arrested development and reduced metabolic rate not immediately referable to environmental conditions. Compare with *dormancy.*

Diastole. Relaxation of the heart muscles allowing the heart to fill with blood. Compare with *systole.*

Dicondylic. A joint having two points of articulation. Compare with *unicondylic.*

Dilated. An expanded or widened portion of a canal.

Dioecious. See *bisexual.*

Direct development. See *ametabolous development.*

Direct wing muscles. The *axillary* and *alar wing muscles* that insert on the wing base and thereby flex or extend the wing, respectively. Compare with *indirect wing muscles.*

Distal. Near or toward the free end of an appendage, being the part most removed from the body as a whole. Compare with *proximal.*

Diurnal. Active during the day. Compare with *crepuscular* and *nocturnal.*

Dormancy. A state of quiescence or inactivity usually referable to temperature, moisture, or other environmental conditions. Compare with *diapause.*

Dorsal. At or belonging to the upper surface, dorsum, or top. Compare with *ventral.*

Dorsal blood vessel. The main circulatory structure of insects consisting of a dilated abdominal *heart* that receives blood from the *hemocoele* and a slender thoracic *aorta* that empties into the head.

Dorsum. The top surface of an organism. Compare with *venter.*

Ecdysial cleavage line. A preformed mid-dorsal line of weakness along the head and thorax where the *cuticle* separates during *ecdysis*; formerly called *epicranial suture.*

Ecdysis. The physical act of shedding the old *cuticle,* or *exuviae,* during the process of *molting.*

Eclosion. Emergence of the adult from the *pupa.*

Economic entomology. The scientific discipline concerned with the study and alleviation of the depredations caused by insects of agricultural and veterinary importance. Compare with *medical entomology.*

Ecosystem. The homeostatic (tending to remain in equilibrium) ecological interaction of the fauna, the flora, and the non-living physical environment of some self-contained place on earth, either large or small. Compare with *community* and *population.*

Ectognathy. Having *exserted* mouthparts not retracted into the head, as is general in pterygotes. Compare with *entognathy.*

Ectoparasite. A *parasite* that lives on or near the body of its host. Compare with *endoparasite.*

Effectors. Muscles or glands that are activated by nerve stimuli. Compare with *receptors.*

Egg. A single cell produced by an *ovary,* containing yolk, surrounded by a shell-like *chorion,* and capable of being fertilized by a sperm cell to form a *zygote,* or 1st-stage embryo.

Ejaculatory duct. The slender, muscular, terminal male duct through which seminal fluid or spermatophores are expelled.

Elevate. To lift or raise a leg, wing, or other appendage with respect to the horizon. Compare with *depress.*

Elytra (sing., *elytron*). The thickened, leathery, or horny fore wings that cover the flight wings of beetles and certain other insects. Compare with *hemelytra* and *tegmina.*

Emigration. To leave, never to return. Compare with *immigration.*

Empodium. See *adhesive organs.*

Encapsulation. The process whereby *hemocytes* form a protective layer around a parasite or other foreign body within the *hemocoele.*

Endocuticle. The inner, flexible, unsclerotized layer of the cuticle subject to digestion during molting. Compare with *exocuticle* and *epicuticle.*

Endoparasite. A parasite that lives within the body cavity or tissues of its host. Compare with *ectoparasite.*

Endopterygotes. Holometabolous insects, *viz.,* ones in which the wings develop as internal wing discs. Compare with *exopterygotes.*

ENTOMOLOGY

Endoskeleton. The internal part of the skeleton including *apodemes, apophyses*, and other structures used for support and muscle attachment.

Energy flow. See *community metabolism.*

Entognathy. Mouthparts retracted within the head, as in the apterygote orders Protura, Collembola, and Diplura. Compare with *ectognathy.*

Entomophagy. See *insectivory.*

Environment. The sum total of conditions affecting organisms in the general place (land, fresh water, or seas) in which they live. Compare with *habitat.*

Environmental factors. See *density-dependent* and *density-independent.*

Epicranial suture. See *ecdysial cleavage line.*

Epicranium. The upper surface of the head from the *frons* backward to the *cervix*, or neck.

Epicuticle. The thin, refractile, outermost layer of the *cuticle*. Compare with *endocuticle* and *exocuticle*.

Epidermis. The single layer of integumentary cells located outside of the *basement membrane* and beneath the *cuticle*, for the secretion of which it is responsible.

Epimeron. The posterior division of a thoracic *pleuron* immediately behind the pleural suture. Compare with *episternum.*

Epipharynx. A lobular organ attached to the inner medial surface of the *labrum* or *clypeus* within the *preoral cavity.*

Epiproct. A dorsal plate covering the *anus*. It arises from the 12th abdominal segment though it appears to stem from the 10th. Compare with *paraprocts*.

Episternum. The anterior division of a thoracic *pleuron* immediately in front of the pleural suture. Compare with *epimeron.*

Eruciform larvae. Moth, butterfly, sawfly, or other caterpillars characterized by a cylindrical body and a well-developed head, thoracic legs, and *larvapods*.

Esophagus. The narrow, tubular portion of digestive tract immediately behind the *pharynx* and in front of the *crop.*

Evagination. An outpocketing or sac-like structure on the outside of the body. Compare with *invagination.*

Eversible. Capable of being everted or turned inside-out.

Exarate pupae. Pupae with free appendages, ones not adherent to the body. Compare with *obtect pupae*.

Excretion. Elimination of nitrogenous and other metabolic wastes to keep constant the body's internal environment.

Exocuticle. The layer of sclerotized, rigid, usually pigmented cuticle sandwiched between the *endocuticle*, beneath, and the *epicuticle*, above, and resistant to molting fluid.

Exopterygotes. Hemimetabolous insects, viz., ones in which the wings develop as external wing pads. Compare with *endopterygotes*.

Exoskeleton. The protective, supportive outer body wall that is cast during *ecdysis*.

Exserted. Protruding or projecting outward from the body.

Extensor. A muscle which straightens or extends an appendage. Compare with *flexor.*

Exuviae (pl., only). The old cast skin of nymphs, larvae, and other juveniles consisting of undigested *exo-* and *epicuticle* discarded following *ecdysis*.

Eyes. See *apposition, superposition, simple,* and *compound eyes.*

Facet. The external surface of an individual visual unit, or *ommatidium*, of a *compound eye*.

Facultative. Able to live and reproduce under varying environmental conditions. Compare with *obligatory.*

Fat body. A network of usually opaque whitish strands, lobes, or masses of closely adherent fat and urate cells within the *hemocoele*, concerned with intermediary metabolism and storage.

Feculae (sing., *fecula*). Waste pellets voided through the *anus*. They contain the combined egestive (*feces*) and excretory (*excreta*) wastes enclosed within a length of discarded *peritrophic membrane*. They are neither feces nor excreta but both and not *frass* (the sawdust left behind by wood-boring insects).

Femur (pl., *femora*). The 3rd, usually stoutest segment of the leg, located between the *trochanter* and the *tibia*.

Fertilization. Penetration of a *spermatozoon*, or sperm, cell through the *micropyle* of an *ovum*, or egg, and fusion of the two to form a new individual called the *zygote*.

Flagellum. The apical, often whip-like 3rd article of an antenna. Compare with *scape* and *pedicel*.

Flexor. A muscle which folds an appendage against itself across a joint. Compare with *extensor.*

Follicles. In males, the testicular tubules from which sperm are produced; in females, cells produced by the germarial mesoderm that enclose developing *oocytes*.

Glossary

Food chain. A simple, linear sequence of foods and feeders involving at least a plant, an herbivore, and a carnivore.
Food web. The complex, interlocking pattern of foods and feeders typical of communities.
Forbivory. A kind of *herbivory* involving the eating of forbs which are non-woody dicots. Compare with *graminivory.*
Fore brain or *protocerebrum.* The largest, most dorsal part of the brain stemming from coalescence of the 1st pair of neural ganglia.
Fore gut. The anterior portion of the alimentary canal from the *mouth, pharynx,* and *crop* through the *proventriculus,* formed by an ectodermal *invagination* called the *stomodaeum.*
Fore legs. The 1st pair of legs, those associated with the *prothorax.*
Fossorial forms. Animals adapted for, or with the habit of, actively digging or burrowing within the substrate. Compare with *interstitial forms.*
Frass. See *feculae.*
Frenulum. A spine from the base of the hind wing of certain Lepidoptera that projects beneath the fore wing, serving to couple the two wings during flight. Compare with *jugum.*
Frons or *front.* The unpaired head sclerite between the *vertex* and the *clypeus* and bearing the median ocellus, if any.
Furca (pl., *furcae*). A forked thoracic *apodeme* arising from pterygotes' *sternum.* Compare with *phragmata* and *pleurademata.*
Furcula. The forked springing organ on the ventral abdominal surface in Collembola.
Galea (pl., *galeae*). The lobular, tongue-like, or flattened lateral process of a *maxilla,* attached to the *stipes.* Compare with *lacinia.*
Gallivory. Dwelling within and taking sustenance from *galls.*
Galls. Abnormal plant growths caused by an external stimulus, usually irritation by an insect larva or other organism.
Ganglion (pl., *ganglia*). A knot-like enlargement of a nerve containing a mass of nerve cells and fibers.
Gaster. See *pedicel.*
Gena (pl., *genae*). The lateral "cheek" region of the head below the *compound eyes* and between the *frons* and the *occiput.*
Genitalia. The external sexual organs associated with the release either of *sperm* or of *ova.*
Genus (pl., *genera*). A group of closely related, similar species. Compare with *species.*
Germarium. The apical portion of an *ovariole* or of a sperm *follicle* from which the germ cells are generated.
Giant fibers. Thick bundles of association neurons that, in certain insects, extend considerable distances without synapsing, suggestive of rapid conduction.
Gills. Thin-walled outpocketings of the body wall or hind gut that facilitate gas exchange in some aquatic insects; they are *tracheal gills* if provided with internal respiratory *tracheae* or *blood gills* if they lack *tracheae.*
Glossae (sing., *glossa*). Paired, inner lobes between the *paraglossae* at the apex of the *labium.*
Gonopore. The external reproductive aperture, either of the *common oviduct* in females or of the *ejaculatory duct* in males.
Graminivory. Eating grasses, sedges, or other monocots. Compare with *forbivory.*
Gregarious insects. Those that live in aggregations but are not truly *social.*
Grubs. See *scarabaeiform larvae.*
Gula. A ventral head sclerite sometimes developed between the *labium* and the *occipital foramen.*
Gut. See *alimentary canal.*
Gynandromorphism. Occurrence of the secondary sexual characteristics of both sexes in one and the same animal which, therefore, is a sex mosaic.
Habitat. The particular place and environmental conditions under which an organism is found. Compare with *environment.*
Halteres. Small, knobbed balancing organs, one on each side of the *metathorax* in Diptera, representing modified hind wings.
Hatching. Egress of the juvenile from the egg. Compare with *eclosion.*
Haustellate or *suctorial.* Mouthparts adapted for sucking, the mandibles either lacking or, if present, not adapted for chewing. Compare with *mandibulate.*
Head. The anterior region of the body bearing *eyes, antennae,* and *mouthparts.*

ENTOMOLOGY

Heart. The posterior, dilated portion of the *dorsal blood vessel* provided with paired *ostia*. Compare with *aorta*.

Heartbeat reversal. Occasional periods of antiperistalsis forcing blood backward. They intervene between the regular periods of peristalsis propelling blood forward into the head.

Hemelytra (sing., *hemelytron*). The fore wings of bugs, usually having a sclerotized, thickened basal and a membranous apical portion. Compare with *tegmina* and *elytra*.

Hemimetabolous development. Pertaining to the incomplete development of those usually terrestrial insects that lack the persistent molting of some adult Ametabola and the pupal stage of Holometabola.. These young, termed *nymphs* or larvae, resemble one another and also the adult except for their smaller size, different body proportions, and incompletely developed genitalia. Compare with *ametabolous* and *holometabolous*.

Hemocoele. The modified body cavity of insects and their arthropod relatives in which the viscera are located and through which blood percolates.

Hemocytes. The blood cells of insects and their arthropod relatives.

Hemolymph. The blood of insects and their arthropod relatives.

Hemopoietic organs. Mitotic larval organs that generate blood cells.

Herbivory. Eating herbaceous, non-woody plants.

Hermaphroditism. Possession of both male and female sex organs in the same individual animal known, therefore, as an *hermaphrodite*.

Hexapoda. An older, alternative name for class Insecta.

Hind brain or *tritocerebrum*. A pair of fused ganglia attached to the underside of the *fore brain* and the back of the *mid brain* which, by *circumesophageal connectives*, lead to the *ventral nerve cord*.

Hind gut. The posterior portion of the alimentary canal from *mid gut* to *anus*, formed by an ectodermal invagination called the *proctodaeum*.

Hind legs. The 3rd pair of legs, those associated with the *metathorax*.

Holometabolous. A type of development with a complete transformation including distinctly separated egg, larval, pupal, and adult stages. See *complete development*.

Host. An animal in, on, or near which a *parasite* lives or a plant on which a phytophagous insect feeds.

Humus. The decomposed or partly decomposed organic matter of soil.

Hyaline hemocytes. See clotting.

Hydrofuge or *hydrophobe*. A surface with water-repellent characteristics.

Hypermetamorphosis. A type of *holometabolous development* having two or more sequential larval types in addition to pupal and adult stages

Hypognathy. The condition of a head whose mouthparts project ventrad. Compare with *opisthognathy* and *prognathy*.

Hypopharynx. A medial, tongue-like sensory structure attached to the internal surface of the *labium* within the *preoral cavity*, near which the ducts of the *salivary glands* empty.

Imaginal discs. Clusters of undifferentiated embryonic cells maintained throughout the larval stage as centers for future development of adult structures.

Imago. The adult, reproductive stage of an insect.

Immigration. Dispersal of individuals into an area, adding to the resident population. Compare with *emigration*.

Incomplete development. See *hemimetabolous development*.

Indirect wing muscles. Muscles that generate the wing beat by sequentially distorting the shape of the thorax. Compare with *direct wing muscles*.

Innervation. The distribution of nerves to body parts.

In-phase locomotion. Simultaneous extension of legs of a pair, as in hurtling a leaper into the air. Compare with *out-of-phase locomotion*.

Inquilines. "House guests"or individuals that live in the nest or abode of another species.

Insectivory. The eating of insects. See *entomophagy*.

Insemination. Completion of the reproductive act by impregnating the female's *ova* with *sperm*.

Instar. The form or structure of an insect between successive molts, as with the 1st instar between hatching and the 1st molt, the 2nd instar until the next molt, *etc*. Compare with *stadium*.

Integument. The outer cellular and cuticular body covering.

Glossary

Integumentary processes. Integumentary outgrowths including *cellular processes* such as *setae*, *claws*, and *spurs*, all with an internal epidermal core, and *cuticular processes* such as nodules, ridges, and similar outgrowths of solid cuticle.

Intersegmental membrane. See *conjunctivae*.

Interstitial forms. Small, weakly moving creatures that occupy soil cavities within which they do not move appreciably. Compare with *fossorial forms*.

Intestine. See *alimentary canal*.

Intima. The internal cuticular lining of the *fore gut, hind gut*, and *tracheae*, continuous with the *cuticle* of the integument.

Invagination. A sac-like infolding or inpocketing of a structure. Compare with *evagination*.

In vitro. Pertaining to a biological reaction taking place experimentally, as in a test tube or other artificial apparatus.

Jet propulsion. The locomotion of dragonfly naiads which, when disturbed, dilate the rectal chamber to accommodate a volume of water which they then forcibly eject through the *anus*, propelling them a distance.

Joint. The articulation or area of flexion between successive segments or parts.

Jugal lobe. A lobe at the base of and behind the wing.

Jugum. A lobe at the base of the fore wing of certain Lepidoptera and Trichoptera that overlaps the hind wings and couples the two organs in flight. Compare with *frenulum*.

Juvenile hormones or *JH*. Hormones produced by the *corpora allata* maintaining juvenile features in immature insects and later controlling aspects of adult physiology and behavior.

Labellum (pl., *labella*). The modified tip of the *labium* in certain Diptera.

Labial glands. Glands that open at the base of the labium, usually concerned with secreting saliva or silk.

Labium. The paired, fused mouthparts of the 6th segment that function as a lower or posterior lip, depending on mouthpart orientation.

Labrum. The flap-like upper or anterior lip, depending on mouthpart orientation, attached to the *clypeus*.

Lacinia (pl., *laciniae*). The hooked or blade-like inner process of a *maxilla*, attached to the *stipes*. Compare with *galea*.

Larvae (sing., *larva*). The juvenile stage of a holometabolous insect before the pupal stage. Compare with *nymphs* and *naiads*.

Larvapods. The abdominal appendages or prolegs of a caterpillar.

Lateral. Related to or attached to the side. Compare with *medial*.

Lateral oviducts. See *oviducts*.

Law of Priority. The taxonomic requirement that, with certain exceptions, the oldest available scientific name be maintained in its original spelling.

Macropterous. Having wings of normal length, not short. Compare with *apterous* and *brachypterous*.

Macrosuccession. The fixed, predictable sequence of biotic communities that occurs in a particular geographic region as a result of the local conditions of climate and soil, usually referred to as *succession*. Compare with *microsuccession*.

Maggots. The worm-like, legless larvae of certain Diptera without a developed head capsule.

Malpighian tubules. Long, slender, blindly ending excretory tubes within the body cavity that void into the *hind gut*.

Mandibles. The paired appendages of the 4th head segment that become jaws in mandibulate insects. Compare with *maxillae*.

Mandibulate. Having jaws adapted for biting or chewing. Compare with *suctorial* or *haustellate*.

Maxillae (sing., *maxilla*). The paired mouthparts of the 5th head segment located between the *mandibles* and the *labium*.

Mechanoreceptors. Sensilla responsive to physical displacement.

Meconium. The accumulated larval excreta of certain holometabolous insects voided upon completion of the alimentary canal shortly after adult emergence.

Medial. Related to or in the midline of the body. Compare with *lateral*.

Median caudal filament. A long terminal process stemming from the *epiproct*.

Medical entomology. The scientific discipline concerned with the study of mankind's insect-transmitted diseases and health problems and their control. Compare with *economic entomology*.

Melanization. A darkening of the insect skin stemming ultimately from the breakdown of tyrosine producing permanent blackish, brownish, or yellowish pigments.

ENTOMOLOGY

Mesenteron. See *mid gut.*
Mesothorax. The middle or 2nd segment of the thorax bearing the mid legs.
Metamere. A somite or primary body segment usually with reference to embryonic stages.
Metamorphosis. The collective, step-wise changes in form that an animal undergoes in its development from hatching to adulthood. See *ametabolous, hemimetabolous, holometabolous.*
Metathorax. The posterior or 3rd segment of the thorax bearing the hind legs.
Microclimate. The local climate of a specific habitat compared to that of the entire geographic region of which it is a part. Compare with *climate.*
Microhabitats. The small, specialized, effectively isolated environmental niches within which most insects live. Compare with *habitats.*
Micron. See *μM.*
Micropterous. With reduced or short wings. Compare with *apterous* and *brachypterous.*
Micropyle. A minute opening in the insect *chorion,* or egg shell, through which *sperm* can enter.
Microsuccession. A lesser succession in nature (such as within a dead body or in feces) differing from a *macrosuccession* in its smaller scale and lack of a climax community.
Mid brain or *deutocerebrum.* Paired ganglia between the *fore* and *hind brains* involved chiefly in smell.
Mid gut. The middle portion of the alimentary canal that forms as a blind sac which, during development, eventually connects with the *fore* and *hind guts.* See *mesenteron.*
Mid legs. The 2nd or middle pair of legs associated with the *mesothorax.*
Mining or *leaf mining.* Excavation, by certain larvae, of burrows within leaves. Compare with *boring.*
Molting. The overall process of cyclical growth, secretion of the *cuticle,* and *ecdysis,* or shedding, of the *exuviae* or old skin.
Molting hormones or *MH.* Hormones secreted by the *prothoracic glands,* involved in molting.
Monoecious. Compare with *bisexual* and *dioecious.*
Monophagy. Restricted to eating a single host, either plant or animal. Compare with *oligophagy* and *polyphagy.*
Morphology. The science that deals with form or structure. Compare with *physiology.*
Mortality. Death, the state of being dead.
Mouth. The anterior opening into the alimentary canal concealed in front by the *mouthparts* and *preoral cavity.*
Mouthparts. See *mandibulate* and *suctorial.*
Myogenic heart beat. Contractions of the heart musculature produced by the muscles themselves in absence of nervous stimulation. Compare with *neurogenic.*
Myology. The science concerned with the arrangement of muscles and muscle groups within the body.
Myoneural junction or *motor end plate.* The junction between a motor nerve and the muscle that it innervates.
Naiads. The aquatic, gill-breathing juveniles of Ephemeroptera, Odonata, and Plecoptera having externally developing wing pads. Compare with *nymphs* and *larvae.*
Natality. Hatching or birth.
Neck or *cervix.* The membranous connection between head and thorax.
Nekton. Insects and other organisms that swim freely within the aquatic environment. Compare with *neuston* and *plankton.*
Neoptera. Those advanced insects capable of flexing the wings on themselves during repose. Compare with *Paleoptera.*
Nephrocytes. Specialized excretory cells massed within the body cavity, especially around the heart, where they are termed *pericardial cells.*
Neurogenic heart beat. Contractions of the heart musculature produced by nervous stimulation. Compare with *myogenic heart beat.*
Neuron. A nerve cell, being one of the functional units of the nervous system.
Neurosecretion. Synthesis and release of hormones by nerve cells.
Neurotransmitters. Substances liberated from an axon that bridge a synapse to initiate the nerve impulse in the next sequential nerve cell or muscle fiber.
Neuston. Usually minute organisms that swim or float on the water surface. Compare with *nekton* and *plankton.*
Nits. The eggs of sucking lice, attached particularly to the host's hair shaft.
Nocturnal. Insects and other organisms that are active at night. Compare with *crepuscular* and *diurnal.*
Nodes. The one/two dorsal protuberances, or scales, atop the *pedicel,* or "wasp waist," in ants.

Glossary

Nodus. A strong cross vein near the middle of the costal wing border in Odonata.
Nomenclature. The naming of organisms so they may be designated with accuracy throughout the scientific world.
Notum. See *alinotum*.
Nymphs. The usually terrestrial juveniles of *hemimetabolous insects*, those having externally developing wing pads and lacking a pupal stage. Compare with *naiads* and *larvae*.
Obligatory. Organisms restricted to particular environmental conditions in absence of which they cannot live and reproduce. Compare with *facultative*.
Obsolete sutures. Fused, non-functional exoskeletal grooves sometimes indicative of lines of articulation along which adjoining segments have united.
Obtect pupae. Pupae in which the appendages are closely appressed to the body. Compare with *exarate pupae*.
Occipital condyles. Points of articulation between the head and prothorax.
Occipital foramen. The opening by which the *neck* attaches to the *prothorax*.
Occiput. The top, rear portion of head between the *vertex* and the *neck*.
Ocelli (sing., *ocellus*). The 2-3 simple eyes of nymphal or adult insects. Compare with *stemmata*.
Oligophagy. Eating a relatively restricted, often closely related group of plants or animals. Compare with *monophagy* and *polyphagy*.
Oligopneustic. A spiracular pattern involving reduced numbers of functional spiracles. Compare with *apneustic* and *polypneustic*.
Ommatidia (sing., *ommatidium*). The individual visual units of a *compound eye*, each externally visible as a *facet*.
Omnivory. Consuming a broad diet including both plant and animal materials. Compare with *carnivory* and *phytophagy*.
Oocytes. Eggs.
Opisthognathy. Mouthparts projecting backward between the fore legs, as in cockroaches. Compare with *hypognathy* and *prognathy*.
Osmoregulation. Regulation of salt and water concentrations within the body cells and blood.
Ostia. Paired, slit-like, segmentally arranged, valvular openings into the insect *heart*.
Out-of-phase locomotion. Walking by alternate use of legs of a pair, as in insects' tripoding. Compare with *in-phase locomotion*.
Ovaries. The reproductive organs of females, responsible for production of *eggs*.
Ovarioles. Tubules that comprise the *ovaries* of insects, their types ranging from *panoistic* (without nutritive cells), to *polytrophic* (with an alternating succession of oocytes and nutritive cells), to *acrotrophic* (with cords that connect the nutritive cells to the descending oocytes).
Oviducts. Paired *lateral oviducts* that lead from the *ovaries* to the *common oviduct* for discharge to the exterior.
Oviparity. The habit of laying eggs from which the young hatch and develop outside the body of the female parent. Compare with *ovoviviparity* and *viviparity*.
Oviposition. Egg laying or deposition.
Ovipositor. The egg-laying apparatus of female insects, usually either 2-3 sets of valves or, in their absence, often just a tubular extension of the abdomen.
Ovoviviparity. Eggs that hatch within the body of the female parent and are liberated to the outside as small, live juveniles. Compare with *oviparity* and *viviparity*.
Ovum. An egg or mature germ cell of a female individual.
Paedogenesis. Production of eggs or young by a larva or other juvenile.
Palatability. The state of a food that is potentially attractive to feeders owing to its physical and chemical characteristics.
Paleoptera. Insects with a direct wing musculature and, hence, incapable of folding the wings backward over the abdomen in repose. Compare with *Neoptera*.
Palpi (sing., *palpus*). The paired, segmented, feeler-like processes of a *maxilla* or the *labium*.
Panoistic. See *ovarioles*.
Paraglossae (sing., *paraglossa*). Paired, outer (lateral) processes that flank the *glossae* at the tip of the *labium*.
Paraprocts. Paired lobes flanking the *anus* that are derived from the 12th sternite though they appear to stem from an earlier segment. Compare with *epiproct*.

ENTOMOLOGY

Parasitism. The relationship between organisms in which one, the *parasite*, usually the smaller of the two, lives on, in, or near the other, the *host*, deriving sustenance from it without killing it. Compare with *parasitoidism* and *predation.*

Parasitoidism. The relationship between organisms in which one, the *parasitoid*, the smaller of the two, lives within the other, the *host*, for a relatively long time, eventually consuming its tissues and killing it. Compare with *parasitism* and *predation.*

Pars intercerebralis. The medial part of the *fore brain* containing the neurosecretory cells responsible for secreting PTTH which is then passed on to the corpora cardiaca.

Parthenogenesis. Production of young from unfertilized eggs. See *unisexual.*

Pedicel. In insects in general, the pedicel is the 2nd subsegment of the antenna between the basal *scape* and the apical *flagellum*; in certain Hymenoptera, however, the pedicel is a "wasp waist" consisting of the constricted 2nd abdominal segment separating the *propodeum* or fused thorax and 1st abdominal segment from the *gaster* or remainder of the abdomen.

Pericardial cavity. That part of the body cavity surrounding the *dorsal blood vessel* and limited ventrally by the *pericardial membrane.*

Pericardial cells. See *nephrocytes.*

Perineural cavity. That part of the body cavity surrounding the *ventral nerve cord* and limited dorsally by the *perineural membrane.*

Periodicity. A daily or circadian gradient of activity. Compare with *diurnal* and *nocturnal.*

Periodism. A seasonal gradient of activity.

Peripheral nervous division. That part of the nervous system consisting of free nerves that radiate between the medially located *brain* and *ventral nerve cord* and the peripheral body wall and associated musculature. Compare with *central* and *sympathetic nervous divisions.*

Periphyton. The assemblage of insects and other fresh-water organisms that cling to rooted plants or brush projecting from the bottom.

Peristalsis. Rhythmic waves of muscular contraction that sweep along the length of a tubular structure carrying with them food, blood, or other contained fluids. Waves sweeping in the opposite direction are termed *antiperistaltic.*

Peritreme. The sclerotic plate encircling a *spiracle.*

Peritrophic membrane. The delicate, tubular protective sheath that surrounds food within the *mid gut* of most insects.

Perivisceral cavity. The body cavity surrounding the digestive tract, reproductive system, *etc.*, between the *pericardial* and *perineural membranes.*

pH. A measure of acidity or alkalinity based on the relative concentration of hydrogen ions in solution, in which a value of pH 7.0 indicates neutrality, lower values down to pH 1.0 acidity, and higher values up to pH 14.00 alkalinity.

Phagocytosis. Destruction of blood particulates by engulfment carried out by certain blood cells.

Phallus. The male copulatory organ, the apical portion of which, the *aedeagus*, bears the *gonopore.*

Pharynx. The anterior, muscular part of the *fore gut* into which the *mouth* opens to lead into the *esophagus.*

Pheromones. Externally secreted chemical substances including sex attractants, alarm substances, *etc.*, produced by one individual that cause a specific reaction in other individuals of the same species.

Photoperiod. The relative number of hours of daylight or of darkness per day.

Phragmata (sing., *phragma*). Paired, plate-like *apodemes* or invaginations of the *terga* that provide attachment for the dorsal longitudinal wing muscles. Contrast with *furcae* and *pleurademata.*

Phylogeny. The evolutionary relationships between an ancestral form and its descendants; *i. e.,* the genealogy of organisms.

Physiology. The science that deals with the vital functions of living organisms. Contrast with *morphology.*

Phytophagy. Eating plants. Contrast with *carnivory.*

Plankton. Submicroscopic or microscopic floating organisms that are largely incapable of moving against the current. Contrast with *nekton* and *neuston.*

Plantulae. See *adhesive organs.*

Plastron. A dense bed of fine, water repellent "hairs" of certain aquatic insects holding an air film against the body across which respiratory gas exchanges take place.

Glossary

Pleurademata (sing., *pleuradema*). Lateral *apodemes* that invaginate from the pleural suture on each side of the thorax. Contrast with *furcae* and *phragmata*.

Pleuron (pl., *pleura*). The plate-like lateral sclerite of a thoracic segment subdivided into an anterior *episternum* and a posterior *epimeron*.

Poikilothermy. So-called *"cold bloodedness,"* the property of having the body temperature rise or fall with the external environmental temperature.

Polyembryony. Production of two or more identical embryos by mitotic division of a single *egg*.

Polymorphism. The simultaneous occurrence of two or more structurally different body forms within the same life stage of a single species.

Polyphagy. Eating many unrelated foods. Contrast with *monophagy* and *oligophagy*.

Polypneustic. Respiration carried out using multiple pairs of functional *spiracles* as in adults of most orders. Contrast with *apneustic* and *oligopneustic*.

Polytrophic. See *ovarioles*.

Pool zone. That part of the running-water environment in which the current is slow, allowing for deposition of gravel, sand, and, in places, silt. Contrast with *riffles*.

Population. An aggregation of potentially or actually interbreeding individuals of the same species occupying a particular area at the same time. Contrast with *community* and *ecosytem*.

Pore canals. Minute, often helically coiled, duct-like spaces that provide a channel for wax secretions from the cellular to the cuticular layers of the integument. Contrast with *dermal gland ducts*.

Posterior. Hind or rear. Contrast with *anterior*.

Posterior intestine. The caudal section of *proctodeum* behind the *anterior intestine,* from which it is usually separated by a rectal valve.

Postmentum. The basal portion of the *labium*, proximad of the labial suture. Contrast with *prementum*.

Postnotum. The dorsal plate behind the scutellum, bearing *phragmata* in winged species.

Postocciput. The extreme posterior section of head between the postoccipital suture and the neck or *occipital foramen*.

Predation or *predatism*. A relationship between organisms in which one, the *predator*, usually the larger, more powerful of the two, attacks or traps, overcomes, eats, and kills the other, the *prey*. Contrast with *parasitism* and *parasitoidism*.

Prementum. The apical portion of the *labium*, distad of the labial suture. Contrast with *postmentum*.

Preoral cavity. The space between the *mouthparts* in front of the *mouth*.

Primarily aquatic. Aquatic insects that live a submerged existence and use dissolved oxygen obtained from the water around them. Contrast with *secondarily aquatic*.

Primary segmentation. The arrangement of body segments such that intersegmental grooves form the internal attachment points for the longitudinal muscles. Contrast with *secondary segmentation*.

Primitively wingless. Arthropod relatives of insects that do not have and presumably never did have functional wings. Contrast with *secondarily wingless*.

Proctodaeum. See *hind gut*.

Prognathy. The condition of a head with the mouthparts projecting forward. Contrast with *hypognathy* and *opisthognathy*.

Prolegs. See *larvapods*.

Pronotum. The dorsal sclerite or notum (*i., e.,* the top) of the prothorax.

Propodeum. See *pedicel*.

Prosternum. The ventral sclerite or sternum (*i. e.*, the bottom) of the prothorax.

Prostomium. See *acron*.

Prothoracic glands. Endocrine glands generally within the prothorax that secrete *molting hormones*.

Prothorax. The anterior or first of the three thoracic segments.

Protocerebrum. See *fore brain*.

Proventriculus. The muscular, most posterior part of the *fore gut*, often provided with a sclerotized armature of grinding dentes controlling entry of food into the *mid gut*.

Proximal. Nearer to the body or to the base of an appendage. Contrast with *distal*.

Pterothorax. The two modified, often fused, wing-bearing segments (*meso-* and *metathorax*) of the *thorax*.

ENTOMOLOGY

Pterygota. The subclass of winged or sometimes secondarily wingless insects of varied development, never ametabolous. Contrast with *Apterygota*.

Ptilinum. A bladder-like, inflatable head sac in certain developing Diptera that is everted to rupture the *puparium* facilitating emergence.

PTTH. Certain neurohormones produced by the brain neurosecretory cells that are passed on to the *corpora cardiaca* to initiate molting by activating the prothoracic glands.

Pulsatile organs. Small, specialized organs whose rhythmic muscular contractions promote blood flow in the head, thorax, and appendages of various insects.

Pulvilli (sing., *pulvillus*). See *adhesive organs*.

Pupa (pl., *pupae*). The non-feeding, comparatively inactive transitional stage between the last larva and the adult of *holometabolous insects*.

Pupal diapause. A temporary rest in development during the pupal stage.

Puparium. A case formed from the hardened last larval skin within which some advanced Diptera develop during the pupal stage.

Pupation. The act of transforming into a pupa.

Pyloric valve. The valve at the anterior extremity of the *hind gut* separating it from the *mid gut*. Compare with *stomodael valve*.

Raptorial. Adapted for seizing and holding prey.

Receptors. Specialized sense organs, or *sensilla*, that perceive energy from internal and external sources and relay it to the *effectors* of the body, *viz.*, muscles and glands.

Rectum. The posterior region of the *hind gut* provided internally with *rectal pads* which are enlarged cells specialized for resorption of water and ions.

Reflex centers. Local nerve ganglia responsible for given reflexes, as with the last abdominal ganglion controlling oviposition, defecation, *etc.*, as opposed to more remote ganglia or *coordinating nerve centers* that exert stimulative or inhibitory influence over the local ganglia.

Repletes. Specialized honey ants that, upon engorging themselves with liquid food, literally become inactive, living "honey casks" that provide food reserves to others of the colony as needed.

Resilin. An insoluble, rubbery, proteinaceous constituent of the insect *cuticle* developed where deformation occurs, for example, at the wing base.

Respiration. The physiological process involving oxygen intake, use of it under enzyme control to release the chemical energy of food, and elimination of the resulting carbon dioxide wastes. It may be termed *tracheal respiration* if facilitated by *tracheae*, *gill respiration* if it takes place within *gills*, or *cutaneous respiration* if it involves gas diffusion through the soft, thin, general body wall.

Resting potential. The nerve potential generated across the external and internal surfaces of neurons or muscle cells based on the dissimilar distribution of sodium and potassium ions. Contrast with *action potential*.

Reticulate venation. The complex network of veins and cross veins typical of the wings of mayflies, dragonflies, and certain other insects.

Retina. The light-sensitive apparatus of an eye.

Rhabdom. The rod-like, light-sensitive core of an *ommatidium* formed of the opposing inner surfaces of sensory cells.

Riffles. That shallow, partly obstructed part of the running-water environment in which the water current flows swiftly. Contrast with *pool zone*.

Ring gland. A composite endocrine gland encircling the *dorsal aorta* in certain advanced Diptera, capable of secreting all three metamorphosis hormones, *viz.*, *PTTH, molting hormones*, and *juvenile hormones*.

Salivary glands. Glands that open into the *preoral cavity* at the base of the *labium* involved in secreting saliva which is a mixture of digestive enzymes, mucus, and, in certain blood-sucking parasites, an anticoagulant.

Scape. The basal segment of an *antenna*. Contrast with *pedicel* and *flagellum*.

Scarabaeiform larvae. Thick-bodied, sluggish grubs with a well-developed head and thoracic legs but without abdominal legs.

Scavenging. Consuming dead or decaying plant or animal materials or animal wastes.

Scientific name. A Latin or Latinized name of a species consisting of a generic name descriptive of the general kind of animal and a specific name descriptive of the particular kind of animal. See *binomen* and *subspecies*.

Glossary

Sclerite. A sclerotized or hardened body-wall plate often bounded by *conjunctivae* that allow for articulation with adjacent plates; if not, it may fuse with other plates by *obsolete sutures.*

Sclerotization. The process of becoming sclerotized or hardened by tanning.

Scolopophorous or *chordotonal organs.* Clusters of stretch receptors between two internal integumentary surfaces.

Sculling. A type of aquatic locomotion involving alternating sidewise abdominal movements that propel the insect forward.

Secondarily aquatic. Aquatic insects that come to the surface to breathe atmospheric oxygen. Contrast with *primarily aquatic.*

Secondarily wingless. Insects that stem from fundamentally winged lines whose development of winglessness is adaptive. Contrast with *primitively wingless.*

Secondary plant substances. Chemical substances produced by plants that are without an apparent metabolic role but are repulsive to herbivorous insects or other animals which have the potential of eating them.

Secondary segmentation. The arrangement of body segments such that the attachment points of the longitudinal muscles lie forward of the true intersegmental line as is typical of adult insects. Contrast with *primary segmentation.*

Segment. A ring or division of the body or of an appendage between articulations. Contrast with *conjunctivae* and *arthrodial membranes.*

Segmented antennae. See *antennae.*

Seminal vesicles. Dilated, sac-like enlargements of the *vasa deferentia* within which the male seminal fluid may be stored until discharged to the female.

Sensilla (sing., *sensillum*). Specialized sense organs capable of detecting external or internal stimuli and transmitting them to the *central nervous division.*

Series. A sample or assortment of individuals of a species, genus, family, or other unit arranged to reveal variability in given characters as a basis for determining classification.

Sessile. Organisms attached to the substrate or vegetation, unable to move.

Setae. Unicellular, hair-like, articulated *integumentary processes* that invest the body.

Simple eyes. See *ocelli* and *stemmata* and compare with *compound eyes.*

Skeletal muscles. Muscles of body movement and posture attached at both ends to integument. Compare with *visceral muscles.*

Social insects. Insects that live together, are variably dependent on one another, and exhibit brood care and division of labor. Compare with *solitary, gregarious*, and *subsocial insects.*

Soil organisms. Insects or other organisms that dwell within soil. Compare with *fossorial* and *interstitial insects.*

Soldiers. A caste developed within a social insect group specialized for protection by enlarged, pincer-like mandibles or other adaptations.

Solitary insects. Insects that live singly and only come together to mate, never aggregating.

Species. A population of individuals similar in structure, function, and behavior and capable of freely interbreeding to produce fertile living offspring. They are designated by a *binomen.* Compare with *genus* and *subspecies.*

Spermatheca. An unpaired, sac-like outpocketing of the female's *copulatory pouch* that receives and often stores *sperm* from the male prior to use in fertilization.

Spermatophore. The sperm-containing gelatinous capsule produced by the accessory glands of males in some species.

Spermatozoon. The mature sperm or germ cell of a male individual.

Spines. Thorn-like, non-articulated, multicellular *integumentary processes*, particularly on the legs. Compare with *spurs.*

Spiracles. The paired external openings into the tracheal system used in breathing.

Spurs. Thorn-like, articulated, multicellular *integumentary processes* at the apex of the tibia. Compare with *spines.*

Stadium (pl., *stadia*). The interval between molts in a developing arthropod. Compare with *instar.*

Stemmata (sing., *stemma*). The simple, lateral eyes of insect larvae. Compare with *ocelli.*

Sternite. A subdivision of the *sternum*, being one of the sclerites of which it is composed. The term is often used interchangeably with *sternum* though the two words are not synonymous.

Sternum. The ventral division or bottom of a thoracic or abdominal segment.

Stigma (pl., *stigmata*). An often-colored, costal thickening of the wing membrane near the apex in some insects.

ENTOMOLOGY

Stipes (pl., *stipites*). The second segment or stalk of a *maxilla* borne on the *cardo*. The stipes bears the movable *palpus, lacinia*, and *galea*.

Stomodael or *cardiac valve*. The cylindrical or funnel-like, valvular invagination from the *proventriculus* into the *ventriculus* preventing regurgitation. Contrast with *pyloric valve*.

Stomodaeum. See *fore gut*.

Stratification. Development of *strata* or vertical gradients of environmental condition within a community. Contrast with *zonation*.

Stridulation. The habit of some insects producing noise by rubbing together roughened structures or surfaces.

Stylopized. The state of being parasitized by female Strepsiptera.

Sub-. A prefix meaning *approximately, nearly, under,* or *below,* for example, subequal (almost but not quite equal) or subocular (beneath the eye). Compare with *supra-*.

Subcoxa. Sclerites attached to the limb basis or *coxa* in primitive arthropods but incorporated into the thoracic wall or *pleura* of insects..

Subesophageal ganglion. The knot-like composite ganglion of the *central nervous division* at the anterior end of the nerve cord, just below the *esophagus*, innervating the *mouthparts, salivary glands,* and neck region.

Subsegments. See *articles*.

Subsocial insects. The relationship between communal individuals in which the parents remain with and care for the young. Compare with *social* and *solitary insects*.

Subspecies. Subdivisions (often geographic races) within a *species* that are not sharply differentiated but intergrade with one another and are capable of interbreeding. They are designated by a *trinomen*. See *scientific name* and compare with *species*.

Succession. See *macrosuccession* and *microsuccession*.

Suctorial or *haustellate*. Having mouthparts adapted for sucking. Compare with *mandibulate*.

Superposition. The image formed by the eyes of dark-adapted insects produced by overlapping points of light perceived through a number of adjacent ommatidia. Compare with *apposition*.

Supra-. A prefix meaning *above, beyond,* or *superior to,* for example, the supraesophageal ganglion. Compare with *sub-*.

Supraesophageal ganglion. See *brain*.

Sutures. Exoskeletal grooves. See *obsolete sutures*. Compare with *conjunctivae*.

Swimming. The propulsion of aquatic insects by *in-phase* or *out-of-phase* use of oar-like legs.

Sympathetic nervous division. That part of the nervous system consisting of nerves to the stomodaeum and related viscera. Compare with *central* and *peripheral nervous divisions*.

Synchronous wing muscles. Wings of relatively slow fliers that beat in direct response to individual contractions of the muscles that drive them. Compare with *asynchronous wing muscles*.

Synonyms. Two or more different names for one and the same thing or the same *taxon*.

Systematics. See *taxonomy*.

Systole. Regular peristaltic contractions of the *dorsal blood vessel* that propel blood forward into the head. Compare with *diastole*.

Taenidia. Spiral thickenings along the internal walls of *tracheae*.

Taiga. The vast circumpolar belt of boreal forest extending southward from the Arctic *tundra* to the temperate region and downward along certain mountain ranges.

Tarsal claws. Paired, multicellular, articulated claws at the apex of the tarsus, derived from the pretarsal subsegment. See *integumentary processes*.

Tarsus (pl., *tarsi*). The distal leg segment, or foot, just beyond the *tibia* usually consisting of one or more subsegments called *tarsomeres* and terminated by the *pretarsus*.

Taxon (pl., *taxa*). A group of presumably related organisms classified together in the same category.

Taxonomy or *systematics*. The scientific study of organisms' *nomenclature*, classification, and evolutionary relationships.

Tegmina (sing., *tegmen*). The leathery, thickened fore wings of Orthoptera and certain related insects. Compare with *elytra* and *hemelytra*.

Tendons. Slender, chitinous bands or straps by which muscles may attach to the internal surface of the exoskeleton.

Teneral. The pale, soft-bodied condition of recently molted insects whose exoskeleton has yet to become sclerotized and darkened.

Glossary

Tentorium. The endoskeleton of the head usually consisting of two pairs of *apodemes* used for support and muscle attachment.
Tergite. A subdivision of the *tergum*, being one of the sclerites of which it is composed. The term is often used interchangeably with *tergum* though the two words are not synonymous.
Tergum (pl., *terga*). The dorsal division or top of a thoracic or an abdominal segment. Compare with *sternum*.
Termitaria. The above-ground nests of certain tropical termite colonies.
Testes (sing., *testis*). The paired male gonads responsible for production of *spermatozoa*.
Thorax. The central body region of an insect between the *head* and the *abdomen*. It bears legs and wings (when present).
Tibia (pl., *tibiae*). The usually long 4th segment of the leg between the *femur* and the *tarsus*.
Tonofibrillae. Cuticular fibrils by which muscles may anchor to the inner surface of the *cuticle*.
Tracheae. The *taenidea*-lined system of branching air tubes that open to the outside at the *spiracles* and terminate internally within the body tissues as *tracheoles* allowing for exchange of gases.
Tracheal gills. The sac-like, flattened or filamentous, thin-walled tracheated evaginations that enable the aquatic insects provided with them to live a submerged existence using dissolved oxygen extracted directly from water.
Tracheolar liquor. A fluid sometimes restricted to the apices of the *tracheoles* but at other times reaching far upward toward the *tracheae*.
Tracheoles. The minute terminal branches or "capillaries" of the respiratory system ending blindly within the internal body tissues.
Transverse. Across or at right angles to the longitudinal axis.
Trinomen. The name given to a subspecies. Compare with *binomen*.
Tritocerebrum. See *hind brain*.
Triungulins. The active 1st instar larvae of Strepsiptera and of certain beetles that undergo *hypermetamorphosis*.
Trivial name. Synonymous with specific name. See *species*.
Trochanter. The typically small 2nd segment of the leg between the *coxa* and the *femur* sometimes fused with the latter.
Trophallaxis. The exchange of regurgitated or egested food among social insects and sometimes their arthropod guests or *inquilines*.
Tundra. Flat or rolling land north of the tree line up to the Arctic zone of perpetual ice and snow and along the margins of Antarctica.
Type. One or more individuals or parts thereof set aside to bear the scientific name of a described subspecies, species, genus, or other taxonomic entity.
µM. One *micron*, a unit of length equal to one thousandth of a millimeter.
Unicondylic. A joint with a single point of articulation. Compare with *dicondylic*.
Unisexual. Parthenogenetic reproduction, a type carried out by females only.
Uric acid. The nitrogenous waste commonly excreted by terrestrial insects.
Vagility. The ability to get around and cross environmental barriers.
Vagina. See *copulatory pouch*.
Variety. An assemblage of ill-fitting individuals atypical of a given species yet insufficiently divergent to be a new species; i. e., an informal taxonomic category inferior to the rank of subspecies.
Vasa deferentia (sing., *vas deferens*). The paired sperm ducts leading from the *testes* and uniting to form the medial *ejaculatory duct*.
Vasa efferentia (sing., *vas efferens*). Short ducts from the testicular *follicles* that connect with the *vas deferens* on their side.
Veins. Chitinous, rod-like thickenings that support the wings of insects especially those extending longitudinally from the wing base to its apex. Compare with *cross veins*.
Venter. The bottom or undersurface of an organism. Compare with *dorsum*.
Ventral. Lower or beneath, pertaining to the body's undersurface.
Ventral diaphragm. See *perineural cavity* and *membrane*.
Ventral glands. Paired ventrolateral head glands that secrete *molting hormones* in relatively unspecialized pterygotes but are absent in advanced pterygotes, being replaced by the apparently homologous *prothoracic glands*.

ENTOMOLOGY

Ventral nerve cord. The paired chain of longitudinal nerves of the *central nervous division* lying along the ventromedial surface of the body and provided with segmentally arranged *ganglia*.

Ventriculus. The "stomach" of an insect involved in digestion and absorption of food molecules. See *mid gut*.

Vertex. The top of the insect head between the *frons, eyes*, and *occiput*.

Vestigial. See *atrophied*.

Visceral muscles. Striated muscles that invest insects' *alimentary canal, Malpighian tubules, dorsal blood vessel*, and certain other internal organs. Compare with *skeletal muscles*.

Visual receptors. See *eyes*.

Viviparity. Bringing forth living young instead of eggs that hatch. Compare with *oviparity* and *ovoviviparity*.

Water of metabolism. Chemically bound water extracted from the breakdown of food.

Weather. The short-term state of the atmosphere as defined by standard meteorological measurements. Compare with *climate*.

Wing-beat frequency. Rate of wing beat per unit of time.

Wing margins. Arbitrary designations according to location, as follows: apical (at the tip), anal (along the trailing edge), axillary (where the wing articulates), and costal (along the leading edge).

Wing muscles. See *asynchronous* and *synchronous*.

Wriggling. The locomotion of certain larval insects that use a whip-like motion of the body, first one way, then the other.

Zonation. Development of *habitats*, zones, or other horizontal gradients of environmental condition within the community. Compare with *stratification*.

Zygote. A fertilized egg representing the 1st stage of a new individual.

Index

Abdomen 1, 7, 11, 23, 37, 40, **44**, 45, 48, 53, 54, 56, 60, 66, 88, 100, 105, 110, 131, 132, 139, 141, 142, 158, 212, 227, 233, 249, 251, 265, 268, 270, 272, 276, 280, 282, 283, 286, 288, 290, 291, 293, 297, 305, 307, 308

Abdominal ganglia 66, 71

Absorption 78, 103, 110, 111, 116, 117

Accessory glands, female 145

Accessory glands, male 95, 147, 152, 225

Acetylcholine/Ach 70

Acrididae/Acridoidea (Orthoptera) 259, 262, **276**, 277, 313, 317

Acron/prostomium 23, 24, 26

Acrotrophic ovarioles 148, 150

Actin 86

Action potential 69, 83

Actions 87

Aculeata (Hymenoptera) 309

Adaptability/adaptiveness 1, 11, 170, 179

Adephaga (Coleoptera) 293, 294, 295

Adipohemocytes 125

Adult stage 97, 98, 171, 251, 272, 290

Aedeagus **49**, 146, **147**, 225, 233, 295

Age groups 170

Aggregations 170, 227, 239, 240, 241, 276

Air 1, 9, 56, 72, 74, 131, 132, 133, 136, 137, 139, 140, 141, 143, 158, 184, 204, 210, 214, 217, 220, 222, 223, 224

Air sacs 133, 135, 136, 139, 140

Air stores 143

Alar sclerites 39, 42, 88, 220

Alderflies (Neuroptera: Megaloptera) 291, 292, 311

Aleyrodidae (Homoptera) 286

Aliform muscles. See dorsal transverse muscles

Alimentary canal/digestive tract 23,79, **104**, 107, 108, 110, 111, 112, 118, 121, 123, 127, 131, 132, 145, 147, 200, 290

Alinotum 39

Allantoic acid 115

Allantoin 115

All or None Principle/Law 83

Alpine rock crawlers. See Grylloblattidae 166

Ambush bugs. (Phymatidae) 194

Ambushing 194

American cockroach (*Periplaneta americana*) 163, 198, 216, 253, **254**, 259, 275

Ametabola 155, 159

Ametabolous/direct development 155

Amino acids **104**, 110, 111, 124

Ammonia 114, 115, 117

Ampullae 112

Amylase 104, 109

Anal lobe/area 280

Anal sphincter 109

Anamorphosis 155, 265

Anisoptera (Odonata) 270, 271, 273

Annulated antennae 29

Anoplura 283, 284, 285

Antennae 1, 7, 23, 25, 27, **29**, 30, 40, 46, 59, 60, 61, 65, 72, 74, 75, 123, 143, 144, 156, 201, 203, 207, 229, 231, 251, 265, 268, 269, 270, 272, 274, 276, 278, 280, 282, 284, 288, 290, 291, 293, 295, 298, 299, 302, 304, 305, 307, 308

Antennal nerves 65, 66

Anterior intestine 109, 111

Anticoagulant 106

Antlions. See Myrmeleontidae 291, 292

Ants. See Formicidae 12, 14, 15, 16, 19, 53, 106, 172, 173, 180, 181, 184, 186, 187, 189, 190, 199, 224, 239, 243, **246**, 247, 248, 249, 291, 297, 307, 309

Anus 23, 44, 45, 49, 51, 104, **109**, 111, 148, 216, 301

Aorta. See dorsal blood vessel 122, 123

Aphididae (Homoptera) 15, 286, 287

Apidae/Apoidea (Hymenoptera) 249, 250

Apneustic 136, 137, 142

339

ENTOMOLOGY

Apocrita (Hymenoptera) 307, 308, 309

Apodemes 26, 40, 51, 61, **62**

Apophyses 61, 62

Aposematic/warning coloration 164, 165

Applied entomology. See economic entomology

Apposition eyes 77,78

Apterygota **10**, 224, 265, 267, 310

Aristae 224

Art and decoration 19, 265

Arthropleona (Collembola) 266, 268

Arthrodial membrane 23, 60

Arthropoda 1

Articles/subsegments 23, 30, 31, 33, 40, 41, **60**, 88, 286

Articulation. See unicondylic, dicondylic, universal 31, 39, 42, 59, **60**, 61, 73, 85, 272

Asilidae (Diptera) 180, 187, 190

Aspiration/suction 129, 199

Assassin bugs. See Reduviidae 159, 194, 262, 288, 289

Asynchronous/fibrillar muscles **84**, **221**, 222

ATP/adenosine triphosphate 86, 228

Attractants/phagostimulants 205

Attractive coloration 164

Atrium 131, 132, 133

Availability **205**, 206, 231

Axillary angle 43

Axillary sclerites 42, 44, 220

Backswimmers. See Notonectidae 143, 182, 215, **288**, 289

Balance of Nature 18, 176

Bark beetles. (Scolytidae) 173, 175, 186

Bark lice. See Psocoptera 281, 283, 310

Basal suture 274, 282

Basement membrane 42, **51**, 75, 76, 82, 104, 105, 113, 133

Bed bugs. See Cimicidae 172, 262, 288

Bee bread 250

Bee flies. (Bombyliidae) 189

Bees. See Apidae 14, 19, 59, 84, 86, 110, 155, 167, 173, 175, 186, 187, 190, 192, 217, 220, 223, 224, 225, 239, 243, **249**, 250, 251, 252, 307, 309

Beeswax 19, 104

Beetles. See Coleoptera 8, 9, 11, 14, 18, 19, 24, 30, 53, 59, 60, 66, 76, 84, 107, 110, 143, 146, 155, 158, 164, 166, 167, 170, 173, 175, 180, 181, 182, 183, 184, 186, 187, 189, 190, 194, 195, 199, 201, 212, 215, 216, 217, 222, 227, 240, 241, 242, 262, 278, **293**, 295, 297, 311

Belostomatidae (Hemiptera) 286, 288, 289

Benthos 182

Bicarbonate 124, 137, 141

Bilateral symmetry 79, 87

Biological clock 163

Biological control 18, 302, 307

Bioluminescence 161, **227**, 295

Biotic potential 168

Bird lice/biting lice/chewing lice. See Mallophaga 30, 155, 233, **283**, 284, 285

Bisexual reproduction 225, 235, 307

Biting mouthparts. See mandibulate mouthparts **30**, 35, 199, 203, 268, 278, 282, 290, 291, 297

Bittacidae (Mecoptera) **297**, 298

Blattidae/Blattodea (Orthoptera) **273**, 275, 276

Blister beetles. See Meloidae 295, 296

Blood/hemolymph 56, 57, 59, 71, 87, 91, 94, 95, 97, 98, 99, 106, 110, 111, 112, 115, 116, 117, 118, 121, 122, 123, **124,** 125, 126, 127, 128, 129, 130, 133, 139, 142, 192, 283, 284, 297, 305

Blood cells. See hemocytes 124, **125**, 129, 130

Blood pressure 56, 115, 126, **129**

Blood volume 118, 121, **129**

Bloodworm larvae. See Chironomidae 142, 182, 303

Blow flies. See Calliphoridae 190, 194, 303

Body regions/tagmata 23, 24, 38

Bombykol 228

Booklice. (Liposcelidae) 212, 283

Boreidae (Mecoptera) 297, 298

Borers. (Buprestidae) 186, 203, **204**, 307

Boring 189, **202**, 203, 291, 295, 301, 304, 307

Bot flies. (Oestridae) 19, 207

Boxelder bug (*Leptocoris trivittatus*) 241

Brachycera (Diptera) 302, 304

Braconidae (Hymenoptera) 187

Brain/supraesophageal ganglion 64

Breathing 71, 88, 125, 135, **138**, 139, 140, 141, 142, 215

Bristletails (Thysanura) 10, 30, 46, 75, 189, **268**, 269

Buccal cavity 105

Buffering 110

Bugs/true bugs. See Hemiptera 13, 15, 20, 34, 35, 53, 59, 66, 84, 101, 143, 146, 155, 159, 172, 181, 187, 189, 194, 199, 212, 214, 215, 225, 262, 286, **288**

Bumble bees. See Apidae 164, 165, **249**, 250, 308

Buoyancy 143, 209

Burrows 174, 183, **184**, 227, 230, 249

Bursa copulatrix **146**, 150, 225

Burying beetles. See Silphidae **242**, 294, 295

Butterflies. See Lepidoptera 11, 14, 18, 19, 34, 35, 57, 59, 60, 84, 104, 111, 119, 146, 155, 158, 163, 164, 167, 180, 187, 189, 190, 199, 217, 220, 221, 222, 227, **299**, 301

Index

Cabbage butterflies. (Pieridae) 300

Caddisflies/caddisworms. See Trichoptera 59, 142, 183, 194, **298**, 299, 300

Caecum/caeca 8, 104, **108**

Caelifera (Orthoptera) 276

Calcium carbonate 23, 115

Calcium oxalate 115

Calliphoridae (Diptera) 259, 303

Calypteres 304

Calyptratae (Diptera) 304

Calyx 146, 150

Camel crickets. See Gryllacrididae 166, 167, 184, **276**

Campaniform organs 72, **73**, 75

Cankerworms. (Geometridae) 186

Cannibal/cannibalism/cannibalistic **194**, 196, 206, 230

Carabidae (Coleoptera) **293**, 294

Carbohydrates 86, **103**, 104, 111, 117, 124

Carbon dioxide 71, 114, **137**, **141**, 142, 151, 176

Carbonic acid 137

Carnivory/carnivore 176, **191**, 193, 195

Carpenter ant (*Camponotus pennsylvanicus*) 187, **248**, 260, 308

Carpet beetles. See Dermestidae 19, **295**, 296

Carrion beetles. See Silphidae 165, 195, 294, **295**

Caste determination 244, **250**

Castes 224, 243, **244**, 245, 246, 248, 251, 274

Caterpillars. See Lepidoptera 11, 18, 21, 24, 57, 66, 75, 79, 104, 106, 109, 111, **156**, 186, 189, 201, 205, 206, 213, 233, 240, 291, 297, **299**, 301

Cat fleas (*Ctenocephalides felis*) 306

Caudal gills 272

Cave crickets. See Gryllacrididae 164, **184**, **276**

Cavernicole/cavernicolous/cavernicoly 184

Cecidomyiidae (Diptera) 203

Cecropia moth (*Hyalophora cecropia*) 99

Cell membrane 63, 68, 83, 91

Cellobiose 103

Cell processes 63

Cellular processes 59

Cement 8, 52, 53, **58**

Central nervous division 63, **64**, 210

Centrolecithal 151

Cerambycidae (Coleoptera) 295, 296

Cercopidae (Homoptera) 286

Cercus/cerci 23, 40, 45, **46**, 59, 60, 72, 265, 266, 268, 269, 270, 272, 274, 278, 280, 282, 286, 290, 291, 293

Cervix. See neck 25, 28, 33, **37**, 233

Chaetotaxy 58

Chalcididae/Chalcidoidea (Hymenoptera) 193, 307, **309**

Chemoreception/chemoreceptors 72, **74**, 207

Chemoreceptive/basiconic setae 74

Chewing lice. See Mallophaga 283

Chironomidae (Diptera) 303

Chitin **53**, 103, 104, 133

Chloride 116, 124

Cholinesterase 70

Chorion 149, 150, **151**

Chrysomelidae (Coleoptera) 295, 296

Chrysopidae (Neuroptera) 291, 292

Cicadellidae (Homoptera) 286, 287

Cicadidae/Cicadoidea (Homoptera) **286**, 287

Cicindelidae (Coleoptera) 293

Cimicidae (Hemiptera) 259, 262, 288, 289

Circadian **163**, 166

Circulation 57, 115, **121**, **126**, 127, 128, 129

Circulatory rate 129

Circumesophageal connectives 65

Claws 40, 41, **59**, 198, 199, 211, 272, 293

Cleavage nuclei 159

Click mechanism **220**, 223

Climate 174, 175, 258

Clothes moths. (Tineidae) 19

Clotting 125, **130**

Clumped distribution 170

Clypeus **27**, 30, 31, 33

Coccidae/Coccoidea (Homoptera) 286, 287, **288**

Cockroaches. See Blattidae 18, 19, 21, 24, 67, 104, 107, 110, 118, 163, 167, 184, 189, 196, 211, 212, 216, 217, 233, 240, 254, 272, **273**, 274, 282

Cocoons **158**, 250

Coefficient of viscosity 209

Coleoptera 14, 76, 190, 193, 204, 224, 262, **293**, 295, 297

Collembola 10, 13, 29, 30, 46, 155, **265**, 266

Colon 109, 111

Color change 164

Color perception 78, 163

Combination coloration 60

Commensalism 172

Common fruit fly (*Drosophila melanogaster*) 167

Common lacewings. See Chrysopidae 30, 194, 234, 235, **291**, 292

Communal insects 239

Community/biotic community 163, 167, **172**, 173, 174, 175, 176, 177, 205

Community metabolism/energy flow 18, **175**, 177

Competition **167**, 168, 172, 290

ENTOMOLOGY

Complete/holometabolous development 13, 14, **155**, **290**

Compound eyes 1, 7, 25, 27, **29**, 30, 65, 75, **76**, 77, 157, 162, 244, 268, 269, 270, 272, 276, 290, 291

Comstock-Needham Scheme 43

Confused flour beetle (*Tribolium confusum*) 167

Connectives. See circumesophageal connectives 65, 66, 68

Connective tissue 104, 105, 122, 123, 145

Copulation 46, 228, 230, 231, **232**

Copulatory pouch **145**, 146, 147, 150, 225, 235

Corixidae (Hemiptera) 288, 289

Cornea 75, 76

Corpora allata 57, 68, 92, 94, **95**, 97, 98, 99, 159

Corpora cardiaca 57, 68, 92, 94, **95**, 97

Corydalidae (Neuroptera) 291, 292

Cottony cushion scale (*Icerya purchasi*) **235**, 236

Courtship behavior 226, 230, **231**, 232

Coxal processes/leg condyles 39

Crane flies. (Tipulidae) 190, 262, 303

Creek/brook 182

Creeping water bugs. (Naucoridae) 214

Cremaster 57

Crepuscular **163**, 299

Cretaceous 275, 283

Crickets. See Gryllidae

Crop 8, 68, 79, 80, 105, **106**, 107, 108, 109, **110**, 111, 206

Cross veins **43**, 44, 274, 283, 290, 293, 302

Crowding 97, 206

Cryptic coloration 164

Cryptonephridial 113, 114

Crystalline cones 75, 76

Culicidae (Diptera) 303

Curculionidae (Coleoptera) 295

Cutaneous respiration **137**, 143

Cuticle 25, 42, 51, **52**, 53, 54, 55, 57, 58, 62, 73, 74, 84, 92, 96, 98, 100, 101, 105, 109, 116, 118, 124, 132, 133, 137, 141, 151, 159, 164, 221, 228

Cuticular processes 58

Cuticulin 52, 53, 54, 55, 56, 133

Cyclorrhapha (Diptera) 302, **304**

Cynipidae/Cynipoidea (Hymenoptera) **203**, 204

Damselflies. See Zygoptera 13, 30, 137, 155, 231, 232, 233, **270**, 271, 272, 273

Danaidae (Lepidoptera) 300

Dance flies. (Empididae) 230

"Dancing" 251, 252

Darkling beetles. (Tenebrionidae) 184, 189, 190

Dead-leaf butterfly. (Nymphalidae) 164

Deamination **114**, 115

Deciduous forests 174, **186**, 187, 189

Decomposers 18, 172, **176**

Deer flies. See Tabanidae 21, 186, 187, 303

Dendrophage/dendrophagy/dendrophagous 199

Density dependent. See environmental factors 168

Density independent. See environmental factors 168

Dermal gland 53

Dermal gland ducts 53

Dermaptera **277**, 279, 282

Dermestidae (Coleoptera) 295, 296

Desert 164, 166, 173, **187**, 188, 189, 302

Desert locust (*Schistocerca gregaria*) 207

Desiccation/desiccate 145, **166**

Deterrents/phagorepellents 205

Deutocerebrum. (mid brain) 65

Diapause 92, 94, 97, 121, **157**, 163, **166**

Diastole 127

Dicondylic articulation 61

Differentiation 97, 98, 100, 113, 114, 116, **158**, **159**

Diffusion 68, 111, 113, 115, 121, 137, **138**, 139, 142

Digestion 55, 109, **110**, 111

Digestive enzymes 106, 111, 194, 196

Digestive tract. See alimentary canal 23, 79, 103, **104**, 105, 106, 107, **109**, 110, 112, 116, 118, 121, 123, 127, 131, 132, 145, 147, 200, 290

Digger wasps. See Sphecidae

Dimorphic/dimorphism 12, 164, **227**, 256, 295

Dioecious bisexual reproduction 168, **225**, 235, 236, 237, 307

Diplura 10, 13, 29, 30, 113, 155, 184, 266, **268**

Diptera 14, 18, 21, 35, 48, 66, 92, 97, 107, 136, 180, 187, 189, 190, 193, 224, 259, 262, 270, 297, 299, **302**, 304, 309, 311

Direct development. See ametabolous development 13, **153**, 155, 156

Disruptive coloration 164

Ditrysia (Lepidoptera) 301

Diurnal 77, **163**, 164, 166, 205

Division of labor 239, 241, **242**, 282

Dobsonflies. See Corydalidae 227, **291**, 292

Domestic/domiciliary 196, 232, 240, 254, 262, 269, 274, 276

Dorsal blood vessel 44, 79, 121, **122**, 123, 125, 128, 131

Dorsal diaphragm. See pericardial membrane 123, **124**, 125, 126, 127

Dorsal transverse/aliform muscles 123, **124**, 127, 128

Double innervation 82

Dragonflies. See Anisoptera **270**, 271, 273

Index

Drinking 116, **167**, 206
Drones/male reproductives 237, **250**, 252
Dung beetles. See Scarabaeidae 294, 295
Dytiscidae (Coleoptera) **293**

Earwigs. See Dermaptera 13, 24, 30, 180, 233, **277**, 278, 279
Economic/applied entomology **19**, 262
Ecdysial cleavage line/epicranial suture **25**, 56, 60
Ecdysis/shedding 51, 55, **56**, 57, 58, 60, 62, 76, 92, 121, 124, 129
Eclosion. See hatching 98, 151, **153**, 169, **234**
Ecosystem 18, **172**, 176
Ectognathy/ectognathous **30**, 268, 270, 272
Ectoparasites. See parasitism 186, **192**, 199, 279, 305
Effectors 64
Egestion/defecation 71, **111**, 116
Egg laying. See oviposition 17, 46, 71, 145, **149**, 150, 198, 201, 203, 217, 227, 230, **233**, 234, 235, 240, 242, 246, 250, 285
Eggs 13, 14, 17, 48, 54, 145, 146, 149, **150**, 151, 153, 163, 179, 192, 201, 203, 233, 234, 235, 236, 237, 239, 240, 241, 242, 244, 245, 246, 249, 250, 251, 274, 275, 279, 283, 284, 304, 305, 307, 309
Ejaculatory duct 147
Elytron/elytra 224, **293**, 294, 295
Embioptera 279, **280**
Emigration 169, **170**
Encapsulation 129
Endocrine system/glands **91**, 92, 94, 96, 98, 100, 151
Endocuticle 52, **53**, 55, 56, 58
Endoparasites. See parasitism 137, **192**, 295
Endopterygota 290
Energy flow. See community metabolism 18, **175**
Ensifera (Orthoptera) 276
Entognathy/entognathous **30**, 265, 268
Entomology **1**, 10, 19, 262
Entomophagy. See insectivory
Environment 14, 18, 51, 52, 70, 72, 100, 103, 112, 116, 145, 151, **161**, 164, 170, 172, 174, 175, 177, **179**, 180, 181, 182, 183, 184, 185, 201, 205, 209, 210, 214, 234, 290, 307
Environmental factors **161**, 163, 165, 167, 168
Enzymes **104**, 106, 109, 110, 111, 124, 165, 194, 196
Ephemeroptera 44, **270**, 271, 273
Epicranial suture. See ecdysial cleavage line **25**, 26, 56, 60
Epicranium **27**, 28, 89
Epicuticle 51, 52, **53**, 54, 55, 56, 58
Epidermis 42, 51, 52, 53, 55, 58, 62, 64, 104, 105, 116, 118, 201, 203
Epipharynx **31**, 33, 34, 35, 305
Epithelium/epithelia **51**, 106, 107, 118, 133, 147, 150, 151
Esophagus 26, 65, 68, 105, **106**
European earwig (*Forficula auricularia*) 241
Excretion **103**, 116, 117, 118, 119
Exocuticle 52, **53**, 54, 56, 58, 59, 60
Exoskeleton **23**, 39, 53, 56, 135, 139, 158, 203
Expiration 139, 140
Exuviae/cast skin **56**, 62, 135, 158
Eyes, compound 1, 7, 25, 27, **29**, 30, 65, 75, **76**, 77, 157, 162, 244, 268, 269, 270, 272, 276, 290, 291
Eyes, simple 1, 26, **29**, 51, 65, 299, 308

Facets 29
Factor ecology 161
Facultative 166, **172**, 184
Fat body 53, 59, 95, 104, 115, **116**, 117, 118, 121, 124, 125, 146, 150, 192
Fatigue 71, 84, **86**, 138, 161
Fats **103**, 111, 124
Feculae/feces **111**, 112, 118, 196, 305
Fecundity 17
Fertilization 145, 192, **225**, 235, 236, 249, 297
Fibrillar muscle 80, 81
Field crickets. See Gryllidae 167, 255, 258, **276**
Fire ant (*Solenopsis xyloni*) 17, 21
Firebrat (*Thermobia domestica*) 156, 166, **269**
Fireflies. See Lampyridae 186, **228**, **295**, 296
Flagellum 30, 32
Flea beetles. See Chrysomelidae 199, 212
Fleahoppers. See Miridae 212, 288, 289
Fleas. See Siphonaptera 19, 20, 21, 29, 34, 40, 58, 75, 139, 177, 184, 195, 212, 224, 233, 297, **305**, 307
Flesh flies. (Sarcophagidae) 187
Flight velocity 222
Flower flies. See Syrphidae 18, 187, 189, 190, 303, 304
Flies/true flies. See Diptera 14, 16, 18, 34, 40, 41, 58, 59, 60, 66, 84, 110, 147, 155, 180, 184, 186, 189, 199, 201, 203, 217, 220, 221, 223, 225, 262, **302**, 304
Flying 14, 39, 40, 84, 86, 125, 127, 135, 196, 210, 217, 220, 222, 227, 229, 270, 290, 291, 297, 307
Follicles (ovaries) 147, 149
Follicles (testes) 147, 149, 151
Food chain 1, 18, 176, 177, 181, 184
Food habits/feeding 11, 12, 14, 23, 24, 35, 99, 104, 112, 117, 118, 121, 156, 166, 176, **191**, 194, **197**, 199, 200, 201, 202, 203, 204, 205, 206, 207, 227, 244, 246, 249, 256, 258, 269, 270, 276, 283, 284, 285, 286, 288, 290,

343

ENTOMOLOGY

295, 301, 304, 305

Food selection 205, 207, **233**

Food web 176

Forbivore/forbivorous/forbivory 199

Forceps 242, 266, 268, 278, 279

Fore brain/protocerebrum 65, 75, 76, 92

Fore gut/stomodaeum 67, **104**, 105, 106, 107, 108, **109**

Fore legs 25, **40**, 184, 194, 195, **210**, 212, 213, 215, 265, 266, 269, 274, 288, 291, 297

Formicidae/Formicoidea (Hymenoptera) **246**, 307, 308

Fossorial forms 183, 184

Frenatae (Lepidoptera) 301

Fritillaries. (Nymphalidae) 186, 187, 190

Frons/front **26**, 29

Frontal ganglion 65, 68

Fructose 103

Fulgoridae/Fulgoroidea (Homoptera) **286**, 287

GABA/aminobutyric acid 70

Gall gnats/midges. See Cecidomyiidae

Gallivory/gallivorous 203

Galls, **203**, 204

Gall wasps. See Cynipidae

Gametes. See germ cells

Ganglia 8, 63, 64, **65**, 66, 67, 71, 91, 92, 95, 96, 97, 128, 141

Gas apertures (oocytes) 151

Gases 121, 124, 139, **141**, 142

Genae 29

Genital aperture/gonopore **46**, 48, 49, 145, 147, 233, 301

Genitalia 1, **47**, **48**, 49, 145, 155, 159, 227, 231, 233, 258, 268, 269, 272, 297, 299

Germ cells/gametes 145, 148, **151**

Giant fibers 67

Giant water bugs. See Belostomatidae

Gills, blood 142

Gills, tracheal 136, **141**, 142, 155, 216, 270

Gliding 217

Glow-worms. (Phengodidae) 228

Glucose 103, 110, 111, 124

Glutamic acid 70

Glycogen 86, **103**, 117, 150, 151

Gonopore. See genital aperture

Graminivores/graminivory/graminivorous 199

Granary weevils. See Curculionidae 18, 19, 167, **295**

Granular hemocytes **125**, 130

Grasshoppers. See Acrididae 13, 15, 18, 24, 30, 41, 84, 110, 111, 112, 140, 155, 163, 164, 166, 171, 175, 176, 177, 187, 189, 199, 205, 206, 210, 212, 216, 221, 233, 236, 272, **276**

Grass moths. (Pyralidae) 187

Grayling butterfly. (Satyridae) 231

Gregarious **239**, 240, 241, 242, 276, 282, 283

Ground beetles. See Carabidae 180, 184, 187, 190, **293**, 294

Grouse locusts. See Tetrigidae 233, **276**

Growth 51, 55, 91, 92, 94, 96, 98, 100, 103, 117, 130, 134, 135, 150, 156, **158**, 164, 166, 167, 168, 170, 180, 187, 188, 228, 246, 250, 256

Gryllacrididae (Orthoptera) 276

Gryllidae/Gryllodea (Orthoptera) 276

Grylloblattidae/Grylloblattoidea (Orthoptera) 276

Gryllotalpidae (Orthoptera) 276

Gula 28, **29**, 33, 286, 293

Gut. See alimentary canal **104**, 107, 110

Gynandromorph/gynandromorphism 237

Gyplure 228, 229

Gypsy moth *(Lymantria dispar)* 186, **229**, 301

Gyrinidae (Coleoptera) 293

Habitat niches. See microhabitats 9, 11, **174**, 182, 217, 227, 272

Habitat occupancy/habitat selection 7, 233, 256

Habitats 174, **179**, 181, 182, 183, 184, 278, 282, 302

Hanging scorpionflies. See Bittacidae 194, **297**, 298

Hatching/eclosion 58, 92, 151, **153**, 169, 234, 283

Haustellum 35, 299

Head 1, 7, 12, 23, **24**, 25, 26, 27, 28, 29, 30, 31, 33, 35, 36, 37, 40, 56, 57, 60, 61, 65, 66, 76, 89, 97, 99, 100, 105, 121, 122, 126, 127, 129, 131, 141, 156, 158, 184, 192, 201, 203, 227, 240, 244, 265, 268, 270, 272, 273, 274, 283, 284, 286, 291, 293, 297, 298, 299, 302, 304, 305, 306, 308

Head, hypognathous/hypognathy 12, **24**, 25, 26, 30, 270, 274, 282, 286, 288, 298, 299

Head, opisthognathous/opisthognathy 12, **25**, 26, 274, 286

Head, prognathous/prognathy 12, **24**, 26, 29, 274, 278, 298

Heart. See dorsal blood vessel

Heartbeat reversal 129

Heat 86, 161, 163, 164, 165, 166, 175, 176, 180, 188, 214

Hemelytron/hemelytra 224, **286**, 289

Hemimetabola 153, **155**, 157, 158, 159

Hemimetabolous development/hemimetaboly. See incomplete development **13**, 270, 272, 290

Hemiptera/Hemipteroidea 15, 155, 193, 224, 259, 262, **282**, **286**, 288

Index

Hemocoele **121**, 125
Hemocytes/blood cells 51, 124, **125**, **129**, 130
Hemolymph. See blood 12, 57, 121, **124**, 129, 133
Hemopoietic organs 125, 129
Herbivore/herbivorous 18, **176**, 198, 204, 205, 309
Hermaphrodites/hermaphroditism **235**, 236, 257
Heteroneura (Lepidoptera) 301
Heteroptera (Hemiptera) 15, 224, **286**, 288, 297
Hexapoda. See Insecta 40
Hind brain/tritocerebrum 65
Hind gut/proctodaeum 104, **109**
Hind legs 40, 143, 194, 195, 210, 212, 213, 215, 230, 274, 297
Hippoboscidae (Diptera) 303
Holocrine secretion 110
Holometabola 116, 153, **155**, 157, 158, 159, 291
Holometabolous development.
   See complete development 13, 14, **155**, 290
Homoneura (Lepidoptera) 301
Homoptera 15, 155, 203, 224, **286**, 288, 297
Honey 19, 248, 249, 250
Honey ants (*Myrmecocystus* spp.) 248
Honey bee (*Apis mellifera*) 16, 19, 21, 107, 110, 237, 243, **250**, 251, 252, 262
Honeycow 173
Honeydew 111, **173**, 246, 248, 249
Hormones 53, **91**, 92, 95, 97, 98, 116, 121, 124
Hornets. See Vespidae 246, 307, 308
Horntails. See Siricidae 186, 203, 306, **307**
Horse flies. See Tabanidae 19, 21, 187, 222, 303
House flies. (Muscidae) 16, 18, 19, 35, 48, 189, 211, 302, 303
Household pests 19
Humus 175, 187, 265, 282
Hunger 206
Hunting 194
Husband cannibalism 230
Hyaline hemocytes/coagulocytes 125, 130
Hydrofuge/non-wettable 143
Hydrophilidae (Coleoptera) 143, 293
Hymenoptera 14, 19, 21, 47, 76, 84, 107, 108, 118, 187, 189, 190, 193, 203, 204, 224, 243, 297, 299, **307**, 309
Hypermetamorphosis 291, 293, 297
Hypocerebral ganglion 68, 97
Hypognathous/hypognathy. See head 12, **24**, 26, 270, 274, 282, 286, 288, 298, 299
Hypopharynx **33**, 34, 35

Ichneumonidae/Ichneumonoidea (Hymenoptera) 187, 190, 193, 307, 309
Ichneumon wasps. See Ichneumonidae 186, 187, 190, 308
Ileum 109
Imaginal discs 159
Immigration 169, 170
Incomplete/hemimetabolous development 13, **155**, 156, 270, 272
Indirect wing muscles 88
Industrial melanism 164
Inertia 209
Infrared 163
Ingestion 12, 109, 110
Inhibition 69, **70**, 71, 72, 228
Injuries and disease 207
Innervation 76, **82**, 109, 128
Inquilines 184
Insect agents of disease 21
Insecta/Hexapoda/insects 1, 40
Insect ectoparasites 192
Insectivores/insectivory/insectivorous 18, 244
Insect livestock pests 19
Insect plant pests 19
Insemination **225**, 230, 232
Inspiration 140
Instars 158, 286
Integument **51**, 52, 58, 118, 137, 280, 284
Intermediary metabolism 110, 111, 117
Intima (digestive tract) 104, 105, 106, 107, 108, 109, 111
Intima (tracheae) 133, 134, 135
Invertase 104, 109
Isoptera **274**, 278

Japygidae (Diplura) 10, 266, **268**, 278
Jet propulsion 216
Johnston's organ 74
Jugatae (Lepidoptera) 301
Jugum/jugal area 42, 43, 223, 224, 272, 282, 290, 301
Jurassic 279, 292, 304
Juvabione. See paper factor 101
Juvenile hormone/JH 92, 94, **95**, 97, 98, 99, 100, 101, 159
Juveniles 13, 29, 95, 96, 101, 128, 136, 137, 142, 155, 156, 159, 183, 198, 224, 245, 270, 280, 291, 295, 298

Katydids. See Tettigoniidae 110, 163, 164, 166, 186, 189, 194, 204, 210, 212, 225, 227, 230, 231, 233, 272, **276**

345

ENTOMOLOGY

Labium 30, 32, **33**, 34, 35, 36, 37, 89, 265, 270, 284, 305
Labral nerves 65
Labrum 27, **30**, 31, 34, 35, 36, 65, 305
Lac 19
Lacewings. See Chrysopidae 30, 194, 234, 235, 290, **291**, 292
Lactic acid 86
Ladybirds. (Coccinellidae) 187, 262
Lakes 142, 179, **181**, 182, 186, 230, 270, 280, 302
Lamellae 53, 59
Lampyridae (Coleoptera) 262, **295**, 296
Lanternflies. See Fulgoridae 286, 287
Larvae 14, 29, 30, 40, 44, 53, 75, 99, 101, 106, 107, 124, 135, 136, 137, 139, 141, 142, 143, **155**, 156, 157, 158, 180, 181, 182, 183, 185, 186, 192, 194, 201, 203, 207, 212, 214, 215, 216, 228, 234, 235, 240, 242, 246, 249, 250, 251, 286, **290**, 291, 292, 293, 295, 297, 298, 299, 302, 304, 305, 307
Larvapods/prolegs 213, 299
Lateral nerves 68, 128
Laws of Thermodynamics 176
Leaf beetles. See Chrysomelidae 295, 296
Leaf bugs. See Miridae 288, 289
Leafhoppers. See Cicadellidae 286, 287
Leaf insects. See Phasmidae 164, 189, **275**
Leaf-mining 201, 304
Leaf miners. (Gracillariidae) 186, 201
Leaping 11, 41, 47, 210, **212**, 213, 280, 305
Legs 1, 7, 10, 11, 23, 25, 38, **40**, 44, 46, 51, 56, 59, 60, 86, 123, 124, 131, 143, 156, 157, 182, 183, 184, 192, 194, 195, 201, 203, 207, 209, 210, 211, 212, 213, 214, 215, 216, 224, 230, 242, 265, 266, 269, 270, 272, 274, 276, 283, 284, 288, 290, 291, 293, 295, 297, 298, 299, 304, 305, 308
Lepidoptera 14, 19, 57, 58, 76, 84, 158, 186, 189, 193, 204, 213, 224, 259, 270, 297, **299**, 301, 309
Lepismatidae (Thysanura) 31, **269**
Life cycle 17, 145, 153, 172, 236, 239, 244, 246, 249, 257, 295
Light 72, 75, 76, 77, 78, 80, **161**, 162, 163, 164, 166, 181, 183, 184, 185, 204, 227, 228, 240, 241, 249, 298
Light organs 228
Liparidae (Lepidoptera) 57, 58, 76, 84, 158, 186, 189, 193, 204, 213, 224, 259, 270, 297
Lipase 104, 109
Literature 19, 259, 297
Locomotion 23, 38, 79, 88, 191, 198, 199, 209, 210, 216, 217, 218, 233, 283
Louse flies. See Hippoboscidae 224, 303
Lucanidae (Coleoptera) 294, 295

Machilidae (Thysanura) 29
Macrolepidoptera 301
Maggots 304
Mallophaga **283**, 284, 285
Malpighian tubules 8, 79, 107, 109, **112**, 113, 114, 115, 116, 118, 121, 124, 159, 272, 282, 290, 309
Maltase 104
Maltose 103
Mandibles 12, 25, 26, 30, **31**, 32, 33, 34, 35, 36, 61, 89, 157, 164, 184, 194, 197, 199, 201, 227, 268, 269, 272, 284, 286, 291, 293, 294, 295, 297, 298, 299, 301, 302, 304, 305
"Mandibulate moths." 301
Mandibulate/biting/chewing mouthparts 1, **30**, 31, 34, 35, 61, 156, 199, 203, 268, 270, 272, 274, 278, 280, 282, 283, 290, 291, 293, 295, 297, 299, 301, 305
Mantidae/Mantodea (Orthoptera) 194,195, **274**, 275
Mantispidae (Neuroptera) 291, 292
Marine 179, 181, 288
Mask 273
Maternal care 227, 279, 281
Maxillae 25, 32, **33**, 34, 35, 36, 284, 286, 301
May beetles. See Scarabaeidae 240, 241, 262, 294, 295
Mayflies. See Ephemeroptera 10, 13, 16, 30, 31, 66, 137, 155, 205, 207, 224, 230, 233, 262, **270**, 271, 273, 280
Mealworms. (Tenebrionidae) 111, 167
Mechanoreceptors 72, 74
Mecoptera 297, 304
Median caudal filament 46, 268, 269, 270, 271, 280
Median nerve 67
Median plates 42
Medical entomology/sanitary entomology 19
Megaloptera (Neuroptera) 291, 292
Megalopygidae (Lepidoptera) 21
Melanin 53, 56
Melanization 56, 57
Meloidae (Coleoptera) 295, 296
Membracidae (Homoptera) 286
Merocrine secretion 110
Mesenteron. See mid gut
Mesothorax 37, 39, 88
Metallic wood borers. (Buprestidae) 186
Metamorphosis 1, 11, **13**, 92, 94, **97**, 98, 101, 108, 118, 121, 129, 130, **153**, 154, 191, 192
Metathorax 7, 37, **39**, 40, 44, 123
Microclimate 166, 185
Microcoryphia/Archeognatha 10, 13, 29, 30, 155, **268**, 269
Microhabitats/habitat niches 9, 11, **174**, 182, 217, 227, 272

Index

Microlepidoptera 190, 301
Micropterygidae (Lepidoptera) 301
Micropyles 151
Microsuccession 173, **175**, 195, 196
Mid brain/deutocerebrum 65
Midges. See Chironomidae 16, 181, 186, 190, 203, 230, 237, 303
Mid gut/mesenteron 104, 106, **107**, 108, 109, **110**, 111, 118
Mid legs 210, 212, 216
Milkweed beetles. See Cerambycidae 164
Mimicry 164
Mining/leaf mining **200**, 201, 202, 301, 304
Miridae (Hemiptera) 288, 289
Mitochondria 63, 80, 86, 113, 114, 228
Mole crickets. See Gryllotalpidae 184, 276
Molting 25, **54**, 55, 56, 57, 58, 62, 92, 94, 95, 96, 97, 98, 99, 100, 101, 118, 124, 129, 135, 155, 192, 244, 268, 269, 295
Molting fluid 56
Molting hormone/MH 57, 92, **96**, 97, 98, 99, 100, 101, 159
Monarch butterfly (*Danaus plexippus*) 199, 262, 300
Monoecious bisexual reproduction. See hermaphroditism 235
Monophage/monophagy/monophagous 203, 204
Monotrysia (Lepidoptera) 148, 301
Mormon cricket (*Anabrus simplex*) 196, 206
Mortality 142, **168**, 169, 170
Mosaic theory of vision 77
Mosquitoes. See Culicidae 16, 20, 21, 34, 35, 59, 101, 162, 181, 186, 190, **192**, 205, 233, 302, 303
Moth flies. (Psychodidae) 136
Moths. See Lepidoptera 11, 14, 34, 35, 59, 71, 106, 119, 146, 155, 163, 186, 187, 189, 199, 203, 220, 223, 227, 236, **299**
Motor end plate. See myoneural junction/neuromuscular junction 70, 82
Motor neurons 63, **64**, 65, 67, 69, 71, 212
Mountain midges. (Deuterophlebiidae) 190
Mountains 186, **189**, 258
Mouth 23, 33, 51, 74, **104**, 105, 297
Mucus 106
Mud-bottom pools 182
Multiple generations 17
Muscle fibers 69, 70, **79**, 80, 81, 82, 83, 86, 122, 126, 212
Muscles 10, 29, 30, 31, 38, 40, 41, 47, 60, 61, 63, 64, 68, 69, 72, **79**, 84, 85, 86, 87, 88, 89, 92, 105, 108, 109, 114, 121, 123, 124, 127, 128, 132, 139, 140, 141, 145, 158, 159, 165, 207, 209, 212, 213, 217, 218, 219, 220, 221, 222, 223, 224

Muscle twitches 83
Mutualism 172
Myogenic heart beat 128
Myology 87
Myoneural junction/motor end plate 82
Myosin 86
Myrmecophiline crickets. See Gryllidae 172, 184
Myrmeleontidae (Neuroptera) 291, 292

Naiads 13, 75, 141, 142, **155**, 171, 181, 182, 183, 216, 270, 272, 280
Natality **168**, 169, 170
Neck/cervix 25, 28, **37**, 89, 233
Nekton 181, **182**, 183
Nematocera (Diptera) **302**, 304, 307
Neoptera 11, 42, 44, **272**, 282, 290
Nephrocytes 118, 121
Nerves **63**, 65, 67, 68, 72, 91, 94, 97, 121, 128, 139
Nerve wings. See Neuroptera
Nervous system 23, 58, 59, **63**, 69, 91, 97, 218, 228
Neural lamella 63, 67, 82
Neurogenic heart beat 128
Neurons **63**, 64, 65, 67, 69, 71, 82, 86, 91, 95, 212
Neuroptera/Neuropteroidea 108, 290, 291, 292
Neurosecretion **91**, 94, 141
Neuston 181, **182**, 183
Nits 234, **284**
Noctuidae (Lepidoptera) 300
Nocturnal 77, 116, **163**, 166, 167, 205, 280, 299
Nodus 270
Notonectidae (Hemiptera) **288**, 289,
Notum. See alinotum 38, **39**, 88, 217, 218, 219, 221
Nutritive cells **146**, 148, 149, 150, 151
Nymphs 13, 75, 101, 143, **155**, 241, 244, 284, 288

Object resemblance 164
Obligatory **172**, 184, 191, 236, 257
Obsolete sutures 26
Occipital condyles 29, 37
Occipital foramen 25, 29
Occiput 26, **28**
Ocellar lobes 65
Ocelli 25, 27, **29**, 65, **75**, 76, 157, 268, 269, 270, 272, 274, 277
Odonata 44, 123, 236, **270**, 272
Oenocytes 51, 53
Oenocytoids 125, 130

ENTOMOLOGY

Oligophage/oligophagous 203

Oligopneustic 135

Ommatidia **76**, 77, 78, 163

Ommochrome pigments 59

Omnivore/omnivory/omnivorous **196**, 249, 269, 273, 276, 279, 309

Oocytes 146, **148**, 149, 150, 236, 250

Oogonia 148, 151

Open circulatory system 121

Opisthognathy/opisthognathous. See head **12**, **25**, 26, 274, 286

Optic lobes 65, 76, 92

Orthoptera/Orthopteroidea 15, 47, 86, 97, 141, 189, 193, 210, 217, 224, 259, **272**, 279, 280, 281, 282

Osmoregulation **116**, 142

Osmotic pressure/osmosis 121, 138

Ostia **122**, 123, 127

Out-of-phase tripoding 210

Ova 145

Ovaries 8, 92, 95, 96, **145**, 146, 235

Ovarioles 8, **145**, 146, 147, 148, 149, 150, 236

Oviduct 8, **145**, 146

Oviparity/oviparous 234

Oviposition/egg laying 17, 46, 71, 145, **149**, 150, 198, 201, 203, 217, 227, 230, **233**, 234, 235, 240, 242, 246, 250, 285

Oviposition site selection 233

Ovipositor 7, 21, **47**, 48, 201, 203, 233, 234, 235, 274, 276, 280, 284, 291, 307, 308, 309

Ovoviviparity/ovoviviparous **234**, 304

Oxygen 9, 71, 86, 104, 134, **137**, 138, 139, 140, **141**, 142, 143, 151, 168, 182, 183, 189, 214, 228

Paedogenesis/paedogenetic 237

Palatability/food attractiveness 205, 206

Paleoptera 10, 11, 44, **270**, 272

Paleozoic 282

Palpi **33**, 34, 35, 36, 46, 72, 74, 75, 206, 207, 268, 269, 283, 293, 302, 304, 305

Panoistic ovarioles 149, 150

Panorpidae/Panorpoidea (Mecoptera) 297, 298

Paper factor/juvabione 101

Paper wasps. See Polistinae

Paranota 288

Parasites/parasitic/parasitism 11, 38, 129, 130, **172**, 176, 177, 184, 187, **191**, 192, 193, 194, 199, 204, 224, 247, 248, 249, 284, 288, 291, 295, 297, 298, 302, 305, 307, 309

Parasitoids/parasitoidism 192, 193

Pars intercerebralis 65, 92, 94

Parthenogenesis/parthenogenetic 235, **236**, 237, 239, 257, 304, 307

Pedicel 30, 32, 146, 149, 246, 247, 308

Pennsylvanian 270

Pentatomidae (Hemiptera) 288, 289

Peppered moth. 164

Pericardial cells 118

Pericardial cavity 124, 126

Pericardial membrane/dorsal diaphragm 123, 124, 125, 126, 127, 129, 146

Perineural cavity 124

Perineural membrane/ventral diaphragm 124, 126

Perineurium 63, 66

Periodicity **163**, 205, 240, 256

Periodism 161, 170, **174**, 256

Peripheral nervous division **63**, 67

Periphyton 181, **182**

Periproct 23, 45

Peristalsis/peristaltic movement 56, 67, 109, **110**, 111, 127, 129, 212

Peritoneum 147

Peritrophic membrane 106, **107**, 108, 111

Perivisceral cavity 124, 126

Permian 272, 281, 283, 286, 292, 295, 297, 298

pH/hydrogen ion concentration 110, 111, 124, 183

Phagocytosis 129

Phallus 48, 49

Phantom midges. (Chaoboridae) 181

Pharynx 65, 105, 109

Phasmidae/Phasmatodea (Orthoptera) 275

Pheromones 228

Phosphogen 86

Photoperiod 97, 163

Photoreceptors. See eyes

Photosynthesis 18, 161, 172, **175**, 181

Phthiraptera 262, **283**, 284, 285

Physical/interference coloration 59, 60

Phytoparasites/phytoparasitism 199

Phytophages/phytophagy/phytophagous 196, **197**, 308

Pigments/pigmentation/chemical coloration 57, **59**, 60, 77, 119, 124, 299

Pine sawyers. See Cerambycidae 164, **295**, 296

Pink bollworm (*Pectinophora gossypiella*) 163

Plague locusts. See Acrididae 167, 207, 217, **240**, 242, 276, 277

Plankton/plankters 181, 183

Planthoppers. See Fulgoridae 212, 286, 287

348

Index

Plantlice. See Aphididae 262, 286, 287
Plasma **124**, 129
Plasmatocytes 51, **125**, 126, 129, 130
Plastron respiration 143, 144
Plate organs 75
Plecoptera **280**, 281, 282
Pleural suture 39
Pleuron/pleura **23**, 38, 39, 44, 88, 131, 270
Poikilothermy/poikilothermic 164
Polistinae (Vespidae) 246, 308
Pollination **18**, 19, 307
Polyembryony 237, 307
Polymorphic/polymorphism **243**, 256, 274, 282
Polyneural innervation 82
Polyphages/polyphagous 203, 204
Polypneustic **135**, 136, 137
Polyphaga (Coleoptera) 293, 294, 295
Polytrophic ovarioles 146, 147, 149
Pomace flies. (Drosophilidae) 163
Ponds 179, **181**, 182, 183, 186, 215, 230, 266, 270, 293, 302
Pool zone 183
Populations 164, **167**, 168, 169, 170, 171, 172, 174, 175, 206, 207, 235, 236, 255, 256
Pore canals 53, 58
Posterior intestine 109, 111
Postgenae 29
Postnotum/postnota 39, 40
Postocciput 26, 28, 37
Potassium 68, 84, 116
Powderpost beetles. (Lyctidae) 186
Praying mantises. See Mantidae 172, 230, **274**
Predacious diving beetles. See Dytiscidae 143, 182, 183, 186, 215, **293**
Predators/predation 9, 111, 163, 164, 168, 169, **172**, 177, 175, 180, 184, 190, **193**, 194, 199, 204, 230, 231, 244, 270, 276, 279, 295, 297, 304
Preoral cavity 30, 33, 65, 74, **104**, 105
Primarily aquatic insects **141**, 142
Primary reproductives. See queens and drones **244**, 274
Primitively wingless insects. See Apterygota **10**, 224, 265, 268, 269
Primordial germ cells 148
Processionary caterpillars. (Thaumetopoeidae) 240
Proctodaeum. See hind gut **104**, 108, **109**, 111, 115, 118
Producers 18, 172, **176**
Prognathous/prognathy. See head
Prohemocytes **125**, 129, 130

Proline 104
Pronotum 240, **272**, 273, 276, 288, 305, 306
Propulsion 210, 216, 223
Prostomium. See acron 23, 24, 26
Protease 104, 109
Proteins 96, 103, **104**, 111, 124, 165
Prothoracic glands 57, 92, 94, **96**, 97, 98, 99, 100, 159
Prothorax **37**, 39, 40, 89, 283, 291
Protocerebrum. See fore brain
Protodonata 272
Protura 10, 13, 29, 30, 40, 45, 46, 155, **265**, 266, 268
Proventriculus 105, 106, **107**, 108, 110
Psocids. See Psocoptera
Psocoptera 281, 282, **283**
Pteridines 59, 119
Pterothorax 40
Pterygota/pterygotes 10, 29, 30, 31, 38, 39, 40, 44, 47, 48, 61, 95, 96, 97, 131, 224, 225, 227, **270**, 272, 282, 290
Ptilinum 303, 304
Pulicidae (Siphonaptera) 306
Pulsatile organs 123, 128
Pulse 128
Pupae 14, 96, 136, 141, **157**, 158, 192, 250, 281, 286, 293, 297, 298, 299, 302, 304, 305
Pupal diapause 94, 157
Puparium 158, 295, 303, 304
Pupation 157, 159
Pyloric valve 109, 111
Pylorus 109
Pyramid of numbers 176
Pyramid of trophic levels 176
Pyruvic acid 86

Queens 237, 244, 246, 248, 249, 250, 252

Rain forests 189
Random distribution 170
Raphidioptera(Neuroptera) 291, 292
Rapids 182
Receptors. See chemoreceptors, mechanoreceptors, visual receptors
Rectal gills 142, 270
Rectal pads 109, 118
Rectal valve 109
Rectal sac 109
Rectum 109, 112, 113, 116, 118, 142
Recurrent nerve 65, 68
Reduviidae (Hemiptera) 288, 289

ENTOMOLOGY

Reflex arcs 70
Reflex center/coordinating center 71, 97
Refractory period 69, 83
Regurgitation 107, 110
Repletes 248
Reproduction 11, 23, 44, 92, 94, 97, 98, 151, **153**, 168, **225**, 235, 237, 244, 245, 304, 307
Resilin **53**, 84, 221
Respiration 121, 136, 137, 138, 141, 142, 143, **175**
Resting potential **68**, 69, 83
Reticulate/net venation 44, 224
Retina 75, 76
Retinula 77
Rhabdom 75, 77
Riffle beetles. (Elmidae) 144
Riffles 182, 183
Ring gland 92, 96, **97**
Rivers 182, 258
Robber flies. See Asilidae 180, 187, 190, 194
Rocky Mountain locust (*Melanoplus spretus*) 16
Roll (in flight) 220
Rotation 89, 220
Round dance 251, 252
Rove beetles. See Staphylinidae 184, 186, 278, **293**, 294
Royal jelly 250
Running **211**, 212, 273, 293

Saliva **106**, 109, 246, 249
Salivary glands **105**, 106, 158, 299
Salts 63, 86, 103, 111, 115, 116, 121, 124, 228
Sand-bottom pools 182, 183
Savanna 186
Sawflies. (Tenthredinidae) 14, 47, 111, 186, 190, 201, 234, 307
Sawyers. See Cerambycidae 164, 186
Scale insects. See Coccidae 19, 40, 155, 186, 189, 224, 227, 236, 286, **288**
Scales **59**, 119, 268, 283, 299
Scape 29, **30**, 32
Scarabaeidae (Coleoptera) 293, 294, **295**
Scavenger/scavenging **194**, 196, 295, 304
Sclerites 29, 37, 38, 39, 40, 42, 44, 53, **60**, 88, 192, 207, 220, 295
Sclerotin **53**, 56
Sclerotization 39, 44, **56**, 57, 60, 290
Scolopophorous/chordotonal organs 72, 74, 223
Scorpionflies. See Panorpidae 225, 227, 230, **297**, 298
Sculling 216

Seaweed flies. (Coelopidae) 180
Secondarily aquatic insects 141, 144
Secondarily wingless insects 11
Secondary reproductives 244
Secondary substances 205
Secretion 53, 55, 58, 95, 98, 108, 110, 113, 115, 116, 143, 150, 230, 250, 288, 299
Segmental ganglia 45, **66**, 141
Segments 1, **23**, 24, 26, 37, 38, 39, 40, 44, 45, 46, 47, 48, **60**, 66, 71, 87, 123, 131, 140, 155, 213, 214, 224, 233, 247, 265, 266, 268, 280, 282, 283, 284, 293, 295, 299, 304, 305, 309
Segmented antennae 29
Seminal vesicles 152
Sense cells 74, 75
Sensilla. See receptors 29, 31, **72**, 74, 223
Sensory neurons **63**, 64, 67, 71
Sensory setae **59**, 72, 74
Setae 19, 21, 51, 58, **59**, 72, 73, 74, 132, 143, 215, 216, 223, 249, 305
Sex attractants 228
Sex cells 159
Sexual color dimorphism 164
Shore flies. (Ephydridae) 180
Sign stimuli 231
Silk 19, 106, 299
Silkworm moth (*Bombyx mori*) 19, 97, 229, 301
Silphidae (Coleoptera) 294, **295**
Silverfish (*Lepisma saccharina*) 10, 97, **269**
Sinking rate 209
Siphonaptera 262, 297, **305**, 307
Siricidae (Hymenoptera) 306
Skeletal muscles **79**, 87
Skippers. See Lepidoptera 299, 300
Slavery/social parasitism 249
Smell/olfaction 29, 72, **74**, 75, 205, 229, 252
Snakeflies. See Raphidioptera 291
Snout beetles. See Curculionidae 187, 190, **295**, 296
Snowfleas. See Collembola 164, 166, 186
Snow scorpionflies. See Boreidae 297, 298
Snowy tree cricket (*Oecanthus fultoni*) 165
Social 16, 19, 184, **239**, **242**, 243, 246, 249, 274, 282, 295, 307
Sodium Pump Mechanism 68
Soil /interstitial organisms 7, 8, 18, 131, 166, **183**, 184, 186, 201, 212, 233, 235, 241, 242, 244, 247, 265, 276, 291, 293, 302
Soldier flies. (Stratiomyidae) 143
Soldiers 12, **244**, 245, 248, 251, 274

Index

Solitary **226**, 239, 240, 246, 249, 274, 283

Sound 12, **72**, **227**, 231, 243

Specialized habits/specialization 11, 196, **197**, 224, 301, 309

Sperm/spermatozoa 145, 149, 150, **151**, 152, **225**, 230, 232, 233, 236, 252

Spermathecae 145

Spermatogenesis 151, 152

Spermatophores **227**, 232

Sphecidae/Sphecoidea. See Hymenoptera 246, 309

Spherule cells 125, 130

Sphincters 79

Spines 41, 58, **59**, 105, 107, 158, 194, 211, 212, 225

Spiracles 1, 44, 67, 71, 131, **132**, **135**, 136, 137, 139, 140, 141, 142, 143, 144, 155

Spittlebugs. See Cercopidae 187, 286

Springs 182

Springtails. See Collembola 10, 29, 46, 48, 75, 95, 113, 118, 137, 166, 180, 183, 184, 186, 189, 199, 212, 232, **265**, 268

Spruce budworm (*Choristoneura fumiferana*) 186

Spurs 41, **59**, 211

Stable flies (*Stomoxys calcitrans*) 35

Stadia/Stadium 56

Stag beetles. See Lucanidae 167, 227, 294, **295**

Staphylinidae (Coleoptera) **293**, 294

Starch 103

Stemmata/lateral ocelli **29**, **75**, 76, 157, 299, 308

Sternum/sterna **23**, 38, **39**, 44, 45, 88, 131, 139, 146, 270, 293, 294

"Sticktights." See Pulicidae 305, 306

Stigma 270, 283, 307

Stigmatic cords 137

Stink bugs. See Pentatomidae 187, 288, 289

Stomachic ganglion 68

Stomodael valve 107, 110

Stomodaeum. See fore gut

Stoneflies. See Plecoptera 13, 40, 155, 166, 170, 233, 262, **280**, 281

Storage 8, 83, 84, 94, 103, 110, 117, 118, 119, 121, 225, 248

Storage excretion 118

Stratification 174, 181

Strepsiptera 193, 295, 298

Stretch. See tension 57, 97, 223

Styli 45, **46**, 265, 268, 269

Stylopidae. See Strepsiptera 193, **295**, 298

Stylopized 297

Subcoxal sclerites 38, 39

Subesophageal ganglion 65, 66, **71**, 92, 95, 97

Subsegments. See articles

Subsocial 239, **241**, 242, 246, 249, 274, 279, 280, 307

Succession 174, 175, 196

Succulence/water content 205

Sucking lice. See Anoplura 34, 155, 192, 233, 234, 283, **284**, 285

Sucrose 103

Suctorial/haustellate mouthparts 1, 30, **34**, 35, 37, 195, 199, 282, 283, 284, 286, 290, 301, 304

Summation **70**, 71, 83

Superposition eyes 77

Surface striding 214

Survival 11, 172

Sutures/conjunctivae/intersegmental membranes 23, 25, 26, 29, 38, 39, 40, 53, 54, 60, 79

Swallowtail butterflies. (Papilionidae) 217

Swarming 246, 249, 252

Swimming 11, 14, 142, 143, **215**, 216

Sympathetic nervous division 65, 66, **67**, 104

Symphypleona (Collembola) 266, 268

Symphyta (Hymenoptera) **307**, 308, 309

Synapse 70

Synaptic vesicles 70

Synchronous wing muscles 221

Syncytium 80

Synonym 261

Syrphidae (Diptera) 303

Systematic Approach 256

Systole 127

Tabanidae (Diptera) 303

Tactile impressions/touch 29, **72**

Taenidia **133**, 134, 135, 139

Taiga 186

Taste/gustation 29, 31, 72, **74**, 75, 205

Tegmen/tegmina 224, **272**, 276, 278, 279, 280

Telson 155, 265

Telsontails. See Protura

Temperature 97, 127, 128, 153, 163, **164**, 165, 166, 167, 168, 181, 183, 184, 185, 189, 204, 210, 305

Tendons 61, 62

Teneral 56

Tension 71, 72, 74, 85, 86, 137, 141, 142, 182, 212, 215

Tent caterpillars. (Lasiocampidae) 240

Tentorium **26**, 28, 30, 60, 62, 65, 89, 265, 269, 286

Tergum/a 23, 38, **39**, 42, **44**, 88

Termites. See Isoptera **243**, **274**, 278

351

Tertiary 292, 304, 307, 309

Testis/testes 146, 147, 235

Tetanus 69, **83**, 84

Tetrigidae (Orthoptera) **276**

Tettigoniidae/Tettigonioidea (Orthoptera) 276

Thermal migrations 166

Thirst 206

Thoracic center 141

Thoracic ganglia 66, 71

Thoracic gills 280

Thorax 1, 7, 10, 23, 26, **37**, 38, 39, 40, 56, 66, 85, 88, 96, 99, 100, 105, 110, 131, 132, 141, 142, 158, 159, 192, 203, 217, 220, 247, 265, 272, 283, 284, 302, 307

Threshold 69, 83, **161**

Thrips. See Thysanoptera 168, 183, 199, 222, **284**, 285, 286

Thysanoptera 282, **284**, 285

Thysanura 10, 13, 29, 155, 232, 268, **269**

Tiger beetles. See Cicindelidae 175, 180, 189, 293

Toad bugs. (Gelastocoridae) 212

Tonofibrillae 58, 61, **62**

Tonus 86

Tormogen cells 51

Tracheae 1, 51, 66, 67, 68, 104, 105, 109, 121, 123, **131**, **133**, 134, 135, 137, 138, 139, 143

Tracheal gills 136, 137, **141**, 142, 155, 216, 270

Tracheoblasts 133

Tracheolar liquor 138

Tracheoles 114, 131, **133**, 134, 137, 138, 141, 145

Trapping 194

Tree crickets. (Gryllidae) 204, 228, 230, **276**

Treehoppers. (Membracidae) 286

Trehalose 103, 110, 117, 124

Triassic 292, 298, 304, 309

Trichogen cells 51

Trichoptera 76, 297, **298**, 299, 300, 301

Tritocerebrum. See hind brain 65, 68

Triungulins 192, **295**

Trophallaxis 106, **243**, 244, 252

Tsetse flies. (Glossinidae) 20

Tumblebugs. See Scarabaeidae **195**, 196, 262, 294, **295**

Tundra 173, **185**, 186

Ultraviolet 78, **162**, 163

Undulation 216

Unicondylic articulation 61

Uniform distribution 170

Unisexual reproduction. See parthenogenesis 235, **236**, 237, 307

Universal articulation 60

Urea 115

Ureter 112

Uric acid **115**, 116, 117, 118, 119, 124

Vacuoles 51, 113, 117

Vagility 170

Vagina. See copulatory pouch

Variation/variability 11, 12, 13, 124, 191, 210, 233, 235, 236, 237, 255, 256, 257, 258, 263

Vasa deferentia 147, 152

Vasa efferentia 147, 152

Veins 1, **43**, 44, 123, 128, 220, 270, 272, 274, 283, 284, 288, 290, 291, 293, 297, 302

Ventral diaphragm. See perineural membrane 124, 126

Ventral glands 97

Ventral nerve cord 26, 64, **66**, 67, 68, 70, 71, 97, 116, 123, 128, 131, 132, 141

Ventral nerve cord connectives 67

Ventral nerve cord ganglia 67, 97

Ventral transverse muscles 124

Ventricular ganglion 68

Ventriculus 106, 107, **108**, 110, 111

Vertex 27

Vespidae/Vespoidea (Hymenoptera) **246**, 308, 309

Vespinae. See Vespidae 246, 308

Vestigial mouthparts 304

Visceral muscles 79, 87

Vision 72, 77, 161, **162**, 249

Visual receptors/eyes 1, 7, 23, 25, 27, **29**, 30, 65, 72, **75**, 76, 77, 157, 162, 184, 192, 201, 244, 268, 269, 270, 272, 274, 276, 284, 288, 290, 291, 295, 308

Vitellogenesis 95

Viviparity/viviparous **234**, 235, 279

Waggle dance 251, 252

Walking 11, 14, 71, 125, **210**, 211, 214, 283

Walkingsticks. See Phasmidae 8, 164, 186, 189, 200, 224, 233, 236, **275**

Warble flies. (Oestridae) 234

Water 8, 18, 56, 103, 109, 111, 114, 115, 116, 118, 121, 124, 129, 131, 132, 136, 137, 139, 141, 142, 143, 144, 145, **166**, 167, 168, 174, 176, 179, 180, 181, 182, 183, 186, 188, 192, 205, 206, 209, 210, 214, 215, 216, 222, 228, 233, 270, 280, 288, 291, 297, 298, 301, 302

Water absorption 111

Water boatmen. See Corixidae 183, 215, **288**, 289

Water measurers. (Hydrometridae) 182

Water-penny beetles. (Psephenidae) 183

Index

Water scavenger beetles. See Hydrophilidae 143, 182, 183, 186, 215, **293**

Waterscorpions. (Nepidae) 194

Water striders. (Gerridae) 179, 182, **215**, 260

Wasps. See Vespidae 7, 14, 19, 21, 106, 110, 155, 175, 184, 186, 187, 189, 190, 203, 220, 221, 222, 223, 227, 237, 243, **246**, 247, 249, 297, 302, 307, **308**, 309

Water of metabolism 116, 167

Wax 52, **53**, 54, **58**, 191, 249, 250, 288, 301

Wax canals 58

Webspinners. See Embioptera 279, **280**, 281

"Wedding gifts" 229

Weevils. See Curculionidae 18, 19, 167, 186, 203, 293, **295**, 296

Wheat stem sawfly (*Cephus cinctus*) 20, 170, **306**

Whirligig beetles. See Gyrinidae 182, 183, **293**

Whiteflies. See Aleyrodidae 262, **286**

Wing beat frequency 220

Wing coupling 224

Wing margin, apical 42, 43, 286

Wing margin, anal 42, 43, 280

Wing margin, costal 42, 43, 220, 270, 283, 290, 307

Wing muscles 10, 40, 88, **218**, 220, 221

Wings 1, 7, 8, **9**, 10, 11, 12, 13, 23, 38, 39, **41**, 44, 51, 56, 59, 71, 84, 88, 124, 131, 143, 155, 157, 158, 159, 164, 192, 209, 210, 217, 218, 219, 220, 221, 222, 223, 224, 227, 228, 231, 244, 245, 246, 248, 249, 251, 270, 272, 274, 276, 278, 279, 280, 282, 283, 284, 286, 288, 290, 291, 293, 295, 297, 298, 299, 301, 302, 304, 307, 308

Wireworms. (Elateridae) 186

Workers 12, 237, 243, 244, 245, 246, 248, 249, 250, 251, 252, **274**

Wriggling 216

Yellow jackets. See Vespidae **246**, 308

Zeugloptera. (Micropterygidae) 301

Zones/zonation 161, 162, **173**, 181, 182, 184, 189, 246, 249, 274, 280, 288, 290, 297, 302, 307

Zoraptera 281, **282**

Zygentoma (Thysanura) 269

Zygoptera (Odonata) **270**, 271, 273

Zygotes 145

About the Author

S. K. (Stan) Gangwere is Professor Emeritus, Wayne State University, Detroit, Michigan 48202, USA. He attended Ohio University, Athens, Ohio, and the University of Michigan, Ann Arbor, Michigan, from the latter of which he received his professional degrees (A. B., M. S., Ph. D.) and was awarded membership in Phi Beta Kappa and several other honorary societies. Over the years he has received travel and research grants from the National Science Foundation, National Geographic Society, National Academy of Sciences, American Philosophical Society, and various other agencies and institutions and has been awarded Fulbright and other visiting lectureships and research scholarships at several Spanish and Argentine institutions. This program of travel and research has resulted in numerous publications on the ecology, biogeography, and behavior of orthopteroid insects, especially of the Iberian, Mediterranean, and Atlantic islands. Numbered among the author's contributions are over fifty peer-reviewed scientific articles as well as many invited review articles, numerous abstracts and other published works, and five previous books/monographs. At Wayne State University, he taught courses in entomology, biogeography, and natural history, served as Director of the university's Northwoods and, later, Fish Lake Biological Station programs, and, for a time, was Chairman of the Department of Biological Sciences. He is past Executive Director of the Orthopterists' Society, past President of the Pan American Acridological Society (precursor to the Orthopterists' Society), and past President of the Michigan Entomological Society (precursor to today's Great Lakes Entomological Society). He has been the principal organizer of five international meetings.